Healing Plants of Nigeria

Traditional Herbal Medicines for Modern Times

Series Editor

Ephraim Philip Lansky

The *Traditional Herbal Medicines for Modern Times* series reports on current developments and discusses key topics relevant to interdisciplinary health sciences research by ethnobiologists, taxonomists, conservationists, agronomists, chemists, clinicians, and toxicologists. The series is relevant to all these scientists and helps them guide business, government agencies, and commerce in the complexities of these matters.

Shengmai San

Robert Kam-Ming Ko

Rasayana: Ayurvedic Herbs for Longevity and Rejuvenation

H.S. Puri

Sho-Saiko-To: Scientific Evaluation and Clinical Applications

edited by Yukio Ogihara and Masaki Aburada

Traditional Medicines for Modern Times: Antidiabetic Plants

edited by Amala Soumyanath

***Bupleurum* Species**: Scientific Evaluation and Clinical Applications

edited by Sheng-Li Pan

Herbal Principles in Cosmetics: Properties and Mechanisms of Action

Bruno Burlando, Luisella Verotta, Laura Cornara and Elisa Bottini-Massa

Medicinal Plants and Malaria: Applications, Trends, and Prospects

Woon-Chien Teng, Ho Han Kiat, Rossarin Suwanarusk and Hwee-Ling Koh

Australian Native Plants: Cultivation and Uses in the Health and Food Industries

edited by Yasmina Sultanbawa and Fazal Sultanbawa

Honey in Traditional and Modern Medicine

edited by Laïd Boukraâ

Complementary and Alternative Medicines in Prostate Cancer: A Comprehensive Approach

edited by K. B. Harikumar

The Genus *Syzygium*: *Syzygium cumini* and Other Underutilised Species

edited by K. N. Nair

***Phyllanthus* Species**: Scientific Evaluation and Medicinal Applications

edited by Ramadasan Kuttan and K.B. Harikumar

Figs: The Genus *Ficus*

Ephraim Philip Lansky and Helena Maaria Paavilainen

Harmal: The Genus *Peganum*

Ephraim Shmaya Lansky, Shifra Lansky and Helena Maaria Paavilainen

Healing Plants of Nigeria: Ethnomedicine and Therapeutic Applications

Anselm Adodo and Maurice M. Iwu

For more information about this series, please visit: https://www.crcpress.com/Traditional-Herbal-Medicines-for-Modern-Times/book-series/CRCTRHEMEMOT

Healing Plants of Nigeria

Ethnomedicine and Therapeutic Applications

by

Anselm Adodo

Maurice M. Iwu

CRC Press is an imprint of the
Taylor & Francis Group, an **informa** business

First edition published 2020
by CRC Press
6000 Broken Sound Parkway NW, Suite 300, Boca Raton, FL 33487-2742
and by CRC Press
2 Park Square, Milton Park, Abingdon, Oxon, OX14 4RN

First issued in paperback 2021

ISBN: 978-1-138-33982-8 (hbk)
ISBN: 978-1-032-24875-2 (pbk)
ISBN: 978-0-429-44092-2 (ebk)

Typeset in Minion
by Deanta Global Publishing Services, Chennai, India

Contents

3. Trees 27

4. Shrubs 57

5. Forbs 75

6. Grasses 101

Part Two Application of Medicinal Plants for Specific Diseases

Authors

ANSELM ADODO

Anselm Adodo is the founder and Director of one of West Africa's largest phytomedicine companies, Pax Herbal Clinic and Research Laboratories, popularly called Paxherbals. He is a prominent advocate of African herbal medicine research, indigenous knowledge systems, rural community development, Africanised economic models, health policy reform and transformation of education in Africa. His doctoral degrees are in Medical Sociology and Anthropology and in Management of Technology and Innovation. Apart from over 500 publications in journals, magazines, national dailies and peer-reviewed journals, Anselm has written more than ten books, which include the best-selling 'Nature Power. Natural Medicine in Tropical Africa'. His 2017 book: 'Integral Community Enterprise in Africa. Communitalism as an Alternative to Capitalism', has received a lot of positive reviews for its indigenised approach to African development. Anselm is an adjunct visiting lecturer at the Institute of African Studies, University of Ibadan, Nigeria, where he teaches African Transformation Studies and Traditional African Medicine. He is an Adjunct Fellow of the Nigerian Institute of Medical Research, Nigeria, a Fellow of the Nigerian Society of Botanists, and a PhD academic supervisor at DaVinci Institute, South Africa. Anselm is the Director of the Africa Centre for Integral Research and Development, Co-founder of Trans4m Communiversity Associates, UK Ltd and the initiator of African Action Research Community.

MAURICE M. IWU

Maurice M. Iwu is the President of Bioresources Development Group (BDG), an independent biosciences research and development organization that cultivates, processes and produces medicines and cosmetics from natural products. He is Chairman of the Boards of Intercedd Health Products Ltd, the Bioresources Institute of Nigeria (BION), and Nature's Emporium healthy living outlets; and a member of the Board of Directors of Neimeth International Pharmaceuticals Plc. He was a Professor of Pharmacognosy at the University of Nigeria Nsukka, and a Senior Research Associate at the Division of Experimental Therapeutics of Walter Reed Army Institute of Research, Washington D.C. He was the Vice President, Research and Development at Tom's of Maine (a personal care manufacturing company in Maine, USA). He attended the School of Pharmacy, University of Bradford, Bradford, UK, where he obtained a Master of Pharmacy Degree and a Ph.D. in Pharmacognosy. He attended the Leadership and Strategy in Pharmaceuticals and Biotech Course (Course 10) at the Harvard Business School, Boston, USA. He was awarded the Doctor of Letters degree (Honoraris Causa) of Imo State University (2009). He is the Chairman of the Herbal Pharmacopeia Review Committee of the Federal Ministry of Health. He is a member of the African Union's African Regional Organization for Standardization (ARSO) Technical Committee on Traditional Medicine and the Chairman of its National Mirror Committee in Nigeria. Prof. Iwu has received many academic and professional honours, among which are the World Health Organization (WHO) Visiting Scholar to the Dyson Perrins Laboratory, University of Oxford (1980), Fulbright Senior Scholar Award (Ohio State University, Columbus Ohio and the Department of Chemistry, Columbia University, New York (1983); Senior Research Scholar Award, U.S. National Research Council, Washington D.C. (1993–1995) and the Richard Schultz International Prize for Ethnobiology (1999). Prof. Iwu has presented over 300 scientific papers, published more than 200 research articles and is the author of four books, including the highly cited *Handbook of African Medicinal Plants* (Taylor and Francis, First Edition 1994; Second Edition 2014) and *Food as Medicine – Functional Food Plants of Africa* (Taylor and Francis, 2016).

Introduction

This book aims to serve as a workbook for students, teachers, and practitioners in the field of ethnobotany and ethnomedicine. It documents the plants that are traditionally used by the local population, the history of their local use, and the traditional beliefs around their use in Nigeria. At a time when so much attention is being given to phytochemical screening of plants, there is a temptation to overlook the philosophy of ethnomedicine and cultural use of plants, thereby losing the link between plants and the community. This research adopts a community-oriented approach to African herbal medicine research and argues for a return to a community-based approach to medicine, wherein the health of the individual is closely aligned with that of the community. Community in this context includes plants, animals and the environment.

There are two approaches to the practice of herbal medicine: a clinic-oriented approach and a community-oriented approach. In the clinic-oriented approach, emphasis is placed on scientific identification, conservation and use of medicinal plants. Laboratory research and screening are carried out to determine plants' chemical composition and biological activities. Great interest is shown in quality control of raw materials and finished products, and in the development of methods for large-scale production of herbal drugs, which are labelled and packaged in the same way as modern drugs. They are also distributed through channels similar to those of modern drugs, that is, through recognised health officials in hospitals, health centres or pharmaceutical supply chains. The government, private companies and non-governmental organisations invest relatively huge sums of money in promoting further research in herbal medicine. In the clinic-oriented approach, minimal interest is shown in the socio-cultural use of the plants (LeGrand and Wondergem, 1990).

In the community-oriented approach, the emphasis is on the crude and local production of herbs to apply simple but effective herbal remedies to common illnesses in the local community. Knowledge of the medicinal uses of herbs is spread among community members to promote self-reliance, and information is given freely on disease prevention and the origin of diseases. No interest is shown in mass production of drugs for transportation to other parts of the country or exportation to other countries. The cultural context of the plants used is taken into account, and local perceptions of health and healing often take precedence over modern diagnostic technology. Simple herbal recipes are used for the treatment of illnesses such as coughs, colds, catarrh, malaria, typhoid and ulcers.

The approaches outlined above are two extremes of the same reality. There is a need to harmonise these two extremes to complement one another, as there seems to be little cooperation between the people working on either side. Scientists, pharmacists and medical doctors who follow the clinic-oriented, exogenous approach tend to sneer at and look down on traditional health practitioners. Meanwhile, traditional healers closely guard their indigenous knowledge and refuse to reveal their formulae and production systems to the so-called professionals. This inevitably leads to herbal medicine practitioners being labelled as "secretive", "esoteric" or "unscientific" (Adodo, 2017).

There is a third approach, which this book endorses: a return to the indigenous way of thinking, though in modern guise, that transcends the dichotomies between man and woman, nature and humans, community and the individual. This is a return to a transdisciplinary, transcultural, transpersonal and transformational mode of knowledge creation – an evolution, so to speak, from the traditional academic confines of University to the much more open "Communiversity", a community of knowledge that emerges from and is fully grounded in nature and community. The uniqueness of this research lies in its open-minded approach to science as a product of the community in which it grows, and as something that should serve the needs of that community. This matters, because, if research is disconnected from the community, it leads to a "dis-eased" research outcome such as a botanist and a taxonomist who, while gaining a technical knowledge of trees and plants, have lost knowledge of the forest, or the astronaut who, while becoming an expert in the study of the moon and the stars, has lost knowledge of the sky.

A study of the history of medicine in Europe shows that medicine and culture are inseparable. In fact, some scholars (Payer, 1996; Lupton, 1994) have demonstrated that medicine is, in fact, as much a culture as it is a science. A study of the British National Health Service (NHS) shows that the NHS is not just about healthcare, but is an integral part of the British culture (Hunter, 2008). The health system in a particular country reflects the health beliefs, illness behaviour, worldview, cultural values and level of development of that country (Lupton, 1994). Medicine does not exist in a vacuum. Medicine exists in a particular context.

Healthcare, health policies and treatment decisions are all contextual issues (Hunter, 2008) and only make sense when studied in their particular context.

The practice of medicine differs from culture to culture (Freidson, 1970). In Britain, doctors do fewer tests and give less medicine and lower doses. Germans use six times the number of heart drugs as the French or English, although the three countries have similar rates of heart disease (Payer, 1996). France's medical community prefers to focus on the "terrain", or the vitality of the body, and chooses gentle therapies, such as natural medicines, nutritional supplements, and the rest, to strengthen and restore the terrain. When they do use pharmaceuticals, they prescribe lower dosages and shorter courses (five days instead of ten). This orientation toward shoring up the terrain has put French scientists and physicians in the lead in fields such as immunotherapy for cancer and AIDS.

In the United States of America, the emphasis is on aggressive treatments using high-tech equipment, with a special focus on surgical interventions. The American attitude is to view disease as a hostile invasion by foreign bodies, whether viruses or bacteria, and then declare war on the "invader". American doctors perform more diagnostic tests than doctors in France, Germany or Britain. They prefer invasive surgery to drug treatment, and, when they use drugs, they are likely to use higher doses and more aggressive drugs. An American woman has two to three times the chance of having a hysterectomy (a surgical procedure to remove a woman's womb) as her counterpart in England, France or Germany. Over 60 percent of hysterectomies in the US are performed on women aged under 44 years of age. Many American doctors consider it routine to perform hysterectomies for women around the age of forty, especially if there is any slight indicator of cancer. In Nigeria, most women, even of menopausal age, would consider a hysterectomy to be a very last resort, while some wouldn't even consider it as an option at all, even under pain of death. Tampering with the womb for any reason is considered a taboo by many Nigerian women. A simplistic explanation would be to regard this perspective as religious naivety. But the matter is far too complex to be explained away by religion alone.

Medical practice is also influenced by the way doctors are remunerated in various countries. Since prices of medical procedures are fixed in advance, the only way a French doctor, for example, can increase his income is by performing more services. Therefore, if a French doctor wants to double his income, he takes out as many appendixes as possible, whereas, if an American doctor wants to double his income, he doubles his fees (Lupton, 1994). American doctors, whose insurance companies would require them to perform a caesarean section after fibroid removal, are more likely to remove the uterus than French doctors, who face no such pressure. In Nigeria, the majority of doctors who work in government hospitals also have their own private clinics. This practice is outlawed in Europe and America as well as in Nigeria. The difference is that, whereas this law is dutifully enforced in other countries, in Nigeria it exists only on paper. This has encouraged doctors to refer many of their patients to their own private clinics. The majority of non-complicated surgeries in Nigeria are therefore done in private clinics owned by government-employed doctors.

Different cultures place particular importance on different organs. The French place more attention on the liver, with its ability to process food and to regulate the body. The Germans focus on the heart. A beautiful woman in French culture has a slim body, is tall, and has straight, smooth legs, clear skin and medium-sized breasts. A beautiful American woman is expected to have large breasts. For the German, the heart is all that matters. In eastern and western Nigeria, being moderately fat is considered beautiful and attractive in a woman. Like Nigerians, the French value fertility and will do everything possible to preserve it. This explains why French doctors would use a hysterosalpingogram (an infertility test that shows whether both fallopian tubes are open and whether the shape of the uterine cavity is normal), because they are afraid of adhesions from surgery that might impair fertility, rather than the D&C (dilation and curettage) procedure that German, British and American doctors use to treat conditions. As mentioned above, American doctors carry out excessive numbers of hysterectomies that would be considered unwarranted in Britain, France and Germany. Where French and German doctors would do lumpectomies (surgical excision of a tumour from the breast with the removal of a minimal amount of surrounding tissue), American doctors are likely to opt for radical mastectomies (surgical removal of one or both breasts).

It is not true that people choose traditional/alternative medicine because they are too poor to afford pharmaceutical medicine (Dean, 2005). Some 39 percent of French doctors and 20 percent of German doctors prescribe alternative medicines. More than 40 percent of British general practitioners refer patients to alternative medicine practitioners, especially homoeopathic doctors, and 45 percent of Dutch physicians consider alternative remedies effective. Recognizing that over 50 percent of its population is using homoeopathy, herbal medicines and other natural healing therapies, Canada established the Office of Natural Health Products in March 1999. The right to choose one's preferred health system is as fundamental as the right to free speech, and must be respected by responsible governments.

Some researchers have observed that the great improvement in health and wellbeing in the developed countries of Europe and America was not as a result of better medication, more hospitals or medical breakthroughs (Nijhawan, 2010; Mendelsohn, 1979; Dean, 2005). Rather, it was a result of disease-prevention habits, such as improved hygiene, good drainage systems and good dietary habits. Evidence from surveys has shown that many prevalent diseases can be prevented and managed through proper health education, healthy dietary habits and physical exercise (World Bank, 2010). What Africa needs is not necessarily more hospitals, more sophisticated diagnostic equipment or more aggressive drugs. Rather, Africa needs more health educationists, more nutritionists, more health counsellors, and more preventive health care specialists. Indeed, prevention is better than cure.

It is no accident that the golden age of modern medicine and drug discovery, from 1900 to 1950, was the period when science was not yet disconnected from nature, when scientists tilled the soil, refined their gifts for the observation of nature and the human body, and were in touch with the natural environment. After many years of wandering in the wilderness of genetic engineering, genetic screening and gene therapy, costing billions of dollars in research with little to show for it, modern science is now looking back toward the African bush. Could it be that the next blockbuster drug is lying somewhere here?

Traditional healers in Africa have become an endangered species. Many have died and more continue to die without passing on their knowledge and expertise to their children or apprentices. The children often show no interest in taking over the healing work from their parents, but rather have migrated to the cities in search of white-collar jobs. Some became fiery evangelists, pastors and Christian fanatics who wage "war" on traditional medicine, which they regard as "pagan", "idolatrous" and "fetishist". As a result, much useful knowledge on healing plants and medicines is being lost, knowledge that appears worthless because it is not properly valued. This book is a contribution to efforts to preserve such indigenous knowledge, especially traditional medical knowledge, in Africa. It is the fruit of over 15 years' research and fieldwork, documenting the healing plants traditionally used for medicine in different parts of Nigeria, especially the southern part of Nigeria, and specifically Edo State. In the belief system of the traditional healers, every plant has a reason for existing. Plants grow for a particular purpose. Every plant is a manifestation of the energy field which is the universe, a mirror reflecting the intensity and nature of the energy field of the particular environment where they grow. Some plants exist to give nourishment to the earth. Some exist to give support to other plants. Some grow to help regulate the exchange of oxygen and carbon dioxide between human beings and plants. Some give information about past, current and coming events: the coming of rain or a drought, or an epidemic. The types of plants growing in a particular place often reflect the needs or problems in that place. Shortly before any epidemic or disease, the plant which has the antidote begins to sprout. A plant growing in a house of happy and fulfilled people is radiant, attractive and healthy, even when little attention is given to it. On arrival in a new place, experienced herbalists and mystics can gauge the prevalent sicknesses and mood simply by observing the kinds of plants growing nearby. For every disease or deficiency, there is always a medicinal plant growing nearby. It is up to human beings to open nature's book of Wisdom and learn how to use these plants. There is a very close link between plants and animals, so close that

several of the plants we know derive their names from this relationship. We have catnip, horsetail, goat weed, bird's eye, cat eye and pigweed, among many others. There are over 400,000 species of plants on this planet. Human beings in different continents use about 100,000 of these plants for medicinal purposes. Of these, only 10,000 are said to have been clinically analyzed and thus recommended for human consumption.

In traditional African communities, plants are believed to speak many languages. Do plants speak? The answer is yes. Plants speak. Animals speak. Human beings speak. Nature speaks. Intercommunication and interconnectivity is the nature of the universe. But what language do plants speak? Such a question reflects how limited we are in our understanding of reality. We often mistake speech for communication. But speech, the vocal sound we produce, is only one of the means of communication, and it is the most limited. In fact, only 30 per cent of our communication is done through speech. The rest is through gestures, facial expressions, and other forms of body language. Failure to always remember this has led to so much misunderstanding in our world. Plants are very sensitive to sound and can pick up sound vibrations long before they reach the human ear. The adult human ear can hear any sound that vibrates within the frequency of 20 Hz to 20,000 Hz, babies can detect sounds as high as 50,000 Hz, while dogs and other animals can hear sounds with a frequency as high as 100,000 Hz. Plants can detect sounds at frequencies of over 100,000 Hz .

There is so much in nature that is outside the range of our human perception and experience. The fact that we do not perceive something does not mean that it does not exist. It only shows how limited we are, and thus should make us humble. In the ancient Yoruba culture, the use of sound to create things was highly developed. Yoruba hunters, warriors and herbalists were believed to have a deep knowledge of sound mysteries. Wars were fought not with guns or knives, but with potent speech. The warring groups stood face to face and engaged each other in a war of chants. The side that was able to utter sounds that evoked higher vibratory frequencies gained control over the other side, who would become weak and helpless and were then captured. It was believed that through the manipulation of sound, trees could be transformed into animals, water into blood, bad into good.

In African societies, the use of creative sound to affect matter came in the form of the pronunciation of names. By uttering the name of a thing, one evokes a response in it. A number of plants and animals are particularly sensitive to sound, such as plants of the *Jatropha* species. Yoruba herbal mystics call *Jatropha* different names depending on what they want to use it for. When they want to use it to attend to a predominantly physical problem, they call it *Lapalapa*. When it is to treat a predominantly spiritual problem, they call it *Iyalode*, and when it is for both physical and spiritual treatment, they call it *Botuje*. By calling it a particular name, they control the vibratory radiation of the plant and condition it to work in the desired way (Adodo, 2003).

Another sound-sensitive plant is *Hypoestes verticillaris*, called Ogbo in Yoruba. Ogbo is used by African mystics and herbalists to cure insomnia, psychosis and cancer, and as an effective help for women in difficult labour. Ogbo is very effective in dissolving fibroid tumours. In fact, the use of Ogbo roots and leaves for treating fibroids is one of the hidden treasures of traditional herbal medicine and is in urgent need of further rigorous scientific investigation.

In traditional African medicine, the shape and peculiarity of a plant is believed to be a reliable source of information about the nature of the plant. In many African societies, and indeed in other parts of the world, people look at the colour and shape as well as the location of a plant to get an insight into its use and importance. This is called the "Doctrine of Signatures" and is based on the belief that plants grow in a specific area because there is a need for them. Herbs that grow on mountains are believed to be good for the respiratory system, including lungs, bronchi, and nostrils, and the nervous system. They are said to correct high blood pressure as well as pneumonia. Herbs that grow in water are regarded as being very medicinal and are almost always edible, since poisonous herbs very rarely grow in water. They are believed to be good for the circulatory system and to help in repairing the liver and kidneys. Herbs that grow in water are also believed to be good for treating all forms of infertility in both men and women. Herbs that grow close to the soil are believed to be good for digestive and circulatory problems. Since they are close to the ground, the mineral content is high, and so they are thought to be good for the bones and blood. Those who suffer from anaemia would find these herbs useful (Engel, 2002). Of course, the doctrine of signatures does not hold true in all cases, but that is not a reason to condemn the whole theory.

From time to time, traditional healers prescribe that certain herbs be harvested only at a particular time of the day. Sometimes they insist that certain plants should be harvested before sunrise, some after sunset. At times they go out themselves in the middle of the night to collect particular herbs. This practice in itself is scientifically correct. During the night, the chemical compounds in many plants, especially trees, settle down to the roots. Therefore, in order to get the best out of these roots, it is advisable to harvest them at sunset, which in tropical countries begins from 5pm. When the sun begins to rise, plants (especially trees) draw up their chemical compounds and distribute them to the leaves and bark. By the middle of the day, these compounds are fully concentrated in the leaves.

For this reason, the best time to harvest the leaves of medicinal plants is at midday, when the sun is about to reach its peak. By late afternoon, the chemical compounds in trees begin their downward journey back to the roots through the stems and barks. For this reason, the best time to collect the bark of a tree is late in the afternoon or in the early evening, before sunset. As a consequence, a traditional healer may explain to a client that harvesting some herbs at 5 pm rather than 10 pm, or at 11 pm instead of 12 midnight will affect the potency of the herbs. The background of our scientific knowledge tells us that this extra detail or explanation is untrue or unnecessary; however, the basic idea of harvesting the root at sunset is well founded. A traditional healer may want to embellish this common-sense science with riddles and rituals simply to increase the mystery surrounding his practice and to preserve his prestige.

The challenge for today's African thinkers, scientists and philosophers is to sift out the fetish and superstitions from our inherited deposits of knowledge without throwing away the truth. This book sets an example of how this could be done, and explains why it should be done.

References

Adodo, A. (2003) *The healing radiance of the soul*. Lagos: Agelex.

Adodo, A. (2017) *Integral community enterprise in Africa: communitalism as an alternative to capitalism*. London: Routledge. Survey. Benin City: University of Benin, Dept. of Sociology & Anthropology Research Series.

Dean, C. (2005) *Death by modern medicine*. New York: Matrix Verite.

Engel, C. (2002) *Wild health: how animals keep themselves well and what we can learn from them*. Boston: Houghton Mifflin Harcourt.

Freidson, E. (1970) *Professional dominance: the social structure of medical care*. New York: Atherton Press.

Hunter, D. (2008) *The health debate*. Bristol, UK: The Policy Press.

Le Grand, A. and Wondergem, P. (1990). *Herbal medicine and health promotion*. Amsterdam: Royal Tropical Institute.

Lupton, D. (1994) *Medicine as culture*. London: Sage Publications.

Mendelsohn, R. (1979) *Confessions of a medical heretic*. New York: McGraw Hill.
Nijhawan, N. (2010) *Modern medicine is killing you*. Washington, DC: L.E.O Publishing.
Payer, L. (1996) *Medicine and culture*. New York: Henry Holt.
World Bank working paper No. 187. (2010) *Improving primary health care delivery in Nigeria. Evidence from four states*. Washington, DC: AHD Series.

Part One

1 The Practice of Medicine in Africa

1.1 Introduction

Africa is home to diverse and innumerable tribes, ethnic and social groups. Some represent huge populations, consisting of millions of people, while others are smaller groups of a few thousand. Some countries have over 100 different ethnic groups (for example Nigeria, with more than 250 ethnic groups), with different culturally based values and belief systems. In all African societies during the early epoch, some 200,000 year ago, the individual at every stage of life had a series of duties and obligations to others in society, as well as a set of rights, namely things that he or she could expect or demand from other individuals. Age and sex were the most important factors determining the extent of a person's rights and obligations. The oldest members of the society were highly respected and usually in positions of authority, and the idea of seniority through age was reflected in the presence of age-graded groups in a great many African societies.

Plants have formed the basis of sophisticated traditional medicine practices that have been used for thousands of years by people in China, India, Nigeria, and many other countries (Farnsworth and Soejarto, 1991). Perhaps as early as Neanderthal man, plants were believed to have healing power. The earliest records of civilization, from the ancient cultures of Africa, China, Egypt, and the Indus Valley reveal evidence in support of the use of herbal medicine (Baqar, 2001). These include the Atharva Veda, which is the basis for Ayurvedic medicine in India (dating back to 2000 years Before The Current Era, BCE), the clay tablets in Mesopotamia (1700 BCE), and the Eber Papyrus in Egypt (1550 BCE). Other famous literature sources on medicinal plants include "De materia medica", written by Dioscorides between CE 60 and 78, and Jing's classic of Chinese traditional medicine, the "Divine Farmer's Materia Medica" (written around 200 CE) (Joy *et al.*, 1998).

1.2 History and Practice of Traditional African Medicine

As with food technology, Africans had evolved their medical technology long before colonialism. According to Andah (1992), from very early times, Africans used plants as curatives and palliatives for various ailments. Successful treatments became formalised, sometimes with prescriptions outlining the correct methods of preparation and dosage. Andah (1992) stated that, in many cases, patients were cured of their physical or psychological ailments. African traditional medicine is ancient and is perhaps the most diverse of all medicinal systems. Africa is considered to be the cradle of humankind, with a rich biological and cultural diversity, and marked regional differences in healing practices. African traditional medicine, in its varied forms, is holistic, involving both the body and the mind. The healer typically diagnoses and treats the psychological basis of an illness before prescribing medicines to treat the symptoms.

Generally speaking, the methods applied by African traditional healers are similar across the continent, although the plants used and the therapeutic values attributed to them are dependent on various factors. Some of these factors are geographical, sociological, and economic, and transcend ethnic, national, and political boundaries. According to Onu (1999), indigenous medical technology in Africa has developed not only drugs and surgical skills for fighting ailments but is also founded on a rational and coherent body of knowledge, which can be used to train specialists in the treatment of various diseases and disorders. Traditional medical practice in Africa has developed since antiquity to the stage of setting bones, treating mental disorders and even conducting relatively complicated operations such as caesarean sections. A picture of indigenous medical practice in Africa is, therefore, a picture of specialists trained in the acquisition of an impressive wealth of knowledge around herbs and other materials of therapeutic value. Through inspiration, observation, and experiments involving trial and error, the medical value of the plant kingdom, minerals, and certain animals was gradually realised and exploited.

Plant-based medicine is so prominent in Africa that a distinct class of practitioners has emerged, with a thorough knowledge of the medicinal properties of plants and the pharmaceutical steps necessary to turn them into drugs. In addition to herbal materials, minerals such as clay, salt, stone, and many other substances have been used as raw ingredients in indigenous pharmaceutical practice. Ubrurhe (2003) drew attention to the lzon ethnic group of the Niger Delta in Nigeria, whose environment does not permit the growth of many herbs, and who, as an alternative, specialise in massage. According to Ubrurhe (2003), this therapeutic system has been employed for the treatment of ailments of the nervous, muscular, and osseous systems, as well as for treating gynaecological problems. The armamentarium of the masseur is the physical manipulation of the muscles, joints, and veins through the bare skin. In most cases, massage treatment may be applied to relax the muscles and veins, as well as to improve circulation of blood. Ubrurhe (2003) concluded that this therapeutic method has spread to the lzon's neighbours, the Urhobo, the lsoko, and the Itsekiri.

Writing on hydrotherapy, Ubrurhe (2003) contended that its curative value is realised by both the practitioner and those who have undergone such treatment, and more recently by scientists. By equalizing the circulation of the blood in all the systems of the body, hydrotherapy helps to increase muscular tone and nerve force, improving nutrition and digestion, and thereby increasing the activity of the respiratory glands. Hydrotherapy facilitates the elimination of broken-down tissue cells and poisonous matters. It involves the use of cold, hot, compressed and steamed vapour baths. Herbs which are added to cold or hot baths are used for the treatment of different diseases and ailments, including fever, headache, rheumatism, and general pains. The hot bath not only makes the skin capillaries relax but also increases the activity of the sweat glands. It has been observed that water increases the absorption of oxygen to about 75%, while about 85% of the carbon dioxide in the body is eliminated through the use and consumption of water. Ubrurhe (2003) also wrote extensively on the practice of "cupping" or blood-letting as a therapeutic method of abstracting impure blood, using abstraction cups or horns. He asserted that this is widely practised in Africa, especially in northern Nigeria, where it has been regarded as an effective treatment for rheumatism and morbid conditions of the blood. Traditional medical practitioners in Africa are adept at performing intricate surgical operations to remove bullets and poisonous arrows from wounded traditional fighters, extricating infected tissues and stitching the wound together, then applying calabash to promote healing. Traditional plants, with anaesthetic properties, are applied before the operation to remove or reduce pain.

Sofowora (1982) observed that, in traditional medicine, burns are treated with herbal preparations which produce a soothing effect. For example, in cases of superficial burns, ointments prepared from papaya juice are applied by Ayurveda practitioners to gradually remove dead tissue. When this process is complete, and the healthy granulation tissue appears, the burn is treated with a herbal medication specially prepared to promote healing. Sofowora (1982) further observed that bone-setting is a specialised area of traditional medicine. It is usually performed without the aid of X-rays; the experienced traditional bonesetter uses his hands and fingers to feel and assess the type and extent of damage to a broken bone. In the case of a broken leg, the patient is made to lie or sit down with the fractured leg lying flat. Herbal dressings are placed on the fracture before planks or sticks are tied around the leg using string or the stem of a climbing plant.

Many traditional medical practitioners have a diverse range of skills and abilities: they may also be psychotherapists and be proficient in faith healing (spiritual healing), apiutic occultism, male and female circumcision, making tribal marks, treating snakebites and whitlows, removing tuberculosis lymphadenitis in the neck, cutting the umbilical cord, piercing earlobes, removing the uvula (uvulectomy), extracting decayed teeth, trephination (drilling a hole in the skull; also known as trepanation), abdominal surgery, preventive medicine, and so on. As far back as 1884, Dr. R.W. Felkin observed a caesarean section among the Bunyoro people of Uganda. Imperato (1977) reported that the patient was first narcotised with herbal preparations. The bleeding vessels were cauterised with red-hot iron rods. Blood was then drained from the abdominal cavity before the uterus was cut open and the baby and placenta carefully removed. The incision in the uterus was then sutured using iron spikes and string made from tree bark. The wound was then covered with herbal pastes, a hot banana leaf and finally a cloth bandage.

According to traditional African healers' understanding, the healing process is holistic (Thorpe, 1993). This implies that the healer deals with the whole person and provides treatment for physical, psychological, spiritual, and social symptoms. Traditional healers do not separate the natural from the spiritual, or the physical from the supernatural. This means that they address health issues from two major perspectives, the physical and the spiritual.

1. **Physical Perspective:** The following are some of the treatment processes used when the sickness is deemed to have physical causes:

    a. ***Prescription of Herbs:*** The traditional healer may prescribe herbs, depending on the kind of symptoms the patient has presented with. These prescriptions come with specific instructions on how to prepare the herb, the dose, and the timeframe for use (Ayim-Aboagye, 1993).

    b. ***Application of Clay and Herbs:*** In some cases, the traditional healer prepares white clay with herbs for the patient to apply onto his or her body for a number of days; this is used particularly in the case of skin diseases. This concept is based on the spiritual belief that the human body is made from dust or the earth, so, in order to heal it, you return it to where it came from. Clay and particular herbs are also sometimes used in preventive rituals (see below), to prevent the spirits behind the illness from attacking the patient (White, 2015).

    c. ***Counselling:*** Sometimes, the patient is advised on how to live his or her life, especially regarding the kind of food they should or should not eat. This is particularly the case when a taboo has been violated. The patient is advised to demonstrate good behavior, should it be felt that the disease occurred as a result of poor behaviour (Sundermeier, 1998).

2. **Spiritual/Non-physical Perspective:** When the cause of the sickness is believed to be more than physical, or is thought to arise from a spiritual problem, the patient is encouraged to seek spiritual counselling and to carry out prayer and other rituals. Some of the healing processes used when the sickness is deemed to have spiritual causes are outlined below.

a. ***Spiritual Protection:*** If the traditional healer perceives the disease to have been caused by an attack by evil spirits, the sufferer may be protected by the use of a talisman, charm, body marks made using moto (spiritually prepared black powder), amulets, or a "spiritual bath" to drive the evil spirits away (see "spiritual cleansing", below). These are rites aimed at eliminating evils or dangers that are seen to have taken root in a family or community (Westerlund, 2006).

b. ***Sacrifices:*** Sacrifices are sometimes offered at the perceived request of the spirits, gods, and ancestors. Sometimes animals are slaughtered or buried alive (Olupona, 2004). Among the Ewes and some other tribes in northern Ghana, dogs or cats are sometimes buried alive at midnight to save the soul of a person at the point of death. It is believed that, because dogs and cats are domestic animals, if someone very close to them is about to die, the animal can offer its life so that the person can live. So, when a dog or cat dies mysteriously, it is often interpreted that they have saved the life of a human. Animal sacrifices are also carried out in some societies as part of spiritual cleansing (see below).

c. ***Spiritual Cleansing:*** In some cases, herbs are prepared for the patient to bathe with at particular times of day for a number of days. Sometimes an animal is slaughtered and the blood poured on the head and feet of the patient as a way of cleansing. This practice is also common among the Ewe communities in Ghana (Westerlund, 2006).

d. ***Appeasing the Gods:*** According to interviews with some traditional priests (diviners) in Kumasi in the Ashanti region of Ghana, if diseases are seen to have been caused by the invocation of a curse or a violation of taboos, the diviner appeases the ancestors, spirits, or gods. This is done, according to the severity of the case, either by sacrificing an animal or by pouring of a libation. In many cases, the affected person would be told to buy the ritual articles needed for the process, such as "spotless" animals (doves, cats, dogs, goats, or fowl), liquor, including schnapps or the traditional akpeteshie, calico (red, white or black), eggs, or cola nuts. Following the rituals, the articles used are sometimes left at a particular place to rot, thrown into a river, as required by the gods or spirits, or placed at a four-way junction, or on the outskirts of the community, depending on the purpose of the ritual.

e. ***Exorcism:*** This is the practice of expelling demons or evil spirits from people or places that are believed to be possessed. Exorcism is usually performed by a person with religious authority, such as a priest or shaman. It was common in ancient societies and was based on the practice of magic. In the ancient Babylonian civilisation, in what is now Iraq, special priests would destroy a clay or wax image of a demon in a ritual meant to destroy the actual demon. The ancient Egyptians and Greeks had similar rites, and many religions in various parts of the world continue the practice of exorcism (Encarta, 2009). For example, it is practised by the Ewes and some Akan tribes in Ghana, where exorcism is mostly carried out by singing, drumming, dancing, spraying powder into the air and onto the body of the possessed, or by striking their body several times with an animal tail until the spirit has been released. During the process, the possessed person rolls and struggles on the ground but becomes stable after the exorcism, usually with a deep sense of relief (White, 2015). Many traditional communities in Ghana are of the view that mental illness is mostly caused by evil spirits and requires exorcism. This approach is common practice, for example, in the Tigari shrines in Ghana.

1.2.1 Types of Healers in African Traditional Health Care

The traditional healer, as defined by the WHO, is a person who is recognised by the community in which he or she lives as competent to provide health care, using plant, animal, and mineral substances, and certain other methods, based on the social, cultural, and religious background, as well as on the knowledge, attributes, and beliefs that are prevalent in the community, regarding physical, mental, and social wellbeing and the causation of disease and disability. The different types of healers in traditional African society are outlined below:

a. ***Traditional Herbalists:*** Herbalists mainly use medicinal plants or parts of such plants, including the whole root, stem, leaves, stem bark, or root bark, flowers, fruits, and seeds. Sometimes they also use animal parts or entire small animals (such as snails, snakes, chameleons, tortoises, lizards, etc.), as well as inorganic residues (including alum, camphor, salt, etc.) and insects (bees, black ants, etc.). Herbal preparations may be offered in the form of (i) a powder which can be swallowed or taken with pap (cold or hot) or any drink; (ii) a powder which is rubbed into incisions made on part of the body with a sharp knife; (iii) a preparation soaked in water or local gin, and decanted as required before drinking (the materials could alternatively be boiled in water, then cooled and strained); (iv) a preparation pounded with native soap and used for bathing; such medicated soaps are commonly used for skin diseases; (v) a paste, pomade, or ointment in a medium of palm oil or shea butter; or (vi) a soup which is eaten by the patient. The herbalist cures mainly with plants which he gathers fresh. When seasonal plants have to be used, these are collected when available and preserved, usually by drying (Ekeopara and Ugoha, 2017). Unlike the bone-setter, the traditional psychiatrist, and the traditional birth attendant, whose duties are well defined and specialised (see below), the herbalist is the general practitioner of traditional medicine. He is expected to be knowledgeable in the various aspects of healing and in the functioning of all the organs of the body. By his wealth of experience and knowledge, he is expected to determine the nature of the patient's illness, treat him and also predict the course of his treatment. In a typical traditional setting, the herbalist combines the role of the present-day doctor with that of the pharmacist and the nurse (Ekeopara and Ugoha, 2017).

b. ***Traditional Birth Attendants:*** The WHO defines a traditional birth attendant (TBA) as a person who assists a mother in childbirth and who initially acquired her skills alone or by working with other birth attendants. TBAs are usually older, experienced women, who primarily see their role as contributing their skills for the good of the community. In northern Nigeria, TBAs are all women, whereas, in some other parts of the country, both males and females take on this role. TBAs occupy a prominent position in Nigeria today, particularly in rural areas; 60–85 percent of births in the country are assisted by TBAs (Ekeopara and Ugoha, 2017). TBAs can diagnose and confirm pregnancy, and determine the position of the growing foetus. They provide prenatal and postnatal care, thereby also successfully carrying out the duties of the modern-day midwife.

   As a result of their wide exposure and experience, many TBAs have been trained to assist in orthodox medical practices at the primary health care level, leading to a reduction in maternal and child mortality and morbidity (Ekeopara and Ugoha, 2017). Highly experienced TBAs have also been known to assist in obstetric and paediatric care, managing simple maternal and childhood illnesses. With experienced TBAs, delivery by caesarean section is uncommon, as it is not usually necessary to seek surgical help.

c. ***Traditional Surgeons:*** The various forms of surgery recognised in traditional medical care include: (i) the cutting of tribal marks – traditional surgeons cut tribal marks into cheeks, bellies, etc. and rub charred herbal products into the wound to encourage the formation of scar tissue; (ii) male and female (clitoridectomy) circumcision – traditional surgeons carry out these simple surgical operations with special knives and scissors, and treat resulting bleeding and wounds with snail body fluid or pastes prepared from plants (these practices are, however, fast dying out in urban areas); and (iii) removal of whitlows – diseased toes or fingers are usually cut open and treated. Other forms of surgery include piercing of earlobes, particularly in young people, to allow the fixing of earrings, and extracting infected teeth before treating gums with herbal medicines prepared in local gin (Ekeopara and Ugoha, 2017).

d. ***Traditional Medicinal Ingredient Dealers:*** These dealers, usually women, are involved in buying and selling the plants, animals, insects, and minerals used in herbal preparations. Some dealers are also involved in preparing herbal concoctions or decoctions (herbal preparations created by boiling herbs in liquid, usually water) for managing or curing febrile conditions in children, or to treat other diseases in women and children; as such, these dealers may qualify to be referred to as traditional healers (Ekeopara and Ugoha, 2017).

e. ***Traditional Psychiatrists:*** The traditional psychiatrist specialises in the treatment of mental disorders. People suffering violent forms of psychosis are restrained using iron chains or wooden shackles. Those who are regarded as being possessed by demons may be called or beaten into submission, and then be given herbal hypnotics or highly sedative herbal potions to bring them to a state of mental, emotional, and psychological calm. The treatment and rehabilitation of people with mental disorders usually takes place over a long period.

f. ***Practitioners of Therapeutic Spiritism:*** These include diviners or fortune tellers, who may be called "seers", "alfas" or "priests". They may use supernatural or mysterious forces; use prayers, citing and singing of incantations, invocations, or rituals associated with the community's religious worship; or prepare sacrificial materials to appease unknown gods. Practitioners are usually consulted on the diagnosis of diseases, their causes, and treatment. With their perceived ability to deal with the unseen and the supernatural, they are usually held in high esteem in the community. They are believed to have extra-sensory perception, to be able to receive telepathic messages and consult oracles, spirit guides etc., and to perform well where other traditional healers and orthodox doctors fail.

It is believed that diseases which are caused by supernatural forces can be readily diagnosed and treated by these practitioners, and that certain medical ingredients have spiritual powers and can be effectively utilised by these practitioners for the good of all. These include ingredients from unusually large trees that are believed to house spirits; plants commonly found in graveyards, like the physic nut; protective plants, such as the wild colocynth or Sodom vine; or even some reproductive herbs, like the sausage tree. Some practitioners read and interpret the sounds made by "magic stones" when they are thrown to the ground. Some read messages in a pool or a glass of water. Others use the throwing of cowries, coins, kola nut seeds, divining rods, keys, or sticks, etc. (Ekeopara and Ugoha, 2017). Diviners seek input from the spiritual world to understand the cause of the illness and prescribe a cure (Asamoah-Gyadu, 2014; Cheetham and Griffiths, 1982). They are understood to play an intermediary role between the spirit world and the physical world; thus, in some cultures they are called "the eyes of the spirits". According to Sundermeier (1998), diviners are believed to be the custodians of theories of healing, and the hope of society. They learn how to cause, cure, and prevent disease, misfortune, infertility, poor crop yields, magic and witchcraft.

1.3 The Rise of Allopathic Medicine in Nigeria

Allopathic medicine, which is also referred to as "orthodox", "Western" or "conventional" medicine, is defined as medicine based on scientific methods and taught in Western medical schools (CSRC, 2005). Recorded European entry into Nigeria began when Portuguese explorers traded with the Benin Empire in 1472; traces of the influence of the Roman Catholic Church also date back to this time, and images of Portuguese soldiers abound in Benin bronzes. The Portuguese were credited with being the first people to bring Western medical care to their traders in outposts, but not to the indigenous African population (Schram, 1966). As the trade in human cargo expanded and accelerated, the high rate of infection with locally endemic diseases to which the previously unexposed European slave traders were subjected – notably malaria, yellow fever and the ubiquitous dysenteric diarrhoea – compelled the proprietors of the slave trade to introduce limited medical facilities for their staff. Throughout this period, healthcare facilities were available only on board the slave ships.

The first practicing doctors of allopathic medicine in Nigeria were the medical missionaries, ship surgeons, medically qualified botanists, and explorers, who sailed into many ports and navigated several large rivers from the 17th century onwards. No hospitals were built on the mainland until the latter part of 19th century, but there were hospitals on offshore islands three centuries earlier. In the mid-19th century, Dr. Williams, a Briton, was credited with carrying out several vaccination sessions and wound-dressings for indigenous populations along the West coast of Africa, including the Niger Delta and up to Lokoja (Schram, 1966). However, orthodox medicine was not formally introduced to Nigeria until the 1860s. The first group of Roman Catholic nuns in Nigeria lived in a convent in Lagos; later, led by Sister Maria of The Assumption, they moved to Abeokuta and worked under Father Francois, founder of the first fully-fledged mainland hospital, the famous Sacred Heart Hospital, in Abeokuta (HERFON, 2006). This was followed by the British colonial government providing formal medical services, with the construction of hospitals and clinics, in Lagos, Calabar, and other coastal trading centres. Following this, a makeshift temporary civil hospital was built in Asaba (now in Delta State) in 1888. A government hospital was also built in Calabar in 1898, as a result of the wide impact of the first hospitals on the indigenous population and the colonial personnel and their families (HERFON, 2006). The role of the Christian missionaries in providing medical and health care services cannot be over-emphasised. Reverend Hope Waddell, an Irish missionary of the United Presbyterian Church of Scotland (UPCS), worked for 30 years in the Calabar area, where he conducted Nigeria's first vaccination programme against smallpox; many of his missionary colleagues acquired skills and training that enabled them to run clinics and dispensaries in and around Calabar (HERFON, 2006).

Henry Townsend and David Hinderer founded the Church Missionary Society (CMS) (Yoruba Mission) and oversaw the mission's activities in Lagos and Abeokuta. In 1864, the Church of England consecrated Reverend Samuel Ajayi Crowther as the Bishop of Western Equatorial Africa. He undertook a mission up the Niger as far as Lokoja, and from there to Calabar by the way of the Cross

River and into South Cameroon. The establishment of various healthcare posts followed in the wake of Crowther's Episcopal missions. The Qua Iboe Mission, founded in 1891, established a number of dispensary and maternity services in southern Nigeria, as did the Baptist Mission. One of the most outstanding legacies of the Baptist Mission is the famous Baptist Hospital in Ogbomosho. A coalition of the Protestant missions built the reputed Iyi Enu Hospital near Onitsha in 1906. The Sudan Interior Mission (SIM), founded in 1893, worked in the core of Nigeria and the predominantly Islamic north. It operated in the early days with two medical stations, in Bida and Pategi (HERFON, 2006).

The religious missions also contributed substantially to the training of nurses and paramedical personnel. A good example is the highly reputed nursing school of the SIM Christian Hospital in Vom. The mission hospitals in Shaki, Ogbomosho, Ilesa, and Eku, among others, performed similarly important health-training roles. The missions also sponsored many of the first-generation Nigerian doctors to undertake professional training in Europe (HERFON, 2006). During the First World War of 1914–1918, substantial numbers of European health care personnel were withdrawn from Nigeria to render professional services for war victims. The Army Medical Corps (AMC) was set up by Lord Lugard in Lokoja, and was the forerunner of government medical services in Nigeria. It was a centralised medical service, which was initially military and later colonial in nature. In 1943, British colonials opened the first orthopaedic centre in Igbobi, Lagos, as a rehabilitation camp for wounded soldiers returning to Nigeria from the Second World War. The Yaba Medical School (YMS) was founded in 1930 to train a cadre of medical assistants, but ceased to exist after the establishment of University College Hospital, Ibadan. The Kano Medical School was inaugurated in 1954. From the 1500s, medical knowledge passed with enslaved Africans to the Americas for many centuries. Indeed, smallpox inoculation was introduced to North America by an enslaved African medical pioneer called Onesimus.

Allopathic medicine and African traditional medicine: a clash of world views. The impact of colonialism in Africa is characterised in ways that range from "fortune" to "agony". Some scholars (such as Curtin, 1989) are of the opinion that the process of modernisation in Africa is intrinsically connected with foreign intervention, particularly in areas of health and democracy. Curtin (1998) argues that the period between 1840 and 1860 was marked by significant and rapid innovation in tropical medicine, notably the discovery of quinine to stem the scourge of malaria in the most endemic region of the world. From this perspective, the institutionalisation of the modern health care system is seen as one of many positive "legacies" of Western encroachment in Africa.

On the contrary, there are those who believe that Western "invasion" was/is a setback in the process of development in Africa (Achebe, 1958; Afisi, 2009), particularly with respect to "modes of knowledge production" (Taiwo, 1993). These scholars mention slavery, capitalism, colonialism and imperialism, neo-colonialism, and all forms of domination and exploitation that were embedded in these epochs, as major stumbling blocks to indigenous African development. Indeed, the current political and socio-economic crises in Africa are often attributed to colonialism. While some critics of colonialism have focused on its economic and political impacts, others have shifted attention to the impact of colonialism on the indigenous knowledge system (Mapara, 2009), and especially knowledge of medicine (Feierman, 2002; Millar, 2004; Paul, 1977). Such arguments underscore the negative impact of colonialism on indigenous medicine. It is explained that the introduction of Western medicine and culture gave rise to a "cultural-ideological clash", that undermined and stigmatised the traditional health care system in Africa. In some extreme cases, traditional medicine was banned outright; for instance, the South African Medical Association outlawed the traditional medical system in South Africa in 1953 (Hassim *et al.*, n.d.). In addition, the Witchcraft Suppression Act of 1957 and the Witchcraft Suppression Amendment Act of 1970 declared traditional medicine unconstitutional, thereby prohibiting practitioners in South Africa (Hassim *et al.*, n.d.).

The banning of traditional medicine was partially based on the belief that the concept of disease and illness in Africa was historically embedded in "witchcraft", a perception which, in Western eyes, reinforces negative ideas of "backwardness" and "superstition", and of Africa as "the dark continent". However, as this book describes and as recent studies have shown, aetiologies of illnesses in Africa are viewed from both natural and supernatural perspectives (Bello, 2006; Erinosho, 1998, 2005, 2006; Jegede, 1996; Oke, 1995). The subjugation of traditional medicine continued in most African countries even after independence, although local efforts were initiated to challenge the condemnation and stigmatisation of traditional medicine in some African communities during and after colonialism. Erinosho (1998, 2006) reported that the first protest against the marginalisation of traditional medicine in Nigeria dates back to 1922, when a group of native healers insisted that their medicine be legally recognised.

Onu (1999) stressed that drugless therapy is the area of African medical practice which has been most misunderstood. He noted that, with the characteristic hypocrisy that goes with such misunderstanding, African traditional medicine has been regarded not only as ineffective, but as a primitive phenomenon, flourishing on ignorance and prelogical thinking, a misunderstanding that persists, even in enlightened circles. Onu (1999) concluded that it persists because of two unresolved issues in traditional medical practice: first, the controversy over the causes of diseases; and second, the fact that indigenous medical practice in Africa is yet to articulate meticulous models for explaining many sensitive aspects of its drugless therapy.

References

Achebe, C. (1958) *Things fall apart*. London:: Heinemann Books Ltd.

Afisi, O.T. (2009) Tracing contemporary Africa's conflict situation to colonialism: a breakdown of communication among natives. *Philosophical Papers and Reviews*, 1(4), 59–66.

Andah, B.W. (1992) *Nigeria's indigenous technology*. Ibadan: Ibadan University Press.

Asamoah-Gyadu, J.K. (2014) Therapeutic strategies in African religions: health, herbal medicines and indigenous Christian spirituality. *Studies in World Christianity*, 20(1), 83. doi:10.3366/swc.2014.0072

Ayim-Aboagye, D. (1993) *The function of myth in Akan healing experience: a psychological inquiry into two traditional Akan healing communities*. Ph.D. thesis, Department of Theology, Uppsala. Uppsala: University Press.

Baqar, S.R. (2001) Anti-spasmodic action of crude methanolic extract. *Journal of Medicinal Plants Research*, 6(3), 461–464.
Bello, R.A. (2006) Integrating the traditional and modern health care system in Nigeria: a policy option for better access to health care delivery. In: H. Saliu, A. Jimoh, and T. Arosanyin (eds.), *The national question and some selected topical issues on Nigeria*. Ibadan: Vantage Publishers.
Cheetham, R.W.S. and Griffiths, J.A. (1982) The traditional healer/diviner as psychotherapist. *South African Medical Journal*, 62, 957–958.
Curtin, P.D. (1989) *Death by migration: Europe's encounter with the tropical world in the eighteenth century*. London: Cambridge University Press.
Curtin, P.D. (1998) *Disease and empire: the health of European troops in the conquest of Africa*. London: Cambridge University Press.
Ekeopara, C.A. and Ugoha, A.M.I. (2017) The contributions of African traditional medicine to Nigeria's health care delivery system. *IOSR Journal of Humanities and Social Science*, 22(5), 32–43. e-ISSN: 2279–0837, p-ISSN: 2279–0845.
Encarta. (2009) *Exorcism*. Microsoft® Student [DVD], Redmond.
Erinosho, O.A. (1998) *Health sociology for universities, colleges and health related institutions*. Ibadan: Sam Bookman.
Erinosho, O.A. (2005) *Sociology for medical, nursing, and allied professions in Nigeria*. Abuja: Bulwark Consult.
Erinosho, O.A. (2006) *Health sociology for universities, colleges and health related institutions*. Ibadan/Abuja: Bulwark Consult. [Reprint].
Farnsworth, N.R. and Soejarto, D.D. (1991) Global importance of medical plants. In: O. Akerele, V. Heywood, and H. Synge (eds.), *Conservation of medical plants*, pp. 200–250. Cambridge: Cambridge University Press.
Feierman, S. (2002) *Traditional medicine in Africa: colonial transformations*. New York: New York Academy of Medicine. 13 March. Reported by Carter, G.M. Foundation for Integrative AIDS Research.
Hassim, A., Heywood, M. and Berger, J. (n.d.) Health and democracy. Accessed at http://www.alp.org.za on 12 January 2018.
HERFON Nigerian Health Review. (2006) Publication of Health Reform Foundation of Nigeria. Lagos: Government Press.
Imperato, P.J. (1977) *African folk medicine: practices and beliefs of the Bambara and other people*. New York: York Press.
Jegede, A.S. (1996) Social epidemiology. In: E.A. Oke and B.E. Owumi (eds.), *Readings in medical sociology*. Ibadan: Resource Development and Management Services (RDMS).
Joy, P.P., Thomas, J., Mathew, S. and Skaria, B.P. (1998) *Medical plants*. Kerala: Kerala Agricultural University Press.
Mapara, J. (2009) Indigenous knowledge systems in Zimbabwe: juxtaposing postcolonial theory. *The Journal of Pan African Studies*, 3(1), 139–155.
Oke, E.A. (1995) Traditional health services: an investigation of providers and the level and pattern of utilization among the Yoruba. *Ibadan Sociological Series*, 5(1), 2–5.
Olupona, J.K. (2004) Owner of the day and regulator of the universe: Ifa divination and healing among the Yoruba of South-Western Nigeria. In: M. Winkelman and P.M. Peeks (eds.), *Divination and healing: potent vision*, pp. 103–117. Tucson: University of Arizona Press.
Onu, A.O. (1999) Social basis of illness: a search for therapeutic meaning. In: A.I. Okpoko (ed.), *Africa's indigenous technology*. Ibadan: Wisdom Publishers Ltd.
Paul, J. (1977) Medicine and imperialism in Morocco. *MERIP Reports*, 60, 3–12.

Schram, R. (1966) *Development of Nigerian health services, 1940–1960.* MD thesis accepted by the Cambridge University, UK, under the title 'A brief history of public health in Nigeria', 1966. Published by the University of Ibadan Press in 1971 under the title, 'A history of the Nigerian health services'.

Sofowora, A. (1982) *Medicinal plants and traditional medicine in Africa.* Ibadan: Spectrum Books.

Sundermeier, T. (1998) *The individual and community in African traditional religions.* Hamburg: LIT Verlag.

Taiwo, O. (1993) Colonialism and its aftermath: the crisis of knowledge production. *Callaloo,* 16(4), 891–908.

Thorpe, S.A. (1993) *African traditional religions.* Pretoria: University of South Africa.

Ubrurhe, J.O. (2003) *Urhobo traditional medicine.* Ibadan: Spectrum Books Limited.

Westerlund, D. (2006) *African indigenous religions and disease causation.* Leiden: Brill.

White, P. (2015) The concept of diseases and healthcare in African traditional religion in Ghana. *HTS Teologiese Studies/Theological Studies,* 71(3), 2762. doi:10.4102/hts.v71i3.2762

2 Medicine, Culture and Health Belief Systems

2.1 The Concept and Meaning of Culture

Culture, as it is usually understood, is the way of life shared by a group of people that claim to share a single origin or descent. It entails a totality of traits and characteristics that are peculiar to a people, to the extent that it marks them out from other peoples or societies. These particular traits include the people's language, dress, music, work, arts, religion, dancing, and so on, as well as their social norms, taboos, and values. It embraces the way people walk and how they talk, the manner in which they treat death and greet the new-born. Edward B. Taylor is reputed to be the scholar who first defined culture, in his work "Primitive Culture" (1871, reprinted in 1958). Taylor saw culture as a complex whole, which includes knowledge, belief, art, morals, law, customs, or any other capabilities and habits acquired by man as a member of society.

Today, there are as many definitions of culture as there are scholars interested in the phenomenon. In an attempt to capture the all-embracing nature of culture, Bello (1991) described it as the totality of the way of life evolved by a people in their attempts to meet the challenge of living in their environment, which gives order and meaning to their social, political, economic, aesthetic, and religious norms, thus distinguishing a people from their neighbours. Indeed, the many definitions of culture all have this single underlying characteristic: the attempt to portray culture as the entire or total way of life of a particular group of people. It can also be said that culture is uniquely human and shared with other people in a society.

Culture is selective in what it absorbs or accepts from other peoples outside a particular cultural group. Culture is passed on from generation to generation, and the acquisition of culture is a result of the socialisation process. That culture is understood to be the way of life of a people presupposes the fact that there can be no people without a culture. To claim that there is a society without a culture would imply that this society has continued to survive without any form of social organisation or institutions, norms, beliefs, and taboos – which would be quite untrue. Some Western scholars, who may be tempted to use their own cultural categories to judge distinctively different people as "primitive", may claim that such people have no history, religion, or even philosophy, but cannot say that they have no culture.

Culture has been classified into material and non-material aspects. While material culture refers to the visible, tactile objects, which are manufactured for the purposes of human survival, non-material culture comprises the norms and social mores of the people, including taboos and beliefs about what is good and bad. While material culture is concrete and takes the form of artefacts and crafts, non-material culture is abstract but has a very pervasive influence on the lives of the people of a particular culture.

Culture is dynamic; it is continually changing. Antia (2005) states that culture is not fixed and permanent; it is always being changed and modified through contacts with and absorption of other peoples' cultures, in a process known as assimilation. As people change their social patterns and institutions, beliefs and values, and even skills and tools of work, then culture has to be an adaptive system. Each element of a culture is related to the whole system. Once an aspect of culture adjusts or shifts in response to changes from within or outside the environment, then other aspects of the culture are affected, whether directly or indirectly.

Some widely held assumptions even among the so-called educated African elite tend to project a narrow concept of culture as outdated practices of witchcraft, voodoo, fetishism and magic. This misconception, we believe, is not widespread but may have arisen from a limited understanding of the meaning of culture. As we shall see, culture generally, and African culture in particular, is like a two-sided coin: it can have negative aspects, but it also has many soul-lifting, life-enhancing and positive dimensions.

This book deals with African culture and draws examples from Nigerian culture. There are, of course, many cultures in Africa; the continent is inhabited by various ethnic nationalities with different languages, modes of dress, food, dancing, and even greeting habits. Obviously, virtually every locality has unique features. Yet, notwithstanding the cultural variations that make each local or regional manifestation of culture unique, the value systems and beliefs of the cultures of traditional African societies are close, and Africans share some dominant traits in their belief systems and have similar values that mark them out from other peoples of the world. Certainly, African cultures differ vastly from the cultures of other regions or continents, justifying our usage of the term "African culture". A Nigerian culture, for instance, would be closer to, say, a Ghanaian culture, with respect to certain cultural parameters, than it would be to the Oriental culture of the Eastern world or to the Western culture of Europe.

2.2 African Traditional Beliefs on Health and Causes of Disease

Traditional African belief systems are characterised by a strong faith in spiritual powers. The word "spiritual", as seen in aspects of the biopsychosocial approach(es) to health care and illness (Mbiti, 1970; Kiev, 1964), is based on beliefs which can be connected to the history and culture of Africa (Mbiti, 1970). The behaviour of Africans is motivated by what they believe, and what they believe is based on what they experience. Although Western medicine and healthcare systems have been introduced into Africa, many African countries still rely on traditional health care (Osei, 2004; Lambo, 1964).

By way of free will, humans make choices based on many factors, including their spiritual beliefs, preferences, knowledge, and perceptions (Anthony, 2003). In Africa, spiritual belief is a major determinant of choice of treatment (Osei, 2004). A belief system based on Creationism (i.e. the religious belief that the Almighty God created the universe and the first man, rather than the scientific conclusion that they were created by natural processes) is strongly held by indigenous Africans (Mbiti, 1970). The continued widespread existence of spiritual belief among Africans suggests its heritability, but, in fact, it is transferred from one generation to another. There are different kinds of beliefs – the belief in the Almighty God and belief in other spirits, for example (Mbiti, 1970; Appiah *et al.*, 2007; Omonzejele, 2008; Kiev, 1964). As this continues to run across generations, spiritual belief in Africa can be described as a form of 'behavioural genetics'.

In Africa, many herbalists prepare their herbs on the basis of spiritual belief. The elite African, who has knowledge about the scientific cause of illness, also incorporates spiritual factors when dealing with illnesses. The current Western model, the biopsychosocial model, considers health as including physical, mental, emotional, and social factors. For the African, however, wellbeing is not just about the healthy functioning of the body system through proper health care and lifestyle but goes beyond scientific causes to include spiritual involvement (Mbiti, 1970; Kiev, 1964).

The modification of the biopsychosocial model to include spiritual factors (McKee and Chappel, 1992) is therefore well suited to the African culture.

For the traditional African, good health consists of mental, physical, spiritual, and emotional stability, not only of oneself, but also of one's family members and community. This integrated view of health is based on the African unitary view of reality (Omonzejele, 2008). Good health is also often understood in terms of the relationship with one's ancestors. For many Africans, it is of paramount importance that ancestors stay healthy so that they can protect the living. Good health is also believed to be a result of appropriate behavior, that is living in accordance with the values and norms of the society. In view of the above, traditional medicine has, at its base, a deep-rooted belief in the interaction between people's spiritual and physical wellbeing (Setswe, 1999). It is imperative to emphasise the significance for health of this perception of the individual as a member of the collective community; as such, good health is dependent on good relations with the community. Mbiti (1990: pp. 108–109) notes that:

> "Only in terms of the other people does the individual become conscious of his own being ... When he suffers, he does not suffer alone but with the corporate group ... Whatever happens to the individual happens to the whole group, and whatever happens to the whole group happens to the individual. The individual can only say: I am because we are, and since we are, therefore I am."

Traditional African beliefs include several ways to explain or understand the causes of disease. The first is the view that disease is often caused by attacks from evil or bad spirits. Some also believe that, if the ancestors are not treated well, they may punish people with disease (Magesa, 1997; Westerlund, 2006). African traditional religion is thus based on maintaining the balance between the visible and invisible world. The maintaining of this balance and harmony is humanity's greatest ethical obligation and determines the quality of life (Magesa, 1997; Westerlund, 2006). Nyamiti (1984: p. 16) pointed out:

> "When ancestors are neglected or forgotten by their relatives they are said to be angry with them and to send them misfortunes as punishment. Their anger is usually appeased through prayers and ritual in the form of food and drinks."

Good behaviour, according to African traditional belief, includes holding and practising the values established by society and culture, participating in religious rituals and practices, and showing proper respect for family, neighbours and the community. Failure to follow these behavioural guidelines often results in good spirits withdrawing their blessing and protection, thereby opening doors to illness, death, drought or other misfortune.

Spell-casting and witchcraft are other reasons used to explain ill-health. There is a view that people with evil powers can punish their enemies or those who are disrespectful to them by causing them to become sick (Olupona, 2004). Furthermore, many traditional African communities believe that certain illnesses which defy scientific treatment can be transmitted through witchcraft and unforeseen forces; these include infertility, attacks by dangerous animals, snakebites, persistent headaches and repeated miscarriages (Obinna, 2012, Thorpe, 1993). In some Ghanaian communities, especially in Akan communities, it is believed that a person can become sick as a result of the invocation of curses in the name of the river deity, Antoa, which is seen as a means of seeking divine justice (White, 2015).

Many traditional healers believe that sickness results from disregarding taboos. Taboos form an important part of African traditional religion. They are behaviours, practices or lifestyles that are forbidden by a community or a group of people (Isiramen, 1998). Taboos encompass social or religious customs prohibiting or restricting a particular practice, or forbidding association with a particular person, place or thing (Westerlund, 2006). Magesa (1997) asserted that taboos exist to ensure that the moral structures of the universe remain undisturbed for the good of humanity. It is believed that, when a person violates any taboo, whether openly or in secret, the consequences are always manifested either in the person(s) concerned, or in the entire community, in the form of diseases, and possibly death. This is what Magesa (1997) termed the "effect of life force", arguing that moral behaviour maintains and enhances one's life force, but disobedience and behaviour disloyal towards traditions passed on by the ancestors will weaken the life force, leading to punishment from the ancestors or spirits in the form of disease and misfortune.

According to Wiredu (1980), among the Akan people of Ghana, morality is based on human welfare. African notions and application of moral precepts have far-reaching implications for how African traditional medicine is practiced. Adherence to moral precepts is an important and integral part of traditional health care in Africa and is subsumed into African ethics.

Patients' individual health beliefs can have a profound impact on their clinical care, as explored further below. They can impede preventive efforts, delay or complicate medical care, and result in the use of folk remedies that can be beneficial or toxic. Culturally based attitudes to seeking treatment and trusting traditional medicines and folk remedies are rooted in core belief systems regarding illness causation, e.g. naturalistic, Ayurvedic, biomedical, etc. Read (1966) observed that, in African systems, there are three groups of illness: trivial or everyday complaints treated by home remedies; "European disease" – that is, disease that responds to Western scientific therapy; and "African disease" – disease that is not likely to be understood or treated successfully by Western medicine. This observation, according to Oke and Owumi (1996), is true for many ethnic groups in Nigeria. Erinosho (1976) and Oke (1995), working among Yorubas, noted that illness aetiology could be traced to three basic factors: natural, supernatural and mystical.

Illness is not only a personal affair; it also arouses a wide variety of feelings in those close to the sufferer as they engage in a search for treatment, which becomes an immediate problem (Onu 1999). According to Onu, for the patient, a serious illness carries with it the underlying fear of death or permanent disability and constitutes a crisis which requires cooperative efforts from both family members and healthcare providers (physical or spiritual).

2.3 Influence of Cultural Beliefs on Health and Illness Behaviour

All cultures have systems of health beliefs to explain what causes illness, how it can be cured or treated, and who should be involved in the process (McLaughlin and Braun, 1998). Research has extended the description of illness beyond actual diseases to include how the sick person, and the members of the family or wider social network, perceive, live with, and respond to symptoms and disability (Kleinman, 1988). Kleinman argued that the illness experience includes categorising and explaining the forms of distress caused by those physiological processes. It therefore follows that any productive effort to improve the wellbeing of patients must include an understanding of their perception of illness and its symptoms.

When patients are diagnosed with an illness, they generally develop an organised pattern of beliefs about their condition. These views are key determinants of behaviour directed at managing illness, and they change in response to shifts in patients' perceptions and ideas about their illness. These illness perceptions or cognitive representations directly influence the individual's emotional response to the illness, as well as their coping behaviour, such as adherence to treatment. Yet, despite their importance, patients' views about their illness or symptoms are rarely sought in medical interviews, and patients tend not to discuss their illness beliefs with doctors. Understanding common perceptions in indigenous communities about the causes of ill-health may help policy makers to design effective integrated primary healthcare strategies to serve these communities.

Illness is commonly described as a condition of pronounced deviation from the normal healthy state. The term is often used to mean disease, but can also refer to a person's perception of their health, regardless of whether they have a disease or not. A person without any disease may feel unhealthy and simply have the perception of having a disease, while another person would experience the same condition as feeling healthy (Vaughn *et al.*, 2009). Johnson (2002) argued that illnesses cannot be investigated solely by the methods of biomedicine because its study ultimately depends directly on phenomenological analysis of experienced suffering through individual self-reports and behaviour, and therefore its presence cannot be objectively established by physical signs.

According to Petrie and Weinman (2012), patients' models of their illnesses share a common structure, made up of beliefs about the cause of the illness, the symptoms that are part of the condition, the consequences for the patient's life, how the illness is controlled or cured, and how long the illness will last. These beliefs are based either on a patient's own medical knowledge and experience, or on the experiences of friends or family members, who have had similar symptoms or diagnoses (Broadbent *et al.*, 2006). Patients with the same illness may have different perceptions of their condition and different emotional reactions to it. Interventions based around changing, inaccurate, or unhelpful perceptions of illness are an important emerging area of health psychology. It is important to note that patients' knowledge of medical concepts and the body is often rudimentary, limiting the accuracy and complexity of the models they build. Nevertheless, as stated above, a patient's perception of their illness can influence their coping ability, compliance with treatment, and functional recovery. Psychological interventions to address negative beliefs and perceptions may facilitate an earlier return to work.

People routinely experience symptoms that may signal illness, such as lethargy, depression, anorexia, sleepiness, hyperalgesia, and inability to concentrate.

Whereas symptoms are critical elements in a person's decision to seek medical attention, the presence of symptoms is not always sufficient to prompt a visit to doctor. Some people seek help for symptoms, whereas others do not. When people seek and perceive symptoms of illness, they often consult others, such as friends, neighbours or family members on whether or not to report the symptoms to a doctor. Some may be advised not do so until their condition worsens. Some deny the symptoms, while others resort to self-medication. Some go to "quacks" (bogus doctors) and may not be properly treated. Mechanic (1968) listed characteristics that determine perceptions of symptoms, as follows: (i) the visibility of the symptoms, that is, how readily apparent they are; (ii) perceived severity of the symptoms; symptom perceived as severe will be more likely to prompt action than will less severe ones; (iii) the extent to which symptoms interfere with personal life, the greater the extent, the more likely an individual will be prompted to seek medical care as quickly as possible; and (iv) the frequency and persistency of the symptoms; conditions that people view as requiring care tend to be those that are severe and continuous, whereas intermittent symptoms are less likely to generate illness behaviour.

Certain biomedical aetiologies maintain that symptoms are the manifestation of bodily malfunction, whereas, in non-orthodox healthcare systems, symptoms are believed to be manifestations of the intrusion of the supernatural (Chipfakacha, 1994). Most cultures support the belief that symptoms are the manifestation of illness, whether it is caused by a pathogen or a spirit invasion. Therefore, in order to effectively treat these illnesses, remedies must be both material (e.g. a herbal remedy) and spiritual (e.g. amulets) (Vaughn *et al.*, 2009). Furthermore, social support has been found to have a significant influence on subjective wellbeing (Kahn *et al.*, 2003). Other research findings also hold that social support seems to exert an influence on health, both directly and indirectly, through certain cognitive mechanisms, coping strategies and health behaviours (Cohen and Wills, 1985).

It can be concluded that, whatever a patient's cultural belief may be, how the patient perceives his/her illness will affect the patient's rate of recovery and response to treatment. It is therefore recommended that health practitioners focus attention on patients' perceptions of their illness symptoms, and provide social and psychological support in order to aid the patient's recovery and improve their response to treatment.

References

Anthony, D.J. (2003) *Psychotherapies in counselling*. Bangalore, India: Anugraha Publications.

Antia, O.R.U. (2005) *Akwa Ibom cultural heritage: its incursion by western culture and its renaissance*. Uyo: Abbny Publishers.

Appiah, P., Appiah, K.A. and Agyeman-Duah, I. (2007) *Proverbs of the Akans: Bu Me Bɛ*. Oxfordshire: Ayebia Clarke Publishing Ltd, p. 336.

Bello, S. (1991) *Culture and decision making in Nigeria*. Lagos: National Council for Arts and Culture.

Broadbent, E., Petrie, K.J., Main, J. and Weinmann, J. (2006) The brief illness perception questionnaire. *Journal of Psychosomatic Research*, 60, 631–637.

Chipfakacha, V. (1994) The role of culture in primary health care. *South African Medical Journal*, 84(12), 860–861.
Cohen, S. and Wills, T.A. (1985) Stress, social support and the buffering hypothesis. *Psychological Bulletin*, 98, 310–357.
Erinosho, O. (1976) *Evolution of modern psychiatric care in Nigeria*. Vol. 1, no. 3, pp. 64–70. Dakar: CODESRIA.
Isiramen, C. (1998) *Philosophy of religion, ethics and early church controversies*. Lago: AB Associate Publishers.
Johnson, R. (2002) The concept of illness behavior: a brief chronological account of four key discoveries. *Veterinary Immunology and Immunopathology*, 87(3–4), 443–450. doi:10.1016/S0165-2427(02)00069-7
Kahn, J.H., Hessling, R.M. and Russell, D.W. (2003) Social support, health, and well-being among the elderly: what is the role of negative affectivity? *Personality and Individual Differences*, 35, 5–17.
Kiev, A. (1964) *Magic, faith and health*. New York, NY: The Free Press.
Kleinman, A. (1988) *The illness narratives: suffering, healing, and the human condition*. New York: Basic Books.
Lambo, T.A. (1964) Patterns of psychiatric care in developing African countries. In: A. Kiev (ed.), *Magic, faith and health*. New York, NY: The Free Press.
Magesa, L. (1997) *African religion: the moral traditions of abundant life*. Maryknoll, NY: Orbis Books.
Mbiti, J.S. (1970) *Concepts of god in Africa*. London: The Camelot Press Ltd, p. 348.
Mbiti, J.S. (1990) *African religions and philosophy*. London: Heinemann.
McKee, D.D. and Chappel, J.N. (1992) Spirituality and medical practice. *The Journal of Family Practice*, 35, 205–208.
McLaughlin, L. and Braun, K. (1998) Asian and Pacific Islander cultural values: considerations for health care decision-making. *Health and Social Work*, 23(2), 116–126.
Mechanic, D. (1968) *Medical Sociology*. New York: Free Press.
Nyamiti, C. (1984) *Christ as our ancestor: christology from an African perspective*. Gweru: Mambo Press.
Obinna, E. (2012) Life is a superior to wealth?: indigenous healers in an African community, Amariri, Nigeria. In: A. Afe, E. Chitando, and B. Bateye (eds.), *African traditions in the study of religion in Africa*. Farnham: Ashgate Publishing, pp. 137–139.
Oke, E.A. (1995) Traditional health services: an investigation of providers and the level and pattern of utilization among the Yoruba. *Ibadan Sociological Series*, Vol. 2 (1), 2–5.
Oke, E.A. and Owumi, B.E. (1996) *Readings in medical sociology*. Ibadan: Adjacent Press.
Olupona, J.K. (2004) Owner of the day and regulator of the universe: Ifa divination and healing among the Yoruba of South-Western Nigeria. In: M. Winkelman and P.M. Peeks (eds.), *Divination and healing: potent vision*, pp. 103–117. Tucson, AZ: University of Arizona Press.
Omonzejele, P.F. (2008) African concepts of health, disease, and treatment: an ethical inquiry. *Explore*, 4(2), 120–123. doi:10.1016/j.explore.2007.12.001
Onu, A.O. (1999) Social basis of illness: a search for therapeutic meaning. In A.I. Okpoko (ed.), *Africa's indigenous technology*. Ibadan: Wisdom Publishers Ltd.
Osei, A.O. (2004) Types of psychiatric illnesses at traditional healing centres in Ghana. *Ghana Medical Journal*, 35, 106–110.
Petrie, K.J. and Weinman, J. (2012) Patients perceptions of their illness: the dynamo of volition in health care. *Current Directions in Psychological Science*, 21(1), 60–65.
Read, M. (1966) *Culture, health and disease, social and cultural influences on health programmes in developing countries*. London:: Tavistock Publishers.
Setswe, G. (1999) The role of traditional healers and primary health care in South Africa. *Health SA Gesondheid*, 4(2), 56–60. doi:10.4102/hsag.v4i2.356

Taylor, E.B. (1871) *Primitive culture: researches into the development of mythology, philosophy, religion, language, art and custom*. 2nd ed. London: John Murray.
Thorpe, S.A. (1993) *African traditional religions*. Pretoria: University of South Africa.
Vaughn, L.M., Jacquez, F. and Baker, R.C. (2009) Cultural health attributions, beliefs, and practices: effects on healthcare and medical education. *The Open Medical Education Journal*, 2, 64–74.
Westerlund, D. (2006) *African indigenous religions and disease causation*. Leiden: Brill.
White, P. (2015) The concept of diseases and health care in African traditional religion in Ghana. *HTS Teologiese Studies/Theological Studies*, 71(3), 2762.
Wiredu, K. (1980) *Philosophy and an African culture*. London: Cambridge University Press.

3 Trees

Trees are woody perennial plants, typically having a single stem or trunk, growing to a considerable height and bearing lateral branches at some distance from the ground. Many of these trees are used by local communities in Nigeria and the rest of Africa for various medicinal purposes. This chapter examines how such trees are traditionally used for medicinal purposes, and also attempts to show the scientific basis of their uses.

3.1 *Azadirachta indica*

Family: Meliaceae

Botanical Name: *Azadirachta indica*

Common Names: Neem tree, dongoyaro

Local Names: ***Edo:*** Ebe-dongoyaro; ***Igbo:*** Ogwu akom; ***Hausa:*** Dongoyaro; ***Yoruba:*** Aforo-oyinbo.

History: The neem tree is mainly cultivated in the Indian subcontinent.

Belief: Neem products are believed by Siddha and Ayurvedic practitioners to be anthelmintic, antifungal, antidiabetic, antibacterial, antiviral, and contraceptive.

Parts Used: Leaves, seeds, roots, and bark

Local Uses: The leaves and stem bark are used in treating malaria. They are also prescribed for treatment of skin diseases (Zillur Rahman and Shamim Jairajpuri, 1996). Vein is used for treating measles and chickenpox (Zillur Rahman and Shamim Jairajpuri, 1996). Neem oil is also used to treat malaria and to balance blood sugar levels (Chopra *et al.*, 1956).

Chemical Constituents: Nimbin, nimbinin, nimbidin, and azadirachtin (Heuzé *et al.*, 2015).

3.2 *Alstonia boonei*

Family: Apocynaceae

Botanical Name: *Alstonia boonei*

Common Names: Stool wood, pattern wood

Local Names: ***Edo:*** Ukhu; ***Igbo:*** Egbu, egbwu-ora; ***Yoruba:*** Ahun; ***Efik***: Ndodo; ***Ijaw***: Egbu; ***Kwale***: Egbu.

History: It is native to tropical West Africa, with a range extending into Ethiopia and Tanzania.

Belief: Africans regard it as a sacred tree, and it is worshipped in the forest, resulting in parts not being cut off the tree.

Parts Used: Leaves, latex, bark, Root

Local Uses: The bark and leaves are used for treating malaria, asthma and pain. The latex is also used as anaesthetic for pain relief; and is used for treating arthritis (Kweio-Okai, 1991)

Chemical Constituents: Tannins, saponins, and alkaloids, such as echitamine, echitamidine (Burkhill, 1985; Arulmozhi *et al.*, 2010; Maiza-Benabdesselam *et al.*, 2007).

Location: Ewu

3.3 *Bombax boenopozens*

Family: Malvaceae

Botanical Name: *Bombax boenopozens*

Common Names: West African bombax, bombax, ceiba, red silk cotton, red-flowered silk cotton tree, Gold Coast bombax

Local Names: ***Edo:*** Obokha, Ogi-ugbogha, Olikharo, Ugbogha; ***Hausa:*** Gujfiyaa, kuryaa; ***Igbo:*** Akpu, Ngara akpu, Akpu-obololo, Atunjaka; ***Yoruba:*** Eso, Ponpola, Olokododo.

History: It is native to western Africa, the Indian subcontinent, and Southeast Asia, as well as subtropical regions of East Asia and northern Australia. It is distinguished from the genus *Ceiba*, by having red rather than white flowers.

Beliefs: In many parts of West Africa forest and areas, specific trees are protected and valued for particular cultural occasions and as historic symbols. Each community has its own traditions associated with sacred areas and, as a result, the species that are found in them vary greatly. In an analysis of traditional African political institutions, Niangoran-Bouah (1983) noted that there are two traditional sacred locations for reunion: sacred groves and "arbresapalabre." The "arbresapalabre" is the venue for political and social meetings, and is the location where elders sit under the big tree and talk until they agree. It is the location where political, judicial, and social decisions are made.

Visser (1975) noted that, among the Dan tribe of the Côte d'Ivoire, there are specific tree species which serve as "arbresapalabre" such as

Microdesmis sp., *Blighia sapida* (also, a symbol of fecundity), *Cordia millenii*, and *Bombax buonopozens*. The bark of *B. buonopozens* is burnt to produce a smoke that is believed to drive away evil spirits called *aliziniin* Dagbani.

Local Uses: Both the flowers and the young fruits are used in making soup.

Chemical Constituents: Alkaloids, flavonoids, tannins, saponins, terpenoids, steroids, phlobatannins, anthraquinones and carbohydrates are all found in *B. buonopozens*, mostly in the root.

3.4 *Borassus aethiopum*

Family: Arecaceae

Botanical Name: *Borassus aethiopum*

Common Names: African fan palm, African palmyra palm, deleb palm, ron palm, toddy palm, black rhun palm, ronier palm.

History: *B. aethiopum* is a species of Borassus palm from Africa. It is widespread across much of tropical Africa, from Senegal to Ethiopia and south to northern South Africa, though it is largely absent from the forested areas of Central Africa and desert regions such as the Sahara and the Namib deserts. This palm also grows in northwest Madagascar and the Comoros.

Belief: They are believed to play a significant role in rituals and oracles as sacred objects in Africa.

Local Uses: The tree has many uses: the fruits are edible, as are the tender roots produced by the young plant. Fibres can be obtained from the leaves, and the wood (which is reputed to be termite-proof) is also used in construction (Reynolds, 1921; Bailey and Bailey, 1976). It is considered to have antidiabetic properties (Pradeep *et al.*, 2015).

Chemical Constituents: Tannins, anthocyanins, saponosides, mucilages, and coumarins (Aguzue *et al.*, 2012)

3.5 *Casuarina equisetifolia*

Family: Casuarinaceae

Botanical Name: *Casuarina equisetifolia*

Common Names: Australian pine tree, she-oak.

History: The plant is native to Burma, Vietnam, Malaysia, New Caledonia, Vanuatu and Australia. It is also found in Madagascar, but it is doubtful if this is within the native range of the species (Boland *et al.*, 2006).

Belief: The legendary miraculous spear Kaumaile came with the hero Tefolaha from the South Pacific island of Nanumea. He fought with it on the islands of Samoa and Tonga. When Tefolaha died, Kaumaile went to his heirs, then to their heirs, and on and on for 23 generations. It is about 1.80 m long and about 880 years old; the tree was cut on Samoa (Sdsee-Speer, 2014)

Local Uses: Extracts of leaves exhibit anticancer properties. The bark is astringent and used to treat stomachache, diarrhoea, dysentery and nervous disorders. The seeds are anthelmintic, antispasmodic and antidiabetic (Ahsan *et al.*, 2009a; Jain and Dam, 1979).

Chemical Constituents: Alkaloids, steroids, carbohydrates, tannins, fixed oils, proteins, triterpenoids, deoxysugars, flavonoids, and cyanogenic and coumarin glycosides.

3.6 *Citrus aurantiifolia*

Family: Rutaceae

Botanical Name: *Citrus aurantiifolia*

Common Names: Key lime, bitter orange, Seville orange, sour orange, bigarade orange, and marmalade orange.

Local Names: ***Yoruba:*** Osan ghanhinghanhin; ***Igbo:*** Olomaoyinbo; ***Hausa:*** Babban lemu.

History: The name "key lime" comes from its association with the Florida Keys, where it is best known as the flavouring ingredient in key lime pie.

Parts Used: Leaves, fruits, peel and oil.

Local Uses: The leaves are used to make a tea to treat bilious headache. Fresh fruit juice is also used to counter diarrhoea. It is used to lighten the skin, to stimulate the digestive system, detoxify the body, cool fever and to treat rheumatic pain. It has an astringent action, clears congested skin and stops bleeding from minor cuts (Aibinu *et al.*, 2006; Nwankwo *et al.*, 2015; Pathan *et al.*, 2012).

Chemical Constituents: Essential oils, and apigenin, hesperetin, kaempferol, nobiletin, quercetin, and rutin (Kawaii *et al.*, 1999; Caristi *et al.*, 2003).

3.7 *Citrus limon*

Family: Rutaceae

Botanical Name: *Citrus limon*

Common Name: Lemon

Local Names: ***Yoruba:*** Osan wewe; ***Igbo:*** Olomankilisi; ***Hausa:*** Lemunoisami.

History: *C. limon,* the lemon tree, is a small tree in the Rutaceae (citrus) family that originated in Asia (probably India and Pakistan)

Belief: There is a belief in Ayurvedic medicine that a cup of hot water with lemon juice in it tonifies and purifies the liver.

Parts Used: Leaves and fruits.

Local Uses: Lemon juice helps to control blood pressure, purifies the blood, reduces a swollen spleen and strengthens the immune system, as it contains vitamins C, B_2, calcium, iron, etc. It protects the body against germs and bacteria. It is used for to treat bronchitis, earache, nose bleeds, hepatitis, gastric ulcers, and menorrhagia. It helps to prevent scurvy, whooping cough, colds, etc. Lemon oil may be used in aromatherapy (Cooke and Ernst, 2000).

Chemical Constituents: Flavonoids, limonoids) and citric acid (Penniston *et al.*, 2008; Rauf *et al.*, 2014).

3.8 *Cussonia barteri*

Family: Araliaceae

Botanical Name: *Cussonia barteri*

Common Name: Octopus cabbage tree

Local Names: ***Benin:*** Evbi-nato; ***Fulani:*** Takandargiwa; ***Hausa:*** Bumarlahi; ***Yoruba:*** Sigosiga, Sigosego.

History: It is commonly found in Africa (Beentje, 1994).

Belief: Because of the likeness of the defoliated tree to deformed limbs, and based on the 'Theory of Signatures', this plant is often used in Africa for the treatment of leprosy. Most commonly, the stem is macerated and taken as a purgative and is also applied externally as a lotion; the leprous sores may also be dressed with the powdered stem bark. The leafy twigs are used with magical rites in the treatment of yellow fever, oedemas, paralysis, and sleeping sickness.

Local Uses: The water in which the leaves have been boiled is purgative and is taken as a remedy for constipation. The pulped up young shoots are eaten as a remedy for diarrhoea.

A decoction of the leaves is used as an eyewash for treating conjunctivitis, and as a massage in cases of epilepsy, in both adults and children. The plant is an emeto-purgative and diuretic, and these cleansing actions upon the body are used in the treatment of various complaints. The plant is also prescribed as a poison antidote and is used in the treatment of fevers (De Villiers *et al.*, 2010).

Chemical Constituents: Cussonosides A and B (two triterpenes), saponins, phenolics, flavonoids and tannins (Cowan, 1999).

Location: Okomu Forest, Nigeria

3.9 *Dacryodes edulis*

Family: Burseraceae

Botanical Name: *Dacryodes edulis*

Common Names: Native pear, African plum, bush butter tree, African pear.

Local Names: ***Edo:*** Orunmwunn; ***Igbo:*** Ube-igbo, ube, Ube-umbu, ube nkputaki; ***Yoruba:*** Elemi; ***Gabon***: atanga.

History: It is a tree native to West Africa, sometimes called Atanga in Gabon (National Research Council, 2008).

Belief: Africans believed that leopard meat can only be eaten if it is first exorcised under a safou tree (*D. edulis*) in a public place. This tradition is based on the belief that leopards, like other mammals which kill and eat humans, are cursed. The fruit of the tree under which the exorcism is carried out are forbidden to women.

Local Uses: It has long been used in the traditional medicine of some African countries to treat various ailments such as wounds, skin diseases, dysentery, and fever (Omonhinmin and Agbara, 2013).

Chemical Constituents: Terpenes, flavonoids, tannins, alkaloids, saponins, palmitic acid, oleic acid and linoleic acid (Ajibesin, 2011)

3.10 *Dialium guineense*

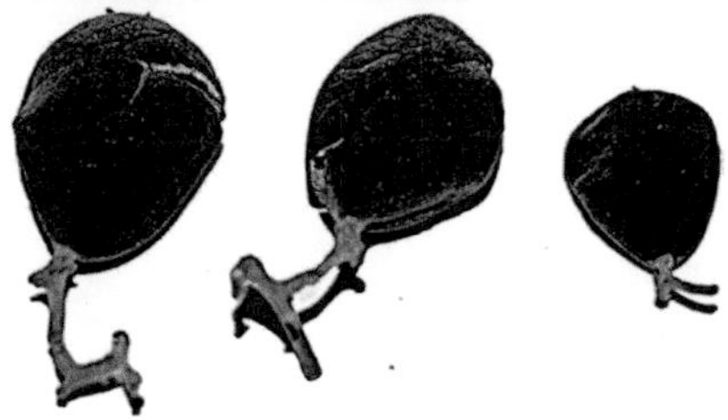

Family: Fabaceae

Botanical Name: *Dialium guineense*

Common Names: Velvet, and black tamarind.

Local Names: ***Yoruba:*** Awin; ***Igbo***: Icheku; ***Hausa:*** Tsamiyar Kurm.

History: It grows in dense forests in Africa along the southern edge of the Sahel. In Togo, it is called atchethewh. The velvet tamarind can be found in West African countries such as Ghana, Sierra Leone, Senegal, Guinea-Bissau and Nigeria.

Parts Used: Whole plant

Local Uses: It is used to treat diarrhoea. An infusion of the leaves and fruits is used to treat fever. Twigs are chewed for cleaning teeth. A decoction and infusion of the bark is used to treat stomachache, toothache, as a gargle and a mouth wash. Root are considered to have aphrodisiac properties.

Chemical Constituents: Tannins, steroids, terpenes, alkaloids, flavonoids, saponin, and phenolics (Utubaku *et al.*, 2017).

3.11 *Desplatsia dewevrei*

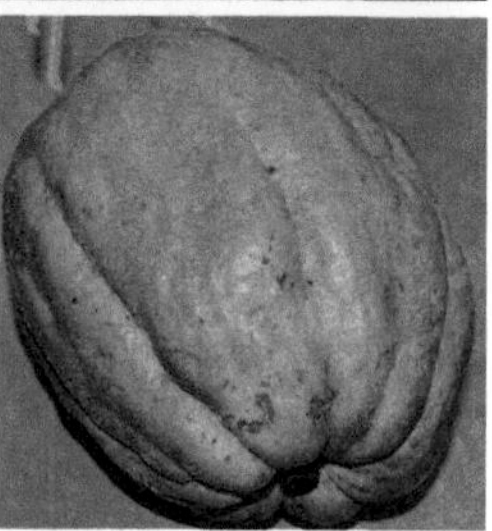

Family: Malvaceae

Botanical Name: *Desplatsia dewevrei*

Common Name: Elephant okra

Local Names: ***Edo:*** Ighiawogho, ikhubeni; ***Yoruba:*** Ila-erin.

Local Uses: The large globose to ellipsoid and angular fibrous juicy ripe fruit of Desplatsia species are boiled to obtain a black dye used for cloth and stains like printers ink (Brink, 2009; Burkill, 2006).

Chemical Constituents: The seed contains fatty acids hexadecanoic acid (palmitic acid), octadecadienoic acid (linoleic acid), and octadecenoic acid (oleic acid).

3.12 *Ficus elastica*

Family: Moraceae

Botanical Name: *Ficus elastica*

Common Name: Rubber fig, rubber bush, rubber tree, rubber plant, Indian rubber bush, Indian rubber tree.

History: *F. elastica* belongs to the genus containing figs, and is native to east India, Nepal, Bhutan, Burma, China (Yunnan), Malaysia, and Indonesia. It has become naturalized in Sri Lanka, the West Indies, and the state of Florida in the United States.

Local Uses: For healing wounds, cuts, and bruises.

Chemical Constituents: Ficus elastic acid, lauric acid, myristic acid, and butyric acid. The latex is used for rubber making (Chang *et al.*, 1998)

3.13 *Entada africana*

Family: Fabiaceae

Botanical Name: *Entada africana*

Local Names: ***Yoruba***: Agurobe, igba oyibo, ogurobe.

History: This plant originates from the south Sahel and Sudanian ecozone savannas, and is found in disturbed places, is fairly common and forms clumps, and is distributed throughout the Sahel to East and South Africa.

Local Uses: Tender young leaves, occasionally harvested from the wild, are cooked, and used in sauces. The plant is commonly used as a traditional medicine within its native range. The leaves have been shown to contain retinol, and the leaves are used to treat stomachache. They are used to make a tonic tea applied externally; the leaves are used for healing wounds. They make a good wound dressing, preventing suppuration. The bark is abortifacient, and the roots are stimulant and tonic in action. Because of their emetic properties, they are said to have antidotal effects against various toxic agents and are used as a fish poison. A fibre obtained from the inner bark is used for making ropes, bands, and storage bins. The bark is a source of tannins. A low-quality gum is obtained from the tree. The bark contains rotenone, which has insecticidal properties (Aubréville, 1950; Berhaut, 1975; Geerling, 1982; Von Maydell, 1983).

Chemical Constituents: Tannins, saponins, and retinol.

3.14 *Entandrophragma angolense*

Family: Meliaceae

Botanical Name: *Entandrophragma angolense*

Common Names: English mountain mahogany, tiama mahogany, tiama, acajou tiama

Local Names: ***Edo:*** Gedunohor; ***Igbo:*** Okeone; ***Yoruba:*** Ijebo

History: *E. angolense* is widespread throughout Africa, occurring from Guinea east to southern Sudan, Uganda and western Kenya, and south to DR Congo and Angola.

Local Uses: The wood, usually traded as 'gedu nohor' or 'tiama', is highly valued for exterior and interior joinery, furniture, cabinet work, veneer, and plywood, and is also used for flooring, interior trim, panelling, stairs, ship building, vehicle bodies, and coffins. It is suitable for light construction, for production of musical instruments, toys, novelties, boxes, crates, carvings, and turnery. Wood that is not suitable for these uses is used as firewood and for charcoal production. The bark is used in traditional medicine.

A bark decoction is drunk to treat fever and the bark is also used, usually as external applications, as an anodyne against stomachache and peptic ulcers, earache, and kidney, rheumatic, or arthritic pains. It is also applied externally to treat ophthalmia, swellings and ulcers. The tree is planted as a roadside tree, and occasionally as a shade tree in banana, coffee, and tea plantations (Burkill, 1985b; Abbiw, 1990).

Chemical Constituents: Tannins, alkaloids, saponins, and cardiac glycoside (Shittu and Akor, 2015).

3.15 *Eucalyptus officinalis*

Family: Myrtaceae.

Botanical Name: *Eucalyptus officinalis*

Common Names: Gum tree

History: Species of the genus *Eucalyptus* dominate the tree flora of Australia, and include *Eucalyptus regnans*, the tallest known flowering plant on Earth. There are more than 700 species of eucalyptus and most are native to Australia; a very small number are found in adjacent areas of New Guinea and Indonesia.

Belief: It is commonly believed that the thirst of the eucalyptus tends to dry up rivers and wells.

Local Uses: *E. officinalis* is a fast-growing evergreen tree native to Australia. As an ingredient in many over-the-counter (OTC) products, eucalyptus oil is used to reduce symptoms of coughs, colds, and congestion. It also features in creams and ointments aimed at relieving muscle and joint pain. The oil that comes from the eucalyptus tree is used as an antiseptic, a perfume, as an ingredient in cosmetics, as a flavoring, in dental preparations, and in industrial solvents (Sadlon and Lamson, 2010).

Chemical Constituents: Eucalyptol, with the leaves also containing flavonoids and tannins (Low *et al.*, 1974; Nagpal *et al.*, 2010).

3.16 *Rothmannia hispida*

Family: Rubiaceae

Botanical Name: *Rothmannia hispida*

Local Names: ***Edo***: Asun, asun-nekwe, asun leghere; ***Yoruba***: Asogbodun.

History: A shrub or small tree growing to 10 m high; a plant of the forest understorey, or in secondary jungle, from Guinea to West Cameroon, and on into Zaïre. On cutting and exposure to the air, the wood takes on a blue colour.

Local Uses: In Nigeria, leaf sap and fruit juice are used to draw black designs on the body and to blacken tattoos; mixed with palm oil, they are applied to the skin to fight fungal infections (Abbiw, 1990).

Chemical Constituents: Monomethyl fumarate, D-mannitol, 4-oxonicotinamide-1-(1'-β-D-ribofuranoside) (Abbiw, 1990).

3.17 *Senna siamea*

Family: Fabaceae

Botanical Name: *Senna siamea*

Common Names: Yellow cassia, Siamese cassia, kassod tree, cassod tree, cassia tree

History: It is native to South and Southeast Asia, although its exact site of origin is unknown.

Parts Used: Flower, root, bark, heartwood

Local Uses: (flower) Food: general; (root and heartwood) Medicines: stomach troubles, venereal diseases; (heart-wood, bark, leaf, fruit) Agri-horticulture: ornamental, cultivated or partially tended, Agri-horticulture:

hedges, markers, Agri-horticulture: biotically active, Agri-horticulture: shade-trees. Products: Building materials (heartwood).

Chemical Constituents: Alkaloids, saponin, phytate, anthraquinones, oxalate, tannins, and phlobatannins (Trease and Evans, 1978; Wheeler and Ferrell, 1971.

3.18 *Spondias mombin*

Family: Anarcadiaceae

Botanical Name: *Spondias mombin*

Common Names: Yellow mombin, hog plum, mombin

Local Names: ***Edo:*** Ogheghe, okhukan; ***Igbo*:** Isikala, Isikere; ***Yoruba*:** Akika, Okikan.

History: It is native to the tropical Americas, including the West Indies. The tree has been naturalized in parts of Africa, India, Bangladesh, Sri Lanka, and Indonesia. It is rarely cultivated except in parts of the Brazilian northeast.

Local Uses: The juice of crushed leaves and a powder of dried leaves is used as a poultice on wounds and inflammations. The gum is employed as an expectorant and to expel tapeworms (Corthout *et al.*, 1994).

Chemical Constituents: Anthraquinones, flavonoids, naphthoquinones, sesquiterpenes, quassinoids, and indole and quinoline alkaloids (e.g. berberine) .

3.19 *Stereospermum kunthianum*

Family: Bignoniaceae

Botanical Name: *Stereospermum kunthianum*

Local Name: ***Hausa*:** Sansami

History: It is an African deciduous shrub or small tree occurring in the Democratic Republic of Congo, Djibouti, Eritrea, Ethiopia, Kenya, Malawi, Senegal, Somalia, Sudan, Tanzania, and Uganda. It is widespread across Africa to as far as the Red Sea, and reaches as far south as Angola, Mozambique, Zambia, and Zimbabwe. There are some 30 species of *Stereospermum* with a Central African and Asian distribution.

Belief: In Zaria, Nigeria, the fruit is sold in local markets for magical uses. It is regarded as a charm to secure riches and to woo wealthy customers.

Local Uses: The pods are chewed with salt to treat coughs and are used in the treatment of ulcers, leprosy, skin eruptions, and venereal diseases, while the stem bark decoction or infusion is used to cure bronchitis, pneumonia, coughs, rheumatic arthritis, and dysentery. The twigs are chewed to clean teeth and to treat toothache. The roots and leaves are used to treat venereal diseases, respiratory diseases, and gastritis (Orwa *et al.*, 2009).

Chemical Constituents: Powdered stem bark contains alkaloids, tannins, phlobatannins, saponins, cardiac glycosides, anthracene derivatives, and reducing sugars.

3.20 *Syzygium samarangense*

Family: Myrtaceae

Botanical Name: *Syzygium samarangense*

Common Names: Java apple, Semarang rose-apple, and wax jambu

History: It is native to an area that includes the Greater Sunda Islands, the Malay Peninsula, and the Andaman and Nicobar Islands, but was introduced in prehistoric times to a wider area and is now widely cultivated in the tropics.

Local Uses: The flowers are astringent and are used in Taiwan to treat fever and diarrhoea.

A root-bark decoction is used to treat dysentery and amenorrhea, and as an abortifacient. Powdered leaves are used to treat cracked tongues. In Hawaii, the juice of salted pounded bark is used to treat wounds, and, in Molucca, a decoction of bark is used to treat thrush.

Chemical Constituents: Reynoutrin, hyperin, myricitrin, quercitrin, quercetin, and guaijaverin

3.21 *Tectona grandis*

Family: Verbanaceae

Botanical Name: *Tectona grandis*

Common Name: Teak

History: *T. grandis* is native to south and southeast Asia, mainly India, Sri Lanka, Indonesia, Malaysia, Thailand, Myanmar, and Bangladesh but it is now naturalised and cultivated in many countries in Africa and the Caribbean.

Local Uses: It is used in the manufacture of outdoor furniture and boat decks. It is also used for chopping boards, indoor flooring, countertops and as a veneer for indoor furnishings. Leaves of the teak tree are used to make Pellakai gatti (jackfruit dumpling), where batter is poured into a teak leaf and steamed.

Chemical Constituents: Seeds contain a fixed oil containing chiefly stearic, palmitic, oleic and linoleic acids. Roots contain lapachol, tectol, dehydrotectol, tectoquinone, β-lapachone, and dehydro-α-lapachone, as well as β-sitosterol.

3.22 *Terminalia superba*

Family: Combretaceae

Botanical Name: *Terminalia superba*

Common Names: Shringlewood, white afara

Local Names: ***Edo:*** Egoyen, egboin nofua, aghoin; ***Yoruba*:** afara.

History: The family Combretaceae is comprised of 20 genera and about 475 species (Thiombiano *et al.*, 2006). Of these, about 200 belong to the genus *Terminalia*, making it the second largest genus of the family after *Combretum*. The family is distributed throughout the tropical and subtropical regions of the world.

Approximately 54 species of *Terminalia* are naturally distributed throughout western, eastern and southern Africa.

Local Uses: *Terminalia* spp. provide economic, medicinal, spiritual and social benefits. The wood of *Terminalia* spp. is highly appreciated as construction timber. It is currently used for light construction, door and window frames, coffin boards, mouldings, beams, rafters, joists, flooring, furniture, carts, tool handles, spindles, shuttles, picker sticks, walking sticks, bowls, boat building, masts, mine props, foundation piles, veneer, and plywood.

3.23 *Terminalia ivorensis*

Family: Combretaceae

Botanical Name: *Terminalia ivorensis*

Common Names: Black afara, framire

Local Names: ***Edo:*** ẹghọẹn-nébì; ***Igbo:*** awunshin-oji; ***Yoruba:*** Afara dudu.

History: It is found in Cameroon, Ivory Coast, Ghana, Guinea, Liberia, Nigeria, and Sierra Leone.

Local Uses: The wood, usually traded as 'framiré' or 'idigbo', is valued for its light weight for construction, door and window frames, joinery, furniture, cabinet work, veneer and plywood. It is suitable for flooring, interior trim, vehicle bodies, sporting goods, boxes, crates, matches, turnery, hardboard, particle board, and pulpwood. It is used locally for house construction, planks, roof shingles, fencing posts, dugout canoes, drums, and mortars. Mixed with other woods, it is suitable for paper making. The wood is also used as firewood and for charcoal production; offcuts are highly valued in Ghana for making charcoal.

Chemical Constituents: Nitrogen, calcium and magnesium.

3.24 *Theobroma cacao*

Family: Malvaceae

Botanical Name: *Theobroma cacao*

Common Name: Cocoa

Local Names: ***Edo:*** Koko; ***Igbo*****:** Koko; ***Hausa*****:** koko; ***Yoruba*****:** koko.

History: This tree is native to the deep tropical regions of Central and South America.

Belief: The Maya believed that the kakaw (cacao) was discovered by the gods in a mountain that also contained other delectable foods to be used by them.

According to Mayan mythology, the Plumed Serpent gave cacao to the Maya after humans were created from maize by the divine grandmother goddess Xmucane. The Maya celebrated an annual festival in April to honor their cacao god, Ek Chuah, an event that included the sacrifice of a dog with cacao-colored markings, additional animal sacrifices, offerings of cacao, feathers, and incense, and an exchange of gifts.

Local Uses: The seeds and cocoa beans are used to make cocoa mass, powder, confectionery, ganache, and chocolate.

Chemical Constituents: Theobromine (an alkaloid), flavanols, proanthocyanidins, and pectin (Decroix *et al.*, 2016).

3.25 *Voacanga africana*

Family: Apocynaceae

Botanical Name: *Voacanga africana*

Common Name: Catapult plant

Local Names: ***Edo:*** Òvìẹn-íbù; ***Igbo*:** Pete pete, akẹte; ***Yoruba*:** Akọdodo; ***Hausa:*** kookiyar birii.

History: *V. africana* is a small tropical African tree that grows to 6 m in height. It has leaves that are up to 30 cm in length, and the tree produces yellow or white flowers, which become fruitswith yellow seeds.

Local Uses: The bark and seeds of the tree are used in Ghana as a poison, a stimulant and an aphrodisiac. The milky latex of the plant is applied to wounds to aid healing in Nigeria and Senegal. A tea made from the leaf is said to be a strengthening potion that relieves fatigue and shortness of breath. It is also used to prevent premature childbirth and to treat painful hernias and menstruation. It is used in many areas of Africa to treat heart problems. The seeds of *Voacanga* spp. are used in Europe due to their high concentration of tabersonine, which. is used as a precursor in the synthesis of vincamine, which is used to treat neural deficiencies in the elderly (Vooglebreinder, 2009).

Chemical Constituents: Iboga alkaloids such as voacangine, voacamine, vobtusine, amataine, akuammidine, tabersonine, coronaridine, and vocangine.

3.26 *Zanthoxylum zanthoxyloides*

Family: Rutaceae

Botanical Name: *Zanthoxylum zanthoxyloides*

Common Name: Fagara

Local Names: ***Edo:*** Ughanghan; ***Hausa***: Fasa kwari; ***Igbo:*** Uko; ***Yoruba***: Ata.

History: It occurs predominantly in Africa, from Guinea east to Ghana.

Local Uses*:* In southern Nigeria, a decoction of the stem bark and roots is taken to treat cancer. Pulped stem bark and root bark is thrown into water to stupefy fish. In West Africa, it is planted as a hedge, as the thorns make it impenetrable. Sheep browse the leaves. The wood is used for manufacturing torches. The timber is yellow, very hard and termite resistant, and used for building purposes, including poles and posts. It also makes good firewood. The roots, young shoots, and twigs are commonly used as chewsticks. The bark or young branches contain much resin, which makes them suitable for ceremonial torches. The spines are thrown into fire to give off a scented smoke. The leaves, which smell like citronella, and the seeds, which taste strongly of cinnamon or pepper, are commonly used to season food. Necklaces are made from the seeds.

Zanthoxyloides also has numerous magico-religious uses, including protection against spirits. It also serves as fetish plant.

Chemical Constituents: Acridone alkaloids, namely 3-hydroxy-1,5,6-trimethoxy-9-acridone; 1,6-dihydroxy- 3- methoxy- 9 - acridone; 3,4,5,7-tetrahydroxy-1- methoxy-10-methyl-9-acridone; and 4- methoxyacridone (Wouatsa *et al.*, 2013).

Location: Ewu

References

Abbiw, D. (1990) *Useful plants of Ghana: West African uses of wild and cultivated plants.* Kew, Richmond, UK: Intermediate Technology Publications/Royal Botanic Gardens, 337 pp.

Aguzue, O.C., Akanji, F.T., Tafida, M.A., Kamal, M.J. and Abdulahi, S.H. (2012) Comparative chemical constituents of some desert fruits in Northern Nigeria. *Applied Science Research*, 4(2), 1061–1064.

Ahsan, M.R., Islam, K.M., Bulbul, I.J., Musaddik, M.A. and Haque, M.E. (2009a) Hepatoprotective activity of methanol extract of some medicinal plants against carbon tetrachloride-induced hepatotoxicity in rats. *European Journal of Scientific Research*, 37, 302–310.

Ahsan, M.R., Islam, K.M., Haque, M.E. and Mossaddik, M.A. (2009b) In vitro antimicrobial screening and toxicity study of some different medicinal plants. *World Journal of Agricultural Sciences*, 5, 617–621.

Aibinu, I., Adenipekun, T., Adelowotan, T., Ogunsanya, T. and Odugbemi, T. (2006) Evaluation of the antimicrobial properties of different parts of *Citrus aurantifolia* (lime fruit) as used locally. *African Journal of Traditional, Complementary and Alternative Medicines*, 4, 185–190.

Ajibesin, K.K. (2011) *Dacryodes edulis* (G. Don) H.J. Lam: a review on its medicinal, phytochemical and economical properties. *Research Journal of Medicinal Plant*, 5(1), 32–41.

Arulmozhi, S., Mazumder, P.M., Narayanan, L.S. and Akurdesai, P. (2010) In vitro antioxidant and free radical scavenging activity of fractions from *Alstonia scholaris* Linn. R.Br. 8 (3) 430–437.

Aubréville, A. (1950) *Flore forestière Soudano-Guinéenne, AOF, Cameroun, AEF.* Paris: Societe d'Edition de Geographic Maritime et Coloniale, 523 pp.

Bailey, L.H. and Bailey, E.Z. (1976) *Hortus Third i–xiv, 1–1290.* New York: Collier & Son Company.

Beentje, H.J. (1994) *Kenya trees, shrubs and lianas.* Nairobi, Kenya: National Museums of Kenya, 722 pp.

Berhaut, J. (1975) *Flore illustrée du Sénégal, vol. 4: Ficoïdées à Légumineuses.* Dakar: Direction des Eaux et Forêts, Ministère du Développement Rural, 525 pp.

Boland, D.J., Brooker, M.I.H., Chippendale, G.M. and McDonald, M.W. (2006) *Forest trees of Australia.* 5th ed. Collingwood, VIC: CSIRO Publishing, p. 82.

Brink, M. (2009) Desplatsia subericarpa Bocq. Record from PROTA4U, in M. Brink and E.G. Achigan-Dako (eds.), *PROTA (Plant Resources of Tropical Africa/Ressources végétales de l'Afrique tropicale)*, Wageningen, Netherlands, viewed 21 July 2015. Available at http://www.prota4u.org/search.asp

Burkill, H.M. (1985a) *The useful plants of tropical west Africa, Vol. 1 (families A-D).* 2nd ed. Kew: Royal Botanic Gardens, pp. 138–140.

Burkill, H.M. (1985b) *The useful plants of tropical west Africa, Vol.* 1. Kew: Royal Botanic Gardens, pp. 211–212.

Burkill, H.M. (1985c) *The useful plants of tropical west Africa, Vol.* 3. Kew: Royal Botanic Gardens.

Burkill, H.M. (1985d) *The useful plants of tropical west Africa, Vol. 4*. 2nd ed. Kew, Richmond, Surrey: Royal Botanical Garden.
Burkill, H.M. (2006) *The useful plants of tropical west Africa, Vol. 5*. 2nd ed. Kew, Richmond, Surrey: Royal Botanical Garden.
Caristi, C., Bellocco, E., Panzera, V., Toscano, G., Vadalà, R. and Leuzzi, U. (2003) Flavonoids detection by HPLC-DAD-MS-MS in lemon juices from Sicilian cultivars. *Journal of Agriculture and Food Chemistry*, 51, 3528–3534.
Chang Siushih, Wu Chengyih and Cao Ziyu. (1998) Moroideae. In: Chang Siushih and Wu Chengyih (eds.), *Fl. Reipubl. Popularis Sin.* 23(1), 1–219.
Chopra, R.N., Nayer, S.L. and Chopra, I.C. (1956) *Glossary of Indian medicinal plants*. New Delhi: CSIR.
Cooke, B. and Ernst, E. (2000) Aromatherapy: a systematic review. *British Journal of General Practice*, 50(455), 493–496.
Corthout, J., Pieters, L.A., Claeys M., Vanden Berghe D.A. and Viletinck, J. (1994) Antibacterial and molluscicidal phenolic acids from Spondias mombin. *Planta Med.* 60(5), 460–463.
Cowan, M.M. (1999) Plant products as antimicrobial agents. *Clinical Microbiology Reviews*, 12, 564–582.
De Villiers, B.J., Van Vuuren, S.F., Van Zyl, R.F. and Van Wyk, B.E. (2010) Antimicrobial and antimalarial activity of *Cussonia* species (Araliaceae). *Journal of Ethnopharmacology*, 129, 189–196.
Decroix, L., Tonoli, C., Soares, D.D., Tagougui, S., Heyman, E. and Meeusen, R. (2016) Acute cocoa flavanol improves cerebral oxygenation without enhancing executive function at rest or after exercise. *Applied Physiology, Nutrition, and Metabolism*, 41(12), 1225–1232.
Geerling, C. (1982) Guide de terrain des ligneux sahéliens et soudano-guinéens (2 eme éd.). Wageningen: Agricultural University, Grasses and Legumes Index (1st ed. 1982), 340 pp. herbpathy.com (Material medical: how to use *Entandrophragma angolense*). Accessed September, 2017.
Heuzé, V., Tran, G., Archimède, H., Bastianelli, D. and Lebas, F. (2015) *Neem (Azadirachta indica)*. Feedipedia, a programme by INRA, CIRAD, AFZ and FAO. Available at https://feedipedia.org/node/182, last updated on October 2, 2015, 15:40.
Jain, S.K. and Dam, N. (1979) Some ethnobotanical notes from Northeastern India. *Economic Botany*, 33, 52–56.
Kawaii, S., Tomono, Y., Katase, E., Ogawa, K. and Yano, M. (1999) Quantitation of flavonoid constituents in Citrus fruits. *Journal of Agricultural and Food Chemistry*, 47, 3565–3571.
Kweio-Okai, G. (1991) Antiinammatory activity of a Ghanaian antiarthritic herbal preparation: II. *Journal of Ethnopharmacol*
Low, D., Rawal, B.D. and Griffin, W.J. (1974) Antibacterial action of the essential oils of some Australian Myrtaceae with special references to the activity of chromatographic fractions of oil of *Eucalyptus citriodora*. *Planta Medica*, 26, 184–185.
Maiza-Benabdesselam, F., Chibane, M., Madani, K., Max, H., and Adach, S. (2007) Determination of isoquinoline alkaloids contents in two Algerian species of Fumaria (*Fumaria capreolata* and *Fumaria bastardi*). *African Journal of Biotechnology*, 6(21), 2487–2492.
Nagpal, N., Shah, G., Arora, M.N., Shri, R. and Arya, Y. (2010) Phytochemical and pharmacological aspects of *Eucalyptus* genus. *International Journal of Pharmaceutical Sciences and Research*, 3, 28–36.
National Research Council. (2008) *Butterfruit. Lost Crops of Africa: Volume III: Fruits*. Johannesburg: National Academies Press.

Nwankwo, I.U., Osaro-Matthew, R.C. and Ekpe, I.N. (2015) Synergistic antibacterial potentials of *Citrus aurantifolia* (Lime) and honey against some bacteria isolated from sputum of patients attending Federal Medical Center, Umuahia. *International Journal of Current Microbiology and Applied Sciences*, 4, 534–544.

Omonhinmin, A.C. and Agbara, I.U. (2013) Assessment of In vivo antioxidant properties of *Dacryodes edulis* and *Ficus exasperata* as anti-malaria plants. *Asian Pacific Journal of Tropical Disease*, 3(4), 294–300.

Orwa, C., Mutua, A., Kindt, R., Jamnadass, R. and Simons, A. (2009) *Agroforestree Database: a tree reference and selection guide version 4.0.* London: Oxford University Press. Available at http://www.worldagroforestry.org/af/treedb/. Accessed March 2018.

Pathan, R., Papi, R., Parveen, P., Tananki, G. and Soujanya, P. (2012) In vitro antimicrobial activity of *Citrus aurantifolia* and its phytochemical screening. *Asian Pacific Journal of Tropical Disease*, 2, 328–331.

Penniston, K.L., Nakada, S.Y., Holmes, R.P. and Assimos, D.G. (2008) Quantitative assessment of citric acid in lemon juice, lime juice, and commercially-available fruit juice products. *Journal of Endourology*, 22(3), 567–570. Accessed 25/9/2017.

Pradeep, G., Anil, K.A., Lakash, M.S. and Singh, G.K. (2015) Antidiabetic and antihyperlipidemic effect of *Borassus flabellifer* in streptozotocin (STZ) induced diabetic rats. *World Journal of Pharmacy and Pharmaceutical Sciences*, 4(1), 1172–1184.

Rauf, A., Uddin, G. and Ali, J. (2014) Phytochemical analysis and radical scavenging profile of juices of *Citrus sinensis, Citrus anrantifolia*, and *Citrus limonum. Organic and Medicinal Chemistry Letters*, 4, 5.

Reynolds, F.J. (1921) *Deleb palm. Collier's new encyclopedia*. New York: P.F.

Sadlon, A.E. and Lamson, D.W. (2010) Immune-modifying and antimicrobial effects of Eucalyptus oil and simple inhalation devices. *Alternative Medicine Review*, 15(1), 33–47.

Shittu, G.A. and Akor, E.S. (2015) Phytochemical screening and antimicrobial activities of the leaf extract of *Entandrophragama angolense. African Journal of Biotechnology*, 14, 3.

Thiombiano, A., Schmidt, M., Kreft, H. and Guinko, S. (2006) Influence du gradient climatique sur la distribution des espèces de Combretaceae auBurkina Faso (Afrique de l'ouest). *Candollea* 61, 189–213.

Trease, M.T. and Evans, S.E. (1978) The phytochemical analysis and antibacterial screening of extracts of *Tetracarpetum conophorum. Journal of Chemical Society of Nigeria*, 26, 57–58.

Utubaku, A.B., Yakubu, O.E. and Okwara, D.U. (2017) Comparative phytochemical analysis of fermented and unfermented seeds of *Dialium giuneense*. Department of Medical Biochemistry, Cross River University of Technology, Calabar, Nigeria and Department of Biochemistry, Federal University Wukari, Nigeria. *Journal of Traditional Medicine and Clinical Naturopathy*, 9(3), 123–110.

Visser, L. (1975) *Plantes médicinales de la Côte d'Ivoire*. No. 75–15. Wageningen, The Netherlands: Mededelingen Landbouwhogeschool.

Von Maydell, J. (1983) *Arbres et Arbustes du Sahel, leurs caractéristiqueset leurs utilisations*. Hambourg: Publié par GTZ, 310 pp.

Voogelbreinder, Snu. (2009) *Garden of Eden: The Shamanic use of psychoactive flora and fauna and the study of consciousness*. Berlin: Snu Voogelbreinder.

Wouatsa, V.N., Mista, L., Kumar, S., Prakash, O., Tchoumbougnang, F and Venkatesh, R. (2013) Aromatase and glycosyl transferase inhibiting acridone alkaloids from fruits of Cameroonian Zanthoxylum species. *Chemistry Central Journal*, 7(1), 125.

Zillur Rahman, S. and Shamim Jairajpuri, M. (1996) *Neem in Unani Medicine*. Neem Research and Development Society of Pesticide Science, India, New Delhi, February 1993, p. 208–219. Edited by N.S. Randhawa and B.S. Parmar. 2nd revised edition.

4 Shrubs

This chapter looks at many shrubs which are used traditionally to treat both physical and spiritual health problems in local communities in Nigeria and other parts of Africa. Shrubs refer to woody plants that are smaller than trees, usually having several stems rather than a single trunk

4.1 *Cajanus cajan*

Family: Fabaceae

Botanical Name: *Cajanus cajan*

Common Name: Pigeon pea

Local Names: ***Edo:*** Olele; ***Hausa*;** Waaken tantabara, Dan-mata; ***Igbo:*** Fio-fio; ***Yoruba*:** Otili.

History: *C. cajan*, more commonly known as pigeon pea, is a drought-tolerant crop, important for small-scale farmers in semi-arid areas, where rainfall is low. Probably native to India, pigeon pea was brought millennia ago to Africa, where different strains were developed. These were brought to the New World in post-Columbian times. Truly wild *Cajanus* has never been found; they exist mostly as remnants of cultivation.

Parts Used: Leaves and seeds

Local Uses: In India and Java, the young leaves are applied to sores. The Indochinese claim that powdered leaves help expel bladder stones. Salted leaf juice is taken for jaundice. In Argentina, the leaf decoction is prized for genital and other skin irritations, especially in females. Floral decoctions are used to treat bronchitis, coughs, and pneumonia. Chinese shops sell dried roots as an alexeritic, anthelminthic, expectorant, sedative, and vulnerary. Leaves are also used to treat toothache, dysentery. sore gums, as a mouthwash and in child delivery. Scorched seeds, added to coffee, are said to alleviate headache and vertigo. Fresh seeds are said to help urinary incontinence in males, whereas immature fruits are believed to be of use for treatment of liver and kidney issues. In recent years, *C. cajan* has also been investigated for the treatment of ischemic necrosis of the caput femoris, aphtha, bedsores and wound healing (Zu *et al.*, 2010).

Chemical Constituents: Cajanin, concajanin, methionine, lysine, and tryptophan (Bressani *et al.*, 1986)

4.2 *Calotropis procera*

Family: Apocynaceae

Botanical Name: *Calotropis procera*

Common Names: Giant milk weed, Sodom apple

Local Names: ***Hausa***: Baabaa ambalee; ***Igbo***: Otokwuru; ***Yoruba***: Bomubomu.

History: The fruit was described by the Roman Jewish historian Josephus, who saw it growing near Sodom; "as well as the ashes growing in their fruits; which fruits have a colour as if they were fit to be eaten, but if you pluck them with your hands, they dissolve into smoke and ashes."

Belief: Some biblical commentators believe that the Sodom apple may have been the poisonous gourd (or poison-tasting gourd) that led to "death in the pot" in 2 Kings 4:38–41.

Parts Used: Leaves, root, bark, and latex

Local Uses: *C. procera* latex has been used to treat leprosy, eczema, inflammation, cutaneous infections, syphilis, malarial and low hectic fevers, and as an abortifacient (Kumar and Basu, 1994; Sharma and Sharma, 2000).

Chemical Constituents: Calotropin, calotoxin, resinols, alkaloids, flavonoids, tannins, saponins, and cardiac glycosides (Mainasara *et al.*, 2011).

Location: Afasho

4.3 *Carpolobia lutea*

Family: Polygalaceae **Botanical Name:** *Carpolobia lutea*

Common Name: Cattle stick

Local Name: ***Edo:*** Aswen; ***Igbo***: Aghba-awa; **A:** Amurejo

History: It is widely distributed in western and central areas of tropical Africa.

Local Uses: The stem is used as chewing stick; the root is also used as chewing stick because of its aphrodisiac potential. Its shrubby and smallish stems give it a practical use as a sweeping material or broom in rural areas among the Ibibio tribes of Akwa Ibom State, Nigeria.

The resilience of the woody stem enhances its patronage by cattle herders as a cane to control their cattle. The decoction of the root is used in locally made alcohol as an aphrodisiac. It is used in the treatment of genitourinary infections, gingivitis and stomach pains (Ettebong and Nwafor, 2009).

Chemical Constituents: Triterpene saponins and polyphenols

4.4 *Clerodendrum splendens*

Family: Lamiaceae

Botanical Name: *Clerodendrum splendens*

Common Names: Flaming glory bower, glory tree

Local Names: ***Igbo***: Afifia omya, Ufuchi; ***Yoruba***: Adabi, Opo-eshi, Araojola.

History: It is a native of tropical western Africa.

Local Uses: The plant is used ethnomedicinally in Ghana for the treatment of vaginal thrush, bruises, wounds, and various skin infections (Irvine, 1961).

Chemical Constituents: Carbohydrates, glycosides, unsaturated sterols, triterpenoids, and flavonoids are reported to be present in the leaves, in addition to the volatile oil in the flowers (Shehata *et al.*, 2001; el-Deeb, 2003)

4.5 *Cochlospermum planchonii*

Family: Cochlospermaceae

Botanical Name: *Cochlospermum planchonii*

Common Name: False cotton

Local Names: ***Edo:*** Gbutu; ***Hausa:*** Zunzunaa; ***Yoruba:*** Oboyo.

History: It originates in western tropical Africa, from Senegal to Sierra Leone, east to Chad and northern Cameroon.

Belief: The Fula of northern Nigeria believe that a leaf infusion bestows magical protection.

Local Uses: The root is a source of a yellow dye, which is used in the Sudan, in Nupe and elsewhere. Hausas in northern Nigeria add indigo to obtain green shades. The root is used in Lagos in the cooking of soup when oil is not available. This explains the anti-diarrhoeal use of *C. planchonii* root in traditional medicine. In addition, tannins have astringent properties, hasten the healing of wounds and of inflamed mucous membranes. Plants with tannins are used for healing of wounds, haemorrhoids and diabetes (Yakubu *et al.*, 2010: Salah *et al.*, 1995).

Chemical Constituents: Prominent among the phytochemicals are saponins, alkaloids, phenolics, and steroids (Mikhail and Musbau, 2013).

4.6 *Hibiscus rosa-sinensis*

Family: Malvaceae

Botanical Name: *Hibiscus rosa-sinensis*

Common Names: China rose, Hawaiian hibiscus

Local Names: ***Igbo:*** Flawa; ***Yoruba:*** Kekeke. **History:** It is native to East Asia.

Belief: It is used in the worship of Devi, and the red variety is especially prominent, playing an important part in tantra

Parts Used: Leaves and flowers

Local Uses: The slimy liquid from leaves washed in water is used to relieve a stomach upset. The flowers are purgative and are used to control menstruation. For mouth sores, leaves are washed clean, cut into pieces and added to 10 ml boiling water for 15 minutes, then filtered when cold, and taken three times daily. Properties include antiaging, anticancer, antimicrobial, and antidiabetic, and it is used for hair care, hair growth (Priya, 2014).

Chemical Constituents: Glycosides, alkaloids, tannins, flavonoids, saponins, and carbohydrates (Udita *et al.*, 2015).

4.7 *Ixora finlaysoniana*

Family: Rubiaceae

Botanical Name: *Ixora finlaysoniana*

Common Names: Siamese white ixora, fragrant ixora

History: *Ixora* is a genus of flowering plants in the Rubiaceae family. It is the only genus in the tribe Ixoreae. It consists of tropical evergreen trees and shrubs and includes around 545 species. Though native to the tropical and subtropical areas throughout the world, its centre of diversity is in tropical Asia. Members of Ixora prefer acidic soil and are suitable choices for bonsai. It is also a popular choice for hedges in parts of South East Asia.

Belief: It is commonly used in Hindu worship, as well as in Ayurveda and Indian folk medicine.

Local Uses: Decoction of roots is used to treat nausea, hiccups, and anorexia.

Chemical Constituents: Root contains an aromatic acrid oil, tannins, and fatty acids. Leaves yield flavonols (kaempferol and quercetin), proanthocyanidins, phenolic acids, and ferulic acid. Flowers contain cyanidin and flavonoids, and a coloured material related to quercitin.

4.8 *Ixora coccinea*

Family: Rubiaceae

Botanical Name: *Ixora coccinea*

Common Names: Flame of the woods, jungle flame, jungle geranium

History: *I. coccinea* is a species of flowering plant in the Rubiaceae family. It is a common flowering shrub native to southern India and Sri Lanka. It has become one of the most popular flowering shrubs in South Florida gardens and landscapes. It is the national flower of Surinam.

Belief: The flowers, leaves, roots, and stem are believed to treat various ailments in the Indian traditional system of medicine, the Ayurveda, and in various folk medicines.

Parts Used: Roots, stems, leaves, and flowers.

Local Uses: Flowers are used to treat dysentery and leucorrhea. Poultice of fresh leaves and stems is used to treat sprains, eczema, boils and contusions. Decoction of the leaves is used to treat typhoid, bronchitis and eczema.

Chemical Constituents: Aromatic acrid oil, tannins, fatty acids. Leaves contain flavonoids, including kaempferol and quercetin, proanthocyanidins, phenolic acids, and ferulic acid.

4.9 *Lawsonia inermis*

Location: Ewu

Family: Lythraceae

Botanical Name: *Lawsonia inermis*

Common Names: Henna, cypress shrub, Egyptian privet

Local Name: ***Edo:*** Lalli; ***Hausa:*** Gwarzo, Lalle, Marandaa; ***Yoruba:*** Lalli.

History: The English name "henna" comes from the Arabic hinnā, loosely pronounced as /ħinna/. The name henna also refers to the dye prepared from the plant and the temporary body art (staining) based on those dyes. Henna has been used since antiquity to dye skin, hair and fingernails, as well as fabrics including silk, wool and leather. Historically, henna was used in the Arabian Peninsula, Indian subcontinent, parts of South East Asia, Carthage, other parts of North Africa and the Horn of Africa. The name is used in other skin and hair dyes, such as black henna and neutral henna, neither of which is derived from the henna plant.

Belief: Some also believe that steaming or warming the henna pattern will darken the stain, either during the time the paste is still on the skin, or after the paste has been removed. It is debatable whether this adds to the colour of the end result. After the stain reaches its peak colour, it remains for a few days, then gradually wears off by way of exfoliation.

Local Uses: Used in Cote d'Ivoire as an ingredient in arrow poison. Bark scales are sometimes used as a fish poison. A decoction of the bark is used as an emmenagogue, and also to treat liver problems and nervous symptoms (Uphof, 1959; Chevallier, 1996).

Chemical Constituents: Coumarins, naphthaquinones (including lawsone), flavonoids, sterols and tannins (Chevallier, 1996).

4.10 *Mimosa pigra*

Location: Okpella

Family: Fabaceae

Botanical Name: *Mimosa pigra*

Common Names: Bashful palm, catclaw mimosa, giant seat, giant sensitive plant, thorny.

Local Names: ***Hausa:*** gwambe, gumbii, kwriyaa, kaidaji, dan kunya; ***Yoruba:*** Ewon, agogo.

History: It is native to the neotropics but has been listed as one of the world's 100 worst invasive species and forms dense, thorny, impenetrable thickets, particularly in wet areas. The genus *Mimosa* (Mimosaceae) contains 400–450 species, which are mostly native to South America. *M. pigra* is a woody invasive shrub that originates from tropical America and has now become widespread throughout the tropics.

Local Use: The plant is used in tropical Africa as a tonic and a treatment for diarrhoea, gonorrhoea and blood poisoning. The leaf is said to contain mimosine; it is purgative and perhaps a tonic. A decoction of the leaves and stems is used to treat thrush in babies and bed-wetting in children. The powdered leaf is taken with water to relieve swelling, and is used in the treatment of asthma, diarrhoea, typhoid fever and genitourinary tract infections (Sonibare and Gbile, 2008).

Chemical Constituents: Saponins, flavonols, and glycosides (Mbatchou *et al.*, 1992).

4.11 *Ocimum kilimandscaricum*

Family: Lamiaceae

Botanical Name: *Ocimum kilimandscaricum*

Local Uses: It is useful to treat coughs, bronchitis, catarrh, foul ulcers and wounds, anorexia and vitiated vata, and glaucoma (Warrier *et al.*, 1995; Khare, 2007).

4.12 *Securinega virosa*

Family: Euphorbiaceae

Botanical Name: *Securinega virosa*

Common Names: Whiteberry bush, snowberry tree

History: It is widely distributed throughout tropical Africa, India, Malaysia, China and Australia. In Nigeria, it is found in virtually all parts of the country. It is found in many parts of Africa including the northeastern Nigeria.

Parts Used: Root, leaves, and bark

Local Uses: The leaves are used as laxatives. Root juice mixed in fat is used as a soothing ointment. The bark is an astringent for treatment of diarrhoea and dysentery. It is an analgesic and it is also used for treatment of abscesses and anaemia.

Chemical Constituents: Alkaloids (securine), tannins and carbohydrates.

4.13 *Senna alata*

Family: Caesalpiniaceae

Botanical Name: *Senna alata*

Common Names: Senna, ringworm plant

Local Names: ***Igbo:*** Ogalu, Ndiuchi; ***Yoruba:*** Asunrun Oyinbo;

History: It is native to Mexico, and can be found in diverse habitats.

Parts Used: Whole plant

Local Uses: The entire plant is used for the treatment of venereal diseases in women. Infusion or decoction of the leaf is used as a mild laxative and a purgative in large doses. For treating ringworm and other fungal infections of the skin, the leaves are ground in a mortar to obtain a kind of "green cotton wool". This is mixed with the same amount of vegetable oil and rubbed on the affected area two or three times a day. A fresh preparation is made every day (Hirt and Bindanda, 2008).

Chemical Constituents: Anthraqunones, azulene, saponins

4.14 *Thevetia neriifolia*

Family: Apocynaceae

Botanical Name: *Thevetia neriifolia*

Common Names: Exile tree, yellow oleander

History: A shrub to 6 m high, native of central and tropical S America from Mexico and the West Indies to Brazil.

Belief: It is believed that the plant's toxicity is comparable to the venom of a rattlesnake.

Parts Used: Leaves, stem, bark, and kernel

Local Uses: All parts of these plants are toxic (especially the seeds) and contain a variety of cardiac glycosides including thevetins A and B. Ingestion of yellow oleander results in nausea, vomiting, abdominal pain, diarrhoea, cardiac dysrhythmias, and hyperkalaemia (Bandara *et al.*, 2010; Roberts *et al.*, 2006).

Chemical Constituents: Nerifolin, the cardenolides thevetin A and B (Bandara *et al.*, 2010).

4.15 *Vernonia amygdalina*

Location: Ewu

Family: Asteraceae

Botanical Name: *Vernonia amygdalina*

Common Name: Bitter leaf

History: A member of the Asteraceae family, it is a small shrub that grows in tropical Africa.

Belief: It is consumed by Hausa women of northern Nigeria in the belief that it enhances sexual attraction

Parts Used: Bark, root, leaves, and fruits.

Local Uses: In a preliminary clinical trial, a decoction of 25 g fresh leaves of *V. amygdalina* was 67% effective in creating an adequate clinical response in African patients with mild falciparum malaria (Challand and Willcox, 2009). Enhancement of the immune system; studies have shown that *V. amygdalina* extracts may strengthen the immune system through regulation of many cytokines (including nuclear factor kappa-light-chain-enhancer of activated B cells (NF-κB), a proinflammatory molecule) (Sweeney *et al.*, 2005).

Chemical Constituents: Sesquiterpene lactones (vernodalin, vernolepin, and vernomygdin) and steroid glycosides (Calixto, 2000; Smith, 2008; Finar, 2008)

References

Bandara, A., Scott, A. Weinstein, J.W. and Michael, E. (2010) A review of the natural history, toxicology, diagnosis and clinical management of *Nerium oleander* and *Thevetia peruviana*. *Toxicon*, 56(3), 273–281.

Bressani, R., Gómez-Brenes, R.A. and Elías, L.G.H. (1986) Nutritional quality of pigeon pea protein, immature and ripe, and its supplementary value for cereals. *Archivos Latinoamericanos de Nutrición*, 36(1), 108–116.

Calixto, J.B. (2000) Efficacy, safety, quality control, marketing and regulatory guidelines for herbal medicines (phytotherapeutic agents). *Brazilian Journal of Medical and Biological Research*, 33, 179–189.

Challand, S., Willcox, M. (2009) A clinical trial of the traditional medicine Vernonia amygdalina in the treatment of uncomplicated malaria. *Journal of Alternative and Complementary Medicine*, 15(11), 1231–1237.

Chevallier, A. (1996) *The encyclopedia of medicinal plants publication*. London: Dorling Kindersley Publishers.

el-Deeb, K.S. (2003) The volatile constituents in the absolute of *Clerodendron splendens* G Don flower oil. *Bulletin of Faculty of Pharmacy, Cairo University*, 41, 259–263.

Ettebong, E. and Nwafor, P. (2009) In vitro antimicrobial activities of extracts of carpolobia lutea root. *Pakistan Journal of Pharmaceutical Sciences*, 22(3), 335–338.

Finar, I.L. (2008) *Organic chemistry, vol. 2: stereochemistry and the chemistry of natural products*. 5th ed. Delhi: Saurabh Printers Pvt Ltd, pp. 956.

Hirt, H.M. and Bindanda, M. (2008) *Natural medicine in the tropics I: Foundation text*. anamed edition. Winnenden, Germany.

Irvine, F.R. (1961) *Woody plants of Ghana*. 1st ed. London: Oxford University Press, p. 74.

Khare, C.P. (2007) *Indian medicinal plants: an illustrated dictionary*. Springer Science & Business Media, p. 445.

Kumar, V.L. and Basu, N. (1994) Anti-inflammatory activity of the latex of *Calotropis procera*. *Journal of Ethnopharmacology*, 44, 123–125.

Mainasara, M.M., Aliero, B.L., Aliero, A.A. and Dahiru, S.S. (2011) Phytochemical and antibacterial properties of Calotropis Procera (Ait) R. Br. (sodom apple) fruit and bark extracts. *International Journal of Modern Botany*, 1(1), 8–11.

Mbatchou, V.C., Ayebila, A.J. and Apea, O.B. (1992) Antibacterial activity of phytochemicals from *Acacia nilotica*, *Entada africana* and *Mimosa pigra* L. on Salmonela typhi. *Journal of Animal and Plant Sciences*, 10(1), 1248–1258.

Mikhail, O.A. and Musbau, A. (2013) Phytochemical and mineral constituents of *Cochlospermum planchonii* (Hook. Ef. x Planch) root. *Bioresearch Bulletin* 33, 98–110.

Ocimum kilimandscharicum Guerke (Lamiaceae): a new distributional record for peninsular India with focus on its economic potential (PDF download available). Available at https://www.research-gate.net/publication/276096752_Ocimum_kilimandscharicum_Guerke_Lamiaceae_A_New:Distributional_Record_for_Peninsular_India_with_Focus_on_its_Economic_Potential [accessed 27 October 2017].

Priya, N. (2014) *Health benefits of Hibiscus rosa sinensis*. Value Food: Nutrition and Health Information Portal. Available at http://www.value-food.info

Roberts, D.M, Southcott, E., Potter, J.M., Roberts, S.M., Eddleston, M. and Buckley, N.A. (2006) Pharmacokinetics of digoxin cross-reacting substances in patients with acute yellow oleander (*Thevetia peruviana*) poisoning, including the effect of activated charcoal. *Therapeutic Drug Monitoring*, 28(6), 784–792.

Salah, W., Miller, N., Pagauga, G., Tybury, G., Bolwell, E., Rice, E. and Evans, C. (1995) Polyphenolic flavonoids as scavenger of aqueous phase radicals and chain breaking antioxidants. *Archives of Biochemistry and Biophysics*, 2, 239–246.

Sharma, P. and Sharma, J.D. (2000) In-vitro schizonticidal screening of *Calotropis procera*. *Fitoterapia*, 71, 77–79.

Shehata, A.H., Yousif, M.F. and Soliman, G.A. (2001) Phytochemical and pharmacological investigation of *Clerodendron splendens* G. Don growing in Egypt. *Egyptian Journal of Biomedical Sciences*, 7, 145–163.

Smith, G.J. (2008) *Organic chemistry*. 2nd ed. New York: McGraw-Hill Publishers, pp. 1175.

Sonibare, M.A. and Gbile, Z.O. (2008) Ethnobotanical survey of anti-asthmatic plants in South Western Nigeria. *African Journal of Traditional, Complementary and Alternative Medicines*, 5(4), 340–345.

Sweeney, C.J., Mehrotra, S., Sadaria, M.R., Kumar, S., Shortle, N.H., Roman, Y., Sheridan, C., Campbell, R.A., Murray, D.J., Badve, S. and Nakshatri, H. (2005) The sesquiterpene lactone parthenolide in combination with docetaxel reduces metastasis and improves survival in a xenograft model of breast cancer. *Molecular Cancer Therapeutics*, 4(6), 1004.

Udita, T., Poonam, Y. and Darshika, N. (2015) Study on phytochemical screening and antibacterial potential of methanolic flower and leaf extracts of *Hibiscus rosa sinensis*. *International Journal of Innovative and Applied Research*, 3(6), 9–14.

Uphof, J.C. Th. (1959) *Dictionary of economic plants*. Weinheim: Weinheim Publisher.

Warrier, P.K., Nambiar, V.P.K. and Ramankutty, C. (1995) In: Varier, P.S. (ed.), *Indian medicinal plants: a compendium of 500 species*, 4, 157–168.

Yakubu, M.T., Akanji, M.A. and Nafiu, M.O. (2010) Anti-diabetic activity of aqueous extract of *Cochlospermum planchonii* root in alloxan-induced diabetic rats. *Cameroon Journal of Experimental Biology*, 6(2), 91–100.

Zu, Y.G., Liu, X.L., Fu, Y.J., Wu, N., Kong, Y. and Wink, M. (2010) Chemical composition of the SFE-CO2 extracts from *Cajanus cajan* (L.) Huth and their antimicrobial activity in vitro and in vivo. *Phytomedicine*, 17, 1095–1101.

5 Forbs

A forb (sometimes spelled phorb) is an herbaceous flowering plant that is not a graminoid (grasses, sedges and rushes). The term is used in biology and in vegetation ecology, especially in relation to grasslands. This chapter takes a foray into the world of forbs, how they are used by traditional healers in treating diseases, the local taboos surrounding these plants, and the scientific basis, if any, of these uses.

5.1 *Acalypha ciliata*

Family: Euphorbiaceae

Botanical Name: *Acalypha ciliata*

Common Name: Copper leaf plant

Local Names: ***Esan:*** Ifoki; ***Igbo:*** Abaleba; ***Yoruba:*** Ẹfiri.

History: It occurs widely in Africa where it is eaten as a vegetable or fed to animals. In West Africa and East Africa, it is used as a medicinal plant.

Local Uses: Leaves are cooked and eaten as a vegetable; it is said to be eaten with okra *(Abelmoschatus esculentus)* or the leaves of cowpea (*Vigna unguiculate*). A decoction of the leaves is drunk as a remedy for female sterility, whereas the mashed leaves are applied as a dressing to sores. It occurs widely in Africa where it is eaten as a vegetable or fed to animals (Goode, 1989). In West Africa and East Africa, it is used as a medicinal plant (Schmelzer and Gurib-Fakim, 2008).

Chemical Constituents: Alkaloids, flavonoids, saponins, and tannins (Odeja, 2017).

5.2 *Acanthospermum hispidum*

Family: Asteraceae

Botanical Name: *Acanthospermum hispidum*

Common Names: Bristly starbur, goat's head, hispid starbur, starbur

History: It is an annual plant in the family Asteraceae, which is native to Central and South America. It is also naturalized in many scattered places in Eurasia, Africa, and North America.

Local Uses: In India, the seeds are ingested orally to treat bed wetting, whereas, in Malaysia, the entire plant is mixed with castor oil and applied to the skin to treat scabies. Corrêa (1926) compiled a basic compendium of medicinal plants in Brazil, and described the use of the roots of *A. hispidum* to treat coughs and bronchitis, while noting that the seeds are toxic to chickens,

Chemical Constituents: Alkaloids, tannins, flavonoids, saponins, and glycosides

5.3 *Aframomum alboviolaceum*

Family: Zingiberaceae

Botanical Name: *Aframomum alboviolaceum*

Common Names: Broad amomum, fruit- or grape-seeded amomum

History: *A. alboviolaceum* is a species of dicotyledonous plants of the family Zingiberaceae, originating in tropical Africa

Local Use: Its leaves are used as a spice and its seeds for stomachic purposes and as a vermifuge

5.4 *Ageratum conyzoides*

Family: Asteraceae

Botanical Name: *Ageratum conyzoides*

Common Name: Billy goat weed

Local Names: ***Edo:*** Ebi-ighedore; ***Igbo:*** Oranjila; ***Yoruba:*** Imi-esu.

History: "*Ageratum*" is derived from the Greek "a geras," meaning nonaging, referring to the longevity of the flowers or the whole plant. The specific epithet "*conyzoides*" is derived from "kónyz," the Greek name for *Inula helenium*, which it resembles (Kissmann and Groth, 1993).

Belief: The plant is believed to have a rank smell, likened in Australia to that of a male goat, hence the common name billy goat weed.

Local Uses: It is used to treat wounds and burns. A decoction of the plant is used as a lotion for scabies (craw-craw) and drunk as a remedy for fever. The juice of the leaves is dropped in the eyes to cure inflammations. As a medicinal plant, *A. conyzoides* is widely used in many traditional cultures, especially as an antidysenteric. It is also used as an insecticide and nematicide (Ming, 1999).

Chemical Constituents: Flavonoids, alkaloids, coumarins, essential oils, and tannins (Jaccoud, 1961).

5.5 *Aloe vera*

Family: Liliaceae

Botanical Name: *Aloe vera*

Synonym: *Aloe barteri*

Common Name: Aloe vera

Local Names: West African Aloe, Aloe

History: From the Mediterranean region, it was carried to the New World in the 16th century by Spanish explorers and missionaries. In the modern era, its clinical use began in the 1930s as a treatment for roentgen dermatitis (Grindlay and Reynolds, 1986).

Belief: The leaves of *A. vera* (syn. *Aloe buettneri*) can be applied externally and is believed to help skin conditions such as burns, wounds, insect bites, Guinea worm sores, and vitiligo.

Local Uses: It is used for the treatment of typhoid fever. It is used to treat skin wounds, burns, scalds, blisters, and insect bites, when applied externally (Burkill, 1995).

Chemical Constituents: Aloe resin is the solid residue obtained by evaporating the latex from the pericyclic cells beneath the plant epidermis (Leung and Foster, 1980). Tannins, polysaccharides, organic acids, enzymes, vitamins, and steroids have been identified (Henry, 1979).

5.6 *Anchomanes difformis*

Family: Araceae

Botanical Name: *Anchomanes difformis*

Common Name: Aroids

Local Names: ***Edo:*** Olikhoror; ***Igbo:*** Oje; ***Yoruba:*** Iwaja.

History: It is commonly found in West Africa, occurring in the forests of Sierra Leone to the western Cameroons.

Belief: It is believed to subserve medicinal properties

Parts Used: Stem and leaves

Local Uses: The leaf tuber is used as a lactation stimulant. The plant is used locally for medicinal purposes, and the tuber is harvested from the wild as an emergency food in times of need. The root, pulped with potter's clay, is applied to maturate abscesses (Burkil, 1985). People with a tendency to rheumatism, arthritis, gout, kidney stones, and hyperacidity should take particular caution if including this plant in their diet (Bown, 1995).

Chemical Constituents: Saponins, tannins, and alkaloids (Oyetayo, 2007)

5.7 *Artemisia annua*

Family: Asteraceae

Botanical Name: *Artemisia annua*

Common Names: Annual wormwood, sweet annie, sweet wormwood

History: It is a common type of wormwood, native to temperate Asia, but naturalized in many countries, including scattered parts of North America.

Belief: It is believed by local people to heal diseases and to eliminate evil.

Parts Used: Whole plant

Local Uses: It is used in the treatment of malaria, fevers, and haemorrhoids. It is used in crafting of aromatics, wreaths, and as flavouring for spirits such as vermouth. Sweet annie (add,) is an aromatic annual herb which is the source of artemisinin and essential oils (Simon *et al.*, 1990).

The secondary plant product artemisinin, known in Chinese folk medicine as qinghasu, is an antimalarial with reduced side effects compared with quinine, chloroquine, or other antimalarials (Klayman, 1985).

Chemical Constituents: The major active constituent of *A. annua*, *Artemisia apiacea*, and *Artemisia lancea* is artemisinin. Derivatives of this compound include arteether, artemether, artemotil, artenimol, artesunate, and dihydroartemisinin, which, along with artemisin, are currently being used to treat drug-resistant and non-drug-resistant malaria (Heide, 2006; Hsu, 2006; Li and Wu, 1998; Phan, 2002; Lommen *et al.*, 2006).

5.8 *Aspilia africana*

Family: Asteraceae

Botanical Name: *Aspilia africana*

Common Names: Haemorrhage plant, African marigold, wild sunflower

Parts Used: Leaves and roots

Local Uses: Historically, *A. africana* was used in Mbaise and most Igbo-speaking parts of Nigeria to prevent conception, suggesting potential contraceptive and anti-fertility properties (Oluyemi *et al.*, 2007). Leaf extract and fractions of *A. africana* extracts effectively arrested bleeding from fresh wounds, inhibited microbial growth of known wound contaminants, and accelerated the wound-healing process.

Chemical Constituents: Alkaloids, tannins, saponins, phlobatannins, terpenoids, flavonoids and cardiac glycosides (Oko and Agiang, 2001).

5.9 *Asystasia gigantica*

Family: Acanthaceae

Botanical Name: *Asystasia gigantica*

Common Names: Creeping foxglove, Chinese violet

Local Names: ***Edo:*** Ebe ogboghiro; ***Yoruba:*** Oseta.

History: It is a species of plant in the Acanthaceae family. It is commonly known as the Chinese violet, coromandel or creeping foxglove. In South Africa, this plant may simply be called asystasia.

Local Uses: In some parts of Africa, the leaves are eaten as a vegetable and used as a herbal remedy in traditional African medicine. The leaves are used in many parts of Nigeria for the management of asthma. The plant is used in ethnomedicine for the treatment of heart pains, stomach pains, rheumatism, and as a vermifuge and an anthelmintic (Chopra, 1933; Kokwaro, 1976). It is also used as an ornamental plant.

Chemical Constituents: Saponins, reducing sugars, steroids, glycosides, flavonoids, and anthraquinones (Hamid *et al.*, 2001).

5.10 *Boerhavia diffusa*

Family: Nyctaginaceae

Botanical Name: *Boerhavia diffusa*

Common Names: Spreading hogweed, red spiderling

History: *B. diffusa* is a species of flowering plant in the four o'clock family, which is commonly known as punarnava (meaning "that which rejuvenates or renews the body" in Ayurveda).

Belief: *B. diffusa* is widely dispersed, occurring throughout India, the Pacific, and the southern United States.

Parts Used: Root, leaves, and seeds.

Local Uses: In arthritis, it helps to reduce inflammation and pain in joints. In indigestion, it acts as a carminative, increases appetite, helps digestion and reduces abdominal pain. It also relieves constipation. In Ayurveda, *B. diffusa* has been classified as a "rasayana" herb, which is said to possess properties like antiageing, re-establishing youth, strengthening life and brain power, and disease prevention, all of which imply that they increase the resistance of the body against any onslaught, in other words, providing hepatoprotection and immunomodulation (Govindarajan *et al.*, 2005).

Chemical Constituents: It contains various categories of secondary metabolites, including flavonoid glycosides, isoflavonoids (rotenoids), steroids (ecdysteroids), alkaloids, and phenolic and lignan glycosides. Recently, a rapid method was developed for quantitative estimation of boeravinones in the plant (Bairwa *et al.*, 2014).

5.11 *Bryophyllum pinnatum*

Family: Crassulaceae

Botanical Name: *Bryophyllum pinnatum*

Common Name: Resurrection plant, African never die, Mother of millions.

Local Names: ***Edo:*** Okekwe; ***Igbo:*** Nkwonkwu, Oda opuo; ***Yoruba:*** Odundun.

History: *B. pinnatum,* also known as the air plant, cathedral bells, life plant, miracle leaf, and Goethe plant, is a succulent plant native to Madagascar, which is a popular houseplant and has become naturalized in tropical and subtropical areas. It is distinctive for the profusion of miniature plantlets that form on the margins of its phylloclades, a trait it has in common with some other members of its family.

Parts Used: Leaves and roots.

Local Uses: A poultice is used for treating intestinal pains. It is used as a diuretic and an anthelminthic. Leaves are chewed with some slices of onions to treat high blood pressure and stroke, and leaves are chewed with some slices of onions and plantain roots to treat convulsion and epilepsy. The pounded leaf is applied to soothe inflammation. *B. pinnatum* has been recorded in Trinidad and Tobago as being used as a traditional treatment for hypertension (Lans, 2006).

Chemical Constituents: The bufadienolide bryophillin A, which showed strong anti-tumor-promoting activity *in vitro,* and bersaldegenin-3-acetate and bryophillin C, which were less active. Bryophillin C also showed insecticidal properties (Supratman *et al.*, 2000, 2001).

5.12 *Caladium bicolor*

Family: Araceae

Botanical Name: *Caladium bicolor*

Common Name: Wild cocoyam

Local Names: ***Igbo***: Ede-ohia, Ede-mmuo; ***Yoruba:*** Eje-jesu, lefunlosun.

History: The genus *Caladium* includes seven species that are native to South America and Central America, and are naturalized in India, parts of Africa, and various tropical islands.

Parts Used: Leaves and bulbs

Local Uses: Extract of fresh leaves is applied to the eyes to cure convulsions. It is also applied to the body to treat skin discolouration. Facial paralysis is treated by crushing the bulbs of the plants and applying to the face. Powdered tubers are employed to treat facial skin blemishes by the French Guianan Palikur. All parts of the leaf are macerated in fresh water for an external bath to remedy numerous maladies of the children of the French Guianan Wayapi. Crushed leaves are used in veterinary medicine to destroy vermin on sores of cattle. (Grenand *et al.*, 1987; Duke, 1985; May, 1982).

Chemical Constituents: Alkaloids, tannins, saponins, and cardiac glycosides (Emmanuel *et al.*, 2015).

5.13 *Carica papaya*

Family: Caricaceae

Botanical Name: *Carica papaya*

Common Name: Pawpaw

Local Names: ***Edo:*** Uhoro; ***Igbo:*** Okwuru bekee; ***Hausa:*** Gwandar masar; ***Yoruba:*** Ibepe.

History: Its origin is in the tropics of the Americas, perhaps from southern Mexico and neighboring Central America

Parts Used: Leaves, latex, roots, fruits, and seeds.

Local Uses: Fruits are applied to treat ringworm, Leaves are ingested to treat malaria. Leaves, mixed with lemongrass and guava leaves, are used in the treatment of hypertension. The fruits are a source of papain, a tenderizer. Leaves are cooked with other plants to reduce fever. In some parts of the world, papaya leaves are made into a tea as a treatment for malaria, but the mechanism is not understood and no treatment method based on these results has been scientifically proven (Titanji *et al.*, 2008).

Chemical Constituents: Carotenoids and polyphenols, as well as benzyl isothiocyanates and benzyl glucosinolates, with levels in skin and pulp increasing during ripening. Papaya seeds also contain the cyanogenic substance prunasin (Rivera-Pastrana *et al.*, 2010; Rossetto *et al.*, 2008; Seigler *et al.*, 2002)

5.14 *Chamaecrista mimosoides*

Family: Fabaceae

Botanical Name: *Chamaecrista mimosoides*

Local Uses: In Japan, young stems and leaves are dried and used as a substitute for tea. An aqueous extract from leaves, stems, and pods, called "hamacha", is a conventional beverage in the San-in district of Japan. In Japan, the raw material is used as a diuretic or antidote in folk remedy (Hemen and Lalita, 2012).

Chemical Constituents: Anthraquinones have been reported from seeds (physcion, physcion-9-antrhone, emodin-9-anthrone, and physcion 10, 10-bianthrone) and aerial parts (chrysophanol, physcion, and emodin). Ethanol extract yielded eight compounds: emodin, luteolin, 1,3-benzenediol, oleanolic acid, (R)-artabotriol, α-L-rhamnose, β-sitosterol and daucosterol (Yamamoto *et al.*, 2000; Zhang *et al.*, 2009).

5.15 *Chromolaena odorata*

Family: Asteraceae

Botanical Name: *Chromolaena odorata*

Common Name: Siam weed

Local Names: ***Edo:*** ebe-awolowo; ***Igbo:*** Igwulube; ***Yoruba:*** Ominira, Akintola.

Parts Used: leaves

Local Uses: Leaf infusion is used to treat fever and diabetes, and to prevent early miscarriage. Crushed fresh leaves are used to treat skin rashes. Leaves are used as a tonic and to aid fertility. Siam weed extract accelerates haemostasis and wound healing (Akah, 1990; Wongkrajang *et al.*, 1990; Phan *et al.*, 1998; Phan *et al.*, 2000).

Chemical Constituents: Alkaloids, essential oils, cardinol-pinene, limonene, oxygenated sesqiterpenoids, flavonoids, oxalates, coumarins, saponins, tannins, and vanillin

5.16 *Citrullus colocynthis*

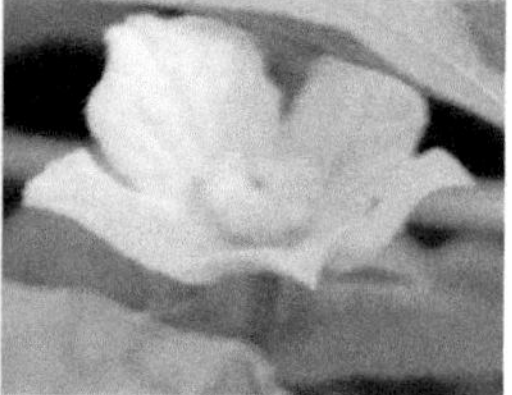

Family: Cucurbitaceae

Botanical Name: *Citrullus colocynthis*

Common Names: Wild gourd, wild melon

Local Names: ***Edo:*** Ikpogi, Ogi; ***Hausa:*** Kwartowa; ***Igbo:*** Elili, Egusi, Ogili; ***Yoruba;*** Egusi baara.

Parts Used: Fruits, leaf

Local Uses: The fruit is eaten in large quantities as a remedy for urinary conditions. The seed shell is powdered, mixed with palm oil and rubbed on the skin to treat fungal infections. The melon shell is used to treat fungal infections on the human skin. Topical *C. colocynthis* application also showed significant efficacy in treatment of patients with painful diabetic neuropathy; the application of a topical formulation of *C. colocynthis* fruit extract can decrease the pain and improve nerve function and quality of life in patients with diabetic neuropathy. This plant has been widely used in folk medicine for centuries. Johann Weyer, in *De praestigiis daemonum* (1563), offered it as a cure for lycanthropy (Mora, 1991).

Chemical Constituents: The oil content of the seeds is 17–19% (w/w), consisting of 67–73% linoleic acid, 10–16% oleic acid, 5–8% stearic acid, and 9–12% palmitic acid. The oil yield is about 400 l/hectare. In addition, the seeds contain high concentrations of arginine, tryptophan, and the sulphur-containing amino acids (Gurudeeban *et al.*, 2010; Schafferman *et al.*, 1998).

5.17 *Cleome viscosa.*

Family: Cleomaceae

Botanical Name: *Cleome viscosa.*

Local Uses: Anti-inflammatory and antipyretic (Tripti *et al.*, 2015; Parimaladevi *et al.*, 2003)

Chemical Constituents: Terpenoids, coumarins, flavonoids, and alkaloids (Jente *et al.*, 1990; Sharaf *et al.*, 1997).

5.18 *Curcuma aeruginosa*

Botanical Name: *Curcuma aeruginosa*

Local Uses: This plant is used to treat a range of conditions, including colic, asthma and cough, obesity, and rheumatism.

Chemical Constituents: Sesquiterpenes, curcumenol, zedoarol, isocur-cum-enol, phytosterol mixtures, including stigmasterol and α-sitosterol (Saad, 2006).

5.19 *Datura stramomium*

Botanical Name: *Datura stramomium*

Local Uses: The Zuni people once used *Datura* as an analgesic to render patients unconscious while broken bones were set. The Chinese also used it as a form of anaesthesia during surgery (Turner, 2009).

Chemical Constituents: All parts of *Datura* plants contain dangerous levels of the tropane alkaloids, atropine, hyoscyamine, and scopolamine, which are classified as deliriants, or anticholinergics. The risk of fatal overdose is high among uninformed users, and many hospitalizations occur amongst recreational users who ingest the plant for its psychoactive effects (Preissel and Hans-Georg, 2002).

5.20 *Dioscorea alata*

Family: Dioscoriaceae

Botanical Name: *Dioscorea alata*

Common Names: water yam or purple yam

History: *D. alata* is native to Southeast Asia, as well as surrounding areas (Taiwan, Ryukyu Islands of Japan, Assam, lowland areas of Nepal, New Guinea, and Christmas Island). It has escaped from its native growth areas and into the wild in many other places.

Local Use: In folk medicine, *D. alata* has been used as a moderate laxative and vermifuge, and to treat fever, gonorrhea, leprosy, tumors, and inflamed haemorrhoids (Wanasundera and Ravindran, 1994).

Chemical Constituents: The colour of the purple varieties is due to various anthocyanin pigments (Moriya *et al.*, 2015).

5.21 *Dioscorea cayenensis*

Family: Dioscoriaceae

Botanical Name: *Dioscorea cayenensis*

Common Name: Yellow yam

5.22 *Euphorbia hirta*

Family: Euphorbiaceae

Botanical Name: *Euphorbia hirta*

Common Names: Asthma plant, cat's hair, hairy spurge

Parts Used: Whole plant

Local Uses: It is used as a remedy for hay fever and catarrh. Leaves increase lactation in nursing mothers. The latex is used to treat conjunctivitis and ulcerated cornea. *E. hirta* is used in the treatment of gastrointestinal disorders (diarrhoea, dysentery, intestinal parasitosis, etc.), bronchial and respiratory diseases (asthma, bronchitis, hay fever, etc.), and conjunctivitis. Hypotensive and tonic properties are also reported from *E. hirta*. The aqueous extract exhibits anxiolytic, analgesic, antipyretic, and anti-inflammatory activities. The stem sap is used in the treatment of eyelid styes and a leaf poultice is used on swellings and boils (Galvez *et al.*, 1993).

Chemical Constituents: Shikimic acid, tinyatoxin, choline, camphol, and quercitol derivatives, containing rhamnose and chtolphenolic acid (Sood *et al.*, 2005).

5.23 *Euphorbia heterophylla*

Family: Euphorbiaceae

Botanical Name: *Euphorbia heterophylla*

Parts Used: Whole plant

Local Uses: The latex is used to treat insect bites, and for the treatment of erysipelas cough, bronchial paroxysmal asthma, hay fever, and catarrh. It is used as an anti-gonorrheal, for treatment of migraine and as a cure for warts. A decoction or infusion of the stems and fresh or dried leaves is taken as a purgative and laxative to treat stomach ache and constipation, and to expel intestinal worms

Chemical Constituents: Flavonoids, such as kaempferol and quercetin (Singh and Kumar, 2013).

5.24 *Hybanthus enneaspermus*

Location: Ewu

Family: Violaceae

Botanical Name: *Hybanthus enneaspermus*

Common Name: Humpback flower

History: This is a plant of coastal savannah, grassland, cultivated fields, barren lands, and roadsides, and is often found growing on open grassland near the seashore. It is found widely in Africa and Madagascar, scattered in India, Sri Lanka, Indochina, South East China, the Philippines, Borneo, East Java, New Guinea, and the northern parts of Australia, and was recently found in the Hengchun Peninsula of Taiwan

Local Uses: In traditional medicines, the whole plant is considered to possess tonic, diuretic and demulcent properties. A decoction of the leaves and tender branches is used to sooth the skin, and these parts of the plant are also made into a cooling liniment for the head. Dried powdered leaves are used to treat asthma.

Chemical Constituents: Alkaloids, terpenoids, saponins, flavonoids, tannins, glycosides, phenols, steroids, and reducing sugars.

5.25 *Jatropha curcas*

Family: Euphorbiaceae

Botanical Name: *Jatropha curcas*

Common Names: Barbados nut, physic nut

Local Names: ***Benin:*** Oru-ebo-, ukpono; ***Igbo:*** Olulu-idu, Owulu-idu, Oluoyibo, Ochocho; ***Hausa:*** Dazugu, Biinida zugaga, Halallamai; ***Yoruba:*** Lapalapa, Botuje, Dodorome, Lakose, Lobotuje, Olobutije, seluju, Serise.

Local Uses: Abortifacient, antibacterial and analgesic.

Chemical Constituents: *J. curcas* contains compounds such as trypsin inhibitors, phytate, saponins and a lectin (Makkar *et al.*, 2008; Martínez-Herrera *et al.*, 2012).

References

Akah, P.A. (1990) Mechanism of hemostatic activity of *Eupatorium odoratum*. *International Journal of Crude Drug Research*, 28(4), 253–256.

Bairwa, K., Srivastava, A. and Jachak, S.M. (2014) Quantitative analysis of boeravinones in the roots of *Boerhaavia diffusa* by UPLC/PDA. *Phytochemical Analysis*, 11, 188–206.

Bown, D. (1995) *Encyclopaedia of herbs and their uses*. London: Dorling Kindersley.
Burkil, H.M. (1985–2004) *The useful plants of west tropical Africa*. Kew: Royal Botanic Gardens Publisher.
Burkill, H.M. (1995) *The useful plants of west tropical Africa*. 2nd ed. Volume 3, Families J–L. Kew, Richmond: Royal Botanic Gardens, 857 pp.
Corrêa, M.P. (1926) *Dicionário das Plantas Úteis do Brasil*. Rio de Janeiro: Editora Nacional.
Chopra, B. (1933) *Further notes on Crustaceae Decapoda in the Indian Museum. V. On Eutrichocheles Modestus (Herbst): Family Axiidae*. http://faunaofindia.nic.in/PDFVolumes/records/035/03/0277-0281.pdf. Accessed June 10, 2016.
Duke, J.A. (1985) *CRC handbook of medicinal herbs*. Boca Raton, FL: CRC Press, 677 pp.
Emmanuel, E.E., Imo, E.J. and Paul, S.T. (2015) Phytochemical composition, antimicrobial and antioxidant activities of leaves and tubers of three *Caladium* species. *International. Journal of Medicinal Plants and Natural Products*, 1(2), 24–30.
Galvez, J., Zarzuelo, A., Crespo, M.E., Lorente, M.D., Ocete, M.A. and Jimenez, J. (1993) Antidiarrhoeal activity of *Euphorbia hirta* extract and isolation of an active flavanoid constituent. *Planta Medica*, 59, 333–336.
Goode, P.M. (1989) *Edible plants of Uganda: the value of wild and cultivated plants as food*. FAO food and nutrition paper 42(1). https://www.worldcat.org/title/edible-plants-of-uganda-the-value-of-wild-and-cultivated-plants-as-food/oclc/19837436. Accessed December 2017.
Govindarajan, R., Vijayakumar, M. and Pushpangadan, P. (2005) Antioxidant approach to disease management and the role of 'Rasayana' herbs of Ayurveda. *Journal of Ethnopharmacology*, 99(2), 165–178.
Grenand, P., Moretti, C. and Jacquemin, H. (1987) *Pharmacopées Traditionnelles en Guyane: Créoles, Palikur, Wayapi*. Paris: Editions de l'ORSTOM, 569 pp.
Grindlay, D. and Reynolds, T. (1986) The aloe vera phenomenon: a review of the properties and modern uses of the leaf parenchyma gel. *Journal of Ethnopharmacology*, 16(2–3), 117–151.
Gurudeeban, S., Satyavani, K. and Ramanathan, T. (2010) Bitter apple (*Citrullus colocynthis*): an overview of chemical composition and biomedical potentials. *Asian Journal of Plant Sciences*, 9(7), 394–401.
Hamid, A.A., Aiyelaagbe, O.O., Ahmed, R.N., Usman, L.A. and Adebayo, S.A. (2001) Preliminary phytochemistry, antibacterial and antifungal properties of extracts of *Asystasia gangetica* Linn T. Anderson grown in Nigeria. https://www.researchgate.net/publication/267416343_. Accessed October 10, 2018.
Heide, L. (2006) Artemisinin in traditional tea preparations of *Artemisia annua*. *Transactions of the Royal Society of Tropical Medicine and Hygiene*, 100(8), 802.
Hemen, D. and Lalita, L. (2012) A review of anthraquinones isolated from Cassia species and their applications. *Indian Journal of Natural Products and Resources*, 3(3), 291–319.
Henry, R. (1979) An updated review of aloe vera. *Cosmetics and Toiletries*, 94, 42–50.
Hsu, E. (2006) The history of qing hao in the Chinese materia medica. *Transactions of the Royal Society of Tropical Medicine and Hygiene*, 100(6), 505–508.
Jaccoud, R.J.S. (1961) Contribuição para o estudo formacognóstico do *Ageratum conyzoides* L. *Revista Brasileira de Farmácia*, 42(11/12), 177–197.
Jente, R., Jakupovic, J. and Olatunji, G.A. (1990) A cembranoid diterpene from *Cleome viscosa*. *Phytochemistry*, 29(2), 666–667.
Kissmann, G. and Groth, D. (1993) *Plantas infestantes e nocivas*. São Paulo: Basf Brasileira.
Klayman, D.L. (1985) Qinghaosu (artemisinin): an antimalarial drug from China. *Science*, 228, 1049–1055.
Kokwaro, J.O. (1976) *Medicinal plants of East Africa*. Nairobi: General Printers Ltd, pp. 12.
Lans, C.A. (2006) Ethnomedicines used in Trinidad and Tobago for urinary problems and diabetes mellitus. *Journal of Ethnobiology and Ethnomedicine*, 2, 45. PMC 1624823. PMID 17040567. doi:10.1186/1746-4269-2-45

Leung, A.Y. and Foster, S. (1980) *Encyclopedia of common natural ingredients used in food, drugs, and cosmetics.* New York: John Wiley & Sons.

Li, Y. and Wu, Y.L. (1998) How Chinese scientists discovered qinghaosu (artemisinin) and developed its derivatives? What are the future perspectives? *Med Trop (Mars),* 58(3 Suppl), 9–12.

Lommen, W.J., Schenk, E., Bouwmeester, H.J. and Verstappen, F.W. (2006) Trichome dynamics and artemisinin accumulation during development and senescence of *Artemisia annua* leaves. *Planta Medica,* 72(4), 336–345.

Makkar, H.P.S., Francis, G. and Becker, K. (2008) Protein concentrate from *Jatropha curcas* screw-pressed seed cake and toxic and antinutritional factors in protein concentrate. *Journal of Science of Food and Agriculture,* 88, 1542–1548.

Martínez-Herrera, J., Jiménez-Martínez, C., Martínez Ayala, A., Garduño-Siciliano, L., Mora-Escobedo, R., Dávila-Ortiz, G., Chamorro-Cevallos, G., Makkar, H.P.S., Francis, G. and Becker, K. (2012) Evaluation of the nutritional quality of nontoxic kernel flour from *Jatropha curcas* L. in rats. *Journal of Food Quality,* 35, 152–158.

May, A.F. (1982) *Surinaams Kruidenboek (Sranan Oso Dresi).* Paramaribo, Surinam: Vaco; and Zutphen, The Netherlands: De Walburg Pers, 80 pp.

Ming, L.C. (1999) Ageratum conyzoides: a tropical source of medicinal and agricultural products. In: J. Janick (ed.), *Perspectives on new crops and new uses.* Alexandria, VA: ASHS Press, pp. 469–473.

Mora, G., Kohl, B.G., Midelfort, E., Bacon, H. (eds.). (1991) Witches, devils, and doctors in the renaissance. In: J. Weyer (ed.), *De praestigiis daemonum.* Translated by John Shea. Binghamton: Medieval and Renaissance Texts and Studies, p. 343.

Moriya, C., Hosoya, T., Agawa, S., Sugiyama, Y., Kozone, I. and Shin-Ya, K. (2015) New acylated anthocyanins from purple yam and their antioxidant activity. *Bioscience, Biotechnology, and Biochemistry,* 79(9), 1484–1492.

Odeja, O., Ogwuche, C.E. and Elemike, E.E. (2017) Clinical phytoscience. *International Journal of Phytomedicine and Phytotherapy,* 2, 12. doi:10.1186/s40816-016-0027-2.

Oko, O.O.K. and Agiang, E.A. (2001) Phytochemical activities of *Aspilia africana* leaf using different extractants. *Indian Journal of Animal Sciences,* 81(8), 814–818.

Oluyemi, K.A., Okwuonu, U.C., Baxter, D.G. and Oyesola Tolulope, O. (2007) Toxic effects of methanolic extract of *Aspilia africana* leaf on the estrous cycle and uterine tissues of wistar rats. *International Journal of Morphology,* 25(3), 609–614.

Oyetayo, V.O. (2007) Comparative studies of the phytochemical and antimicrobial properties of the leaf, stem and tuber of *Anchomanes difformis. Journal of Pharmacology and Toxicology,* 2, 407–410.

Parimaladevi, B., Boominathan, R. and Mandal, S.C. (2003) Evaluation of antipyretic potential of *Cleome viscosa* Linn. (Capparidaceae) extract in rats. *Journal of Ethnopharmacology,* 87(1):11–13.

Phan, T.T., Hughes, M.A. and Cherry, G.W. (1998) Enhanced proliferation of fibroblasts and endothelial cells treated with an extract of the leaves of *Chromolaena odorata* (Eupolin), an herbal remedy for treating wounds. *Plastic and Reconstructive Surgery,* 101(3), 756–765.

Phan, T.T., Allen, J., Hughes, M.A., Cherry, G. and Wojnarowska, F. (2000) Upregulation of adhesion complex proteins and fibronectin by human keratinocytes treated with an aqueous extract from the leaves of *Chromolaena odorata* (Eupolin). *European Journal of Dermatology,* 10(7), 522–527.

Phan, V.T. (2002) [Artemisinine and artesunate in the treatment of malaria in Vietnam (1984–1999)]. *Bulletin de la Société de Pathologie Exotique,* 95(2), 86–88.

Preissel, U. and Preissel, H.-G. (2002) *Brugmansia and Datura: Angel's trumpets and thorn apples.* Ontario: Firefly Books, pp. 124–125.

Rivera-Pastrana, D.M., Yahia, E.M. and González-Aguilar, G.A. (2010) Phenolic and carotenoid profiles of papaya fruit (*Carica papaya* L.) and their contents under low temperature storage. *Journal of the Science of Food and Agriculture*, 90(14), 2358–2365.

Rossetto, M.R., Oliveira do Nascimento, J.R., Purgatto, E., Fabi, J.P., Lajolo, F.M. and Cordenunsi, B.R. (2008) Benzylglucosinolate, enzylisothiocyanate, and myrosinase activity in papaya fruit during development and ripening. *Journal of Agricultural and Food Chemistry*, 56(20), 9592–9599.

Saad, M.S. (2006) *Phytochemical constituents and biological activity of Curcuma aeruginosa Roxb., C. ochrorhiza Val. and Andrographis asculata nees.* Masters thesis, Universiti Putra Malaysia.

Schafferman, D., Beharav, A., Shabelsky, E. and Yaniv, Z. (1998) Evaluation of *Citrullus colocynthis*, a desert plant native in Israel, as a potential source of edible oil. *Journal of Arid Environments*, 40(4), 431–443.

Schmelzer, G.H. and Gurib-Fakim, A. (2008) Plant resources of tropical Africa 11(1). Medicinal plants 1. Wageningen, Netherlands: PROTA Foundation / Leiden, Netherlands: Backhuys Publishers / Wageningen, Netherlands: CTA, 791 pp.

Seigler, D.S., Pauli, G.F., Nahrstedt, A. and Leen, R. (2002) Cyanogenic allosides and glucosides from *Passiflora edulis* and *Carica papaya*. *Phytochemistry*, 60(8), 873–882.

Sharaf, M., El-Ansari, M.A. and Saleh, N.A.M. (1997) Flavonoids of four Cleome and three Capparis species. *Biochemical Systematics and Ecology*, 25(2), 161–166.

Simon, K.V., Charles, D.J., Wood, K.V. and Heinstein, P. (1990) Germplasm variation in artemisinin content of *Artemisia annua* L. using an alternative from crude plant extract. *Journal of Natural Products*, 3, 157–160.

Singh, G. and Kumar, P. (2013) Phytochemical study and screening for antimicrobial activity of flavonoids of *Euphorbia hirta*. *International Journal of Applied and Basic Medical Research*, 3(2), 111–116.

Sood, S.K., Bhardwaj, R. and Lakhanpal, T.N. (2005) *Ethnic Indian Plants in cure of diabetes*. New Delhi: Scientific Publishers.

Supratman, U., Fujita, T., Akiyama, K. and Hayashi, H. (2000) New insecticidal bufadienolide, bryophyllin C, from *Kalanchoe pinnata*. *Bioscience, Biotechnology, and Biochemistry*, 64(6), 1310–1312.

Supratman, U., Fujita, T., Akiyama, K., Hayashi, H., Murakami, A., Sakai, H., Koshimizu, K. and Ohigashi, H. (2001) Anti-tumor promoting activity of bufadienolides from *Kalanchoe pinnata* and *K. daigremontiana* x *tubiflora*. *Bioscience, Biotechnology, and Biochemistry*, 65(4), 947–949. PMID 11388478. doi:10.1271/ bbb.65.947

Titanji, V.P., Zofou, D. and Ngemenya, M.N. (2008) The antimalarial potential of medicinal plants used for the treatment of malaria in Cameroonian folk medicine. *African Journal of Traditional, Complementary and Alternative Medicines*, 5(3), 302–321.

Tripti, J., Neeraj, K. and Preeti, K. (2015) A review on *Cleome viscosa*: anendogenous herb of Uttarakhand. *International Journal of Pharma Research and Review*, 4(7), 25–31.

Turner, M.W. (2009) *Remarkable plants of Texas: uncommon accounts of our common natives*. Texas: University of Texas Press, p. 209. ISBN 978-0-292-71851-7.

Wanasundera, J.P. and Ravindran, G. (1994) Nutritional assessment of yam (*Dioscorea alata*) tubers. *Plant Foods for Human Nutrition*, 46(1), 33–39.

Wongkrajang, Y., Muagklum, S., Peungvicha, P., Jaiarj, P. and Opartkiattikul, N. (1990) *Eupatorium odoratum* Linn: an enhancer of hemostasis. *Mahidol University Journal of Pharmaceutical Sciences*, 17, 9–13.

Yamamoto, M., Shimura, S., Itoh, Y., Ohsaka, T., Egawa, M. and Inoue, S. (2000) Anti-obesity effects of lipase inhibitor CT-II, an extract from edible herbs, Nomame Herba, on rats fed a high-fat diet. *International Journal of Obesity*, 24(6), 758–764.

Zhang, W., Zhang, J. and Li, R. (2009) Inhibitory effect of ethanol extract of *Cassia mimosoides* Linn. on dimethylnitrosamine-induced hepatic fibrosis in rats. *Traditional Chinese Drug Research and Clinical Pharmacology*, 23, 189–202.

6 Grasses

Grasses are mostly herbaceous monocotyledonous plants (apart from *Bambusa* spp.) with jointed stems and sheathed leaves. They are usually upright, cylindrical, with alternating leaves. They have roots, which, in some cases, are modified into rhizomes or stolons. Their inflorescences are differentiated into a panicle, a spike, or a raceme.

6.1 *Axonopus compressus*

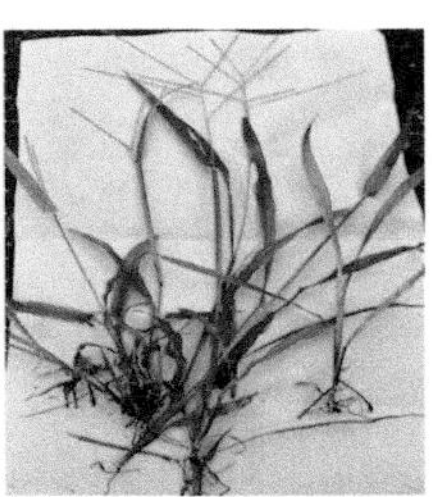

Family: Poaceae

Botanical Name: *Axonopus compressus*

Common Names: Broadleaf carpet grass, blanket grass

Local Name: ***Yoruba:*** Idi.

History: Blanket grass (*Axonopus compressus*) is a robust creeping perennial grass that forms dense mats. Leaves generally reach up to 15 cm long and flowering culms are up to 30–45 cm high. It is shallow rooted, shortly rhizomatous, with slender elongate and branched stolons that root at the nodes. Leaf blades are shiny, flat, folded, lanceolate, 4–15 cm long and 2.5–15 mm broad. Flowering culms are erect and laterally compressed. They bear racemose panicles. There are generally two to three racemes, although up to five is possible. The two upper racemes are paired and borne on a slender peduncle; they are generally one sided. The secondary racemes usually remain hidden in the sheath.

Local Uses: Blanket grass is used as green forage by traditional rabbit raisers in Central Java during both the wet and the dry seasons (Prawirodigdo, 1985).

Chemical Constituents: Crude protein is generally low. A high content of non-structural carbohydrates has also been reported, which may explain the high *in vitro* digestibility observed in some cases (Samarakoon *et al.*, 1990).

6.2 *Brachiaria nigropedata*

Family: Poaceae

Botanical Name: *Brachiaria nigropedata*

History: *B. nigropedata* is a perennial grass belonging to the grass family (Poaceae). It is native to southern Africa, namely the tropical regions of

South Africa and East Africa. *B. nigropedata* is used as a fodder grass in Namibia.

Local Uses: Ashes from the whole plant are used as a treatment against snake bite Brachiaria, Urochloa (Gramineae-Paniceae) in Malesia. The rhizomes act as a diuretic. A paste made from the rhizomes is used in the treatment of kidney problems.

6.3 *Chloris pilosa*

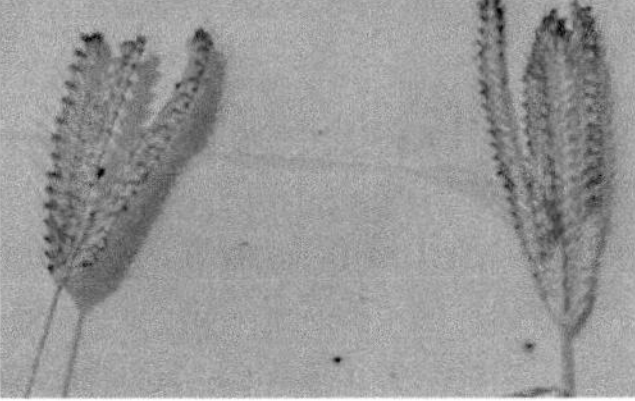

Family: Poaceae

Botanical Name: *Chloris pilosa*

Common Names: Windmill grass, finger grass.

Local Name: ***Yoruba:*** Eéran.

History: *Chloris* is a widespread genus of plants in the grass family, known generally as windmill grass or finger grass. The genus is found worldwide, but especially in the tropical and subtropical regions, and more often in the southern Hemisphere. The species are variable in morphology, but, in general, the plants are less than 0.5 m in height. They bear inflorescences shaped like umbels, with several plumes lined with rows of spikelets.

Chemical Constituents: Aqueous extract of leaves yielded phytosterols, flavonoids, tannins, phenols, carbohydrates, proteins and amino acids.

6.4 *Dactyloctenium aegyptium*

Family: Poaceae

Botanical Name: *Dactyloctenium aegyptium*

Common Names: Crowfoot grass, Indian wheat, coast button grass

Local Name: ***Hausa:*** Kutukku.

History: Coast button grass is a glaucous annual with culms up to 70 cm high, not stoloniferous, but often rooting from the lower nodes. The plant can form a mat with short underground stems. The seed is sometimes harvested from the wild for food, but generally only in times of scarcity. The plant also has local medicinal uses (Ruffo *et al.*, 2002).

Local Uses: The whole plant is used in a decoction to treat lumbago. An infusion of the leaves, mixed with the seeds of pigeon pea, *Cajanus cajan*, is used to accelerate childbirth. A decoction of the leaves, combined with *Scoparia dulcis*, is used as a remedy for dysentery (DeFilipps *et al.*, 1984).

Chemical Constituents: The plant is rich in cyanogenic glucosides at certain stages of growth.

6.5 *Digitaria* spp.

Family: Poaceae

Botanical Name: *Digitaria* spp.

History: *Digitaria* is an annual, growing to 0.5 m (1ft 8in) at a fast rate.

It is hardy to zone (UK) 7. It is in flower from August to September. The flowers are hermaphrodite (i.e. have both male and female organs) and are pollinated by wind. It is suitable forlight (sandy) and medium (loamy) soils and can grow on acid, neutral or basic (alkaline) soils. It cannot grow in the shade, but prefers moist soil.

Local Uses: A decoction of the plant is used in the treatment of gonorrhoea. A folk remedy for cataracts and debility, it is also said to be emetic.

Chemical Constituents: Crabgrass harvest may potentially yield more than 15% crude protein and 60% total digestible nutrients. Studies on chemical composition have yielded (on % dry matter, DM basis) crude protein 12.0–19.2%, total condensed tannins 0.12–0.20%, soluble sugars 5.7–6.3%, NDF44.2–52.3%, OMD 71.2–76.4%. Study on digestibility (IVDMD) and metabolizing energy study yielded 59.3% and 7.99% at early bloom, respectively, and 42.6% and 5.52% at maturity, respectively (Orskov, 1982).

6.6 *Eragrostis tenella*

Family: Poaceae

Botanical Name: *Eragrostis tenella*

Common Name: Lovegrass

History: *E. tenella* is a small densely tufted annual grass, of variable size, usually not much more than 50 cm high. Culms glabrous, spindly, with nodes at the base, may be ramified or not. Leaves are up to 10 cm long. Inflorescence usually has many slender spreading branches. (Galinato *et al.*, 1999).

Local Uses: Seed is sometimes eaten as a cereal, where it is said to be nutritious. The seed is small and fiddly to utilize – it is most commonly seen as a famine food, used when nothing better is available (Burkil 1985)

6.7 *Echonochloa* spp.

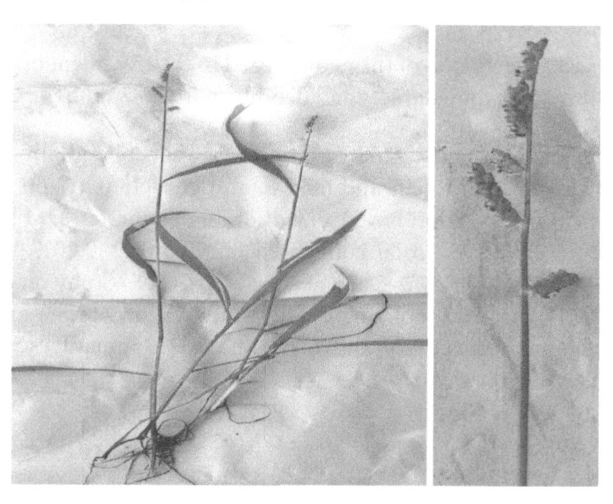

Family: Poaceae

Botanical Name: *Echonochloa* spp.

History: Barnyard millet is an annual plant that can succeed in a wide range of environments from the temperate zone to the tropics. It can be found at elevations up to 2,500 m. It grows best in areas where annual daytime temperatures are within the range 17–30°C but can tolerate 2–40°C. It prefers a mean annual rainfall in the range 700–1,100 mm, but tolerates 310–2,500 mm. An easily grown plant, it is adapted to nearly all types of wet places, and is often a common weed in paddy fields, roadsides, cultivated areas, and fallow fields.

Local Uses: Reported to be preventative and tonic, barnyard grass is a folk remedy for treating carbuncles, haemorrhages, sores, spleen trouble, cancer, and wounds. (Duke, 1983).

Chemical Constituents: Nitrate (Duke, 1983).

6.8 *Eragrostis tremula*

Family: Poaceae

Botanical Name: *Eragrostis tremula*

History: *E. tremula* is a loosely clump-forming plant, an annual to perennial grass with erect, usually unbranched culms, up to 100 cm tall. These culms have attractive trembling panicles. The plant is harvested from the wild for local use as a food and source of material (Burkil 1985).

Local Uses: The culms, bundled together, are used as hand-brooms for indoor use. The culms may also be used for thatching and are woven together to make mats and cordage.

6.9 *Imperata cylindrica*

Family: Poaceae

Botanical Name: *Imperata cylindrica*

Common Names: Spear grass, blady grass, Congo grass

Local Name: ***Igbo:*** Atta; ***Yoruba:*** Ekan.

History: Congo grass is a perennial grass growing to around 120 cm tall and forming extensive clumps by means of its aggressively spreading rhizomes. Considered to be a noxious weed in many areas of the tropics, it is sometimes harvested from the wild for its wide range of edible, medicinal and other uses. It is sometimes used in soil stabilization schemes.

Local Uses: The flowers and the roots are antibacterial, diuretic, febrifuge, sialagogue, styptic, and tonic. The flowers are used in the treatment of haemorrhages, wounds etc. They are decocted and used to treat urinary tract infections, fevers, thirst, etc.

Chemical Constituents: The closely related plant (*I. cylindrica* var. *koenigii*) contains the triterpenoids arundoin, cylindrin and fernenol (Nishimoto *et al.*, 1968).

6.10 *Oplismenus burmannii*

Family: Poaceae

Botanical Name: *Oplismenus burmannii*

Common Names: Basketgrass, wavyleaf basketgrass

History: It can be found in Florida and in Hawaii but is native to Zimbabwe.

Local Uses: Variegated forms have been cultivated as house plants in Europe. Locally occurring species in Australia have been used for revegetation and reclamation in shady or wet areas, though some can be invasive. Some have been promoted as local native plants for wildlife gardens, and as a lawn grass; they are edible to livestock (Scholz, 1981).

Chemical Constituents: Hytosterol glucosides, acylated phytosterol glucosides, 3α-clionasterol glucosides, Acylated 3β-clionasterol glucosides.

6.11 *Panicum maximum*

Family: Poaceae

Botanical Name: *Panicum maximum*

Common Name: Guinea grass

History: Guinea grass is East African in origin but is now widely cultivated throughout the tropics and subtropics as a forage. It was introduced from Africa into the West Indies before 1756, but for production of bird seed, rather than as a forage. It reached Singapore in 1876 and the Philippines in 1907 and is now widely distributed throughout South-East Asia.

Local Uses: Guinea grass is a palatable and high-quality tropical grass, used as forage for ruminants in grazed pastures or in cut-and-carry systems. Guinea grass forage is also dried and ground for use in mixtures with legumes as leaf meal, mainly for non-ruminants such as chickens and pigs. It can be conserved as hay or ensiled. It is also used as medicine for heartburn by the Malays under the name "berita" (Van Oudtshoorn, 1999).

Chemical Constituents: The forage is reported to contain 5.9 g protein, 1.6 g lipid, 81.9 g total carbohydrate, 35.7 g fibre, 10.6 g ash, 2090 mg Ca, and 590 mg P.

6.12 *Panicum latifolium*

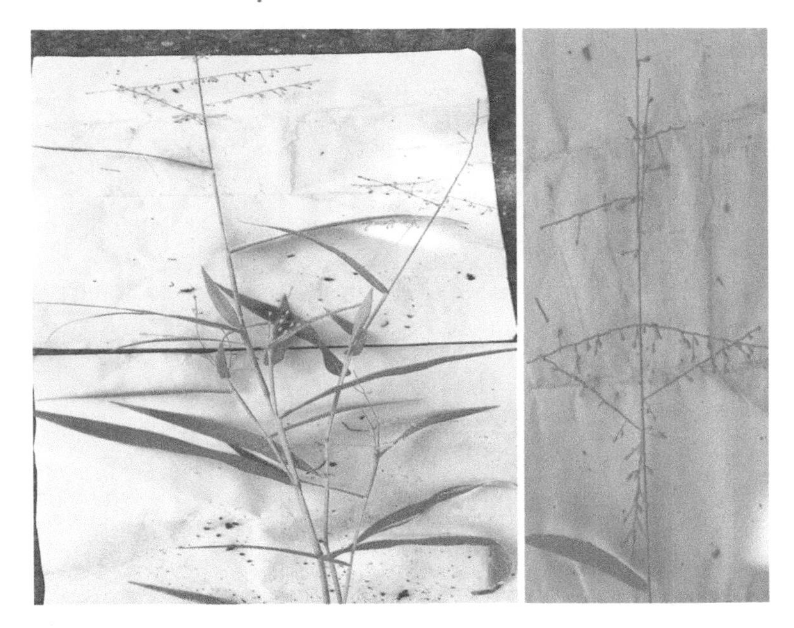

Family: Poaceae

Botanical Name: *Panicum latifolium*

Common Name: Panicgrass

History: *Panicum* (panicgrass) is a large genus of about 450 species of grasses native throughout the tropical regions of the world, with a few species extending into the northern temperate zone. They are often large, annual or perennial grasses, growing to 1–3 m tall. "Panicum". Natural Resources Conservation Service PLANTS Database.

Local Uses: The smoke of the burning plant is used to fumigate wounds and as a disinfectant in the treatment of smallpox and throat infections.

6.13 *Paspalum scrobiculatum*

Family: Poaceae

Botanical Name: *Paspalum scrobiculatum*

Common Names: Kodo millet, cow grass, rice grass, ditch millet.

History: *P. scrobiculatum,* kodo millet, is an annual grain that is grown primarily in India, but also in the Philippines, Indonesia, Vietnam, Thailand, and in West Africa, where it originated. It is grown as a minor crop in most of these areas, with the exception of the Deccan plateau in India where it is grown as a major food source. It is a very hardy crop that is drought tolerant and can survive on marginal soils where other crops may not survive, and can supply 450–900 kg of grain per hectare. Kodo millet has considerable potential to provide nourishing food to subsistence farmers in Africa and elsewhere.

Local Uses: Widely cultivated as a minor millet in Africa and Asia, especially India (Senthivel *et al.*, 1994; Anon., 1996; Ramasamy *et al.*, 1996). Also used for forage (Bisset *et al.*, 1974; Kitamura and Nada, 1986) and as a feed supplement (Kapoor *et al.*, 1987). In India, it has been used as a substrate for mushroom production and for medicinal purposes.

Chemical Constituents: The grain is composed of 11% of protein, providing 9 g/100 g consumed. It is an excellent source of fibre at 10 g/100 g (37–38%), as opposed to rice, which provides 0.2 g/100 g, and wheat, which provides 1.2 g/100 g. An adequate fibre source helps combat the feeling of hunger. Kodo millet contains 66.6 g of carbohydrates and 353 kcal per 100 g of grain, comparable to other millets. It also contains 3.6 g of fat per 100 g. It provides minimal amounts of iron, at 0.5 mg/100 mg, and minimal amounts of calcium, at 27/100 mg. Kodo millets also contain high amounts of polyphenols, a group of antioxidant compound.

6.14 *Pennisetum purpureum*

Family: Poaceae

Botanical Name: *Pennisetum purpureum*

Common Names: Napier grass, elephant grass, Uganda grass

History: A species of perennial tropical grass native to the African grasslands. Historically, this wild species has been used primarily for grazing

Local Uses: It is more affordable for farmers than insecticide use. In addition to this, Napier grass improves soil fertility, and protects arid land from soil erosion. It is also utilized for firebreaks, windbreaks, in paper pulp production and, most recently, to produce bio-oil, biogas and charcoal.

Chemical Constituents: Trace of alkaloids have been detected in the leaves

6.15 *Saccharum officinarum*

Family: Poaceae

Botanical Name: *Saccharum officinarum*

Common Name: Sugarcane

Local Names: ***Benin*:** Ukhure; ***Hausa*:** Reke.

History: It originated in Southeast Asia and is now cultivated in tropical and subtropical countries worldwide for the production of sugar and other products.

Local Uses: Portions of the stem of this and several other species of sugarcane have been used from ancient times for chewing to extract the sweet juice. *S. officinarum* and its hybrids are grown for the production of sugar, ethanol, and other industrial uses in tropical and subtropical regions around the world. The stems and the byproducts of the sugar industry are used for feeding to livestock. As its specific name (*officinarum*, "of dispensaries") implies, it is also used in traditional medicine, being applied both internally and externally.

Chemical Constituents: Contains sucrose (glucose and fructose) in juice, cellulose (polymer of many glucose units) in the fibre.

6.16 *Setaria longiseta*

Family: Poaceae

Botanical Name: *Setaria longiseta*

History: Three species of *Setaria* have been domesticated and used as staple crops throughout foxtail millet *(Setaria italica)*, korali *(Setaria pumila)* in India, and, before the full domestication of maize, *Setaria macrostachya* in Mexico. Several species are still cultivated today as food or as animal fodder, such as foxtail millet *(S. italica)* and korali *(S. pumila)*, while others are considered to be invasive weeds. *Setaria viridis* is currently being developed as a genetic model system for bioenergy grasses.

Local Uses: In Kenya, *S. longiseta* has increasingly been grown as a forage grass in recent times.

6.17 *Setaria pumila*

Family: Poaceae

Botanical Name: *Setaria pumila*

History: It is native to Europe and Africa, but it is known throughout the world as a common weed. It grows in lawns, sidewalks, roadsides, cultivated fields, and many other places.

Local Uses: The grass can be made into a good hay. In Lesotho, sheaves of grain are tied together using rope made from culms of *S. pumila* that are twisted together. In some areas, this grass plays an important role in stabilising bare soil to protect it from erosion. It can also be eaten as a sweet or savoury food in all the ways that rice is used, or ground into a powder and made into porridge, cakes, puddings.

6.18 *Sorghum bicolor*

Family: Poaceae

Botanical Name: *Sorghum bicolor*

History: Sorghum originated in northern Africa and is now cultivated widely in tropical and subtropical regions.

Local Uses: Sorghum is the world's fifth most important cereal crop after rice, wheat, maize and barley. It is also used for making a traditional corn broom. Sorghum is one of a number of grains used as wheat substitutes in gluten-free recipes and products. It is used in feed and pasturage for livestock.

Chemical Constituents: Hydroxybenzoic acids, gallic acid, flavones, apigenin, flavanones, naringen flavonols.

6.19 *Sporobolus indicus*

Family: Poaceae

Botanical Name: *Sporobolus indicus*

History: This bunch grass is native to temperate and tropical areas of the Americas. It can be found in more regions, as well as on many Pacific Islands, as an introduced species and a common weed of disturbed habitats

Local Uses: It is considered to be an antifertility drug in some countries. A fibre is obtained from the leaves. The tough culms are used for making hats and other items that can be woven from straw

6.20 *Tridax procumbens*

Family: Poaceae

Botanical Name: *Tridax procumbens*

Common Names: Coatbuttons or tridax daisy

History: It is native to the tropical Americas, but it has been introduced to tropical, subtropical, and mild temperate regions worldwide.

6.21 *Zea mays*

Family: Poaceae

Botanical Name: *Zea mays*

Common Name: Maize

Local Names: ***Yoruba:*** Agbado; ***Igbo:*** Oka; ***Hausa:*** Masara.

History: Maize is an erect, robust, usually unbranched annual plant. It can grow up to 6 m tall but is usually around 2 m, with cultivars that can range from around 1 m up to 3 m or more in height. A common crop around the world, providing a range of foods including popcorn and sweet corn, it can be ground into a flour. First domesticated in the Americas around 4,000 BC, its cultivation has since spread to most parts of the world.

Local Uses: The mature kernel can be dried and used whole or ground into a flour. It has a very mild flavour and is used especially as a thickening agent in foods, such as custards. The dried kernel of certain varieties can be heated in an oven, when they burst to make popcorn. The kernels can also be sprouted and used in making uncooked breads and cereals. The fresh succulent 'silks' (the styles of the female flowers, also attached to the mature cobs) can also be eaten. The pollen is used as an ingredient of soups. The kernel is diuretic and a mild stimulant. It provides a good emollient poultice for ulcers, swellings and rheumatic pains, and is widely used in the treatment of cancer, tumours and warts.

Chemical Constituents: The maize protein is rich in proline, glutamine, leucine and/or alanine.

References

A Barefoot Doctors Manual. Running Press. ISBN 0-914294-92-X. Bisset *et al.*, 1974; Kitamura and Nada, 1986; Su and Lin, 1994; Compere *et al.*, 1995.

Anon. (1996) *Lost crops of Africa. Volume 1: grains.* Washington, DC: National Academy Press.

Bisset, W.J. and Marlowe, G.W.C. (1974) Productivity and dynamics of two Siratro based pastures in the Burnett coastal foothills of south east Queensland. *Tropical Grasslands*, 8(1), 17–24.

Burkill, H.M. (1985) Entry for *Lasiurus hirsutus* (Forssk.) Boiss. [family POACEAE]. In: *The useful plants of west tropical Africa.* 2nd edition. Kew, UK: Royal Botanic Gardens.

DeFilipps, R.A., Maina, S.L. and Crepin, J. (1984) *Medicinal plants of the Guianas.* Vol. 2004, p. 20013-7012. Washington, DC: Department of Botany, National Museum of Natural History, Smithsonian Institution.

Duke, J.A. (1983) *Handbook of energy crops.* New CROPS web site, Purdue University.

Duke, J.A. and Ayensu, E.S. (1985a) *Medicinal plants of China.* 2 Vols. Algonac, MI: Reference Publications Inc. ISBN 0-917266-20-4.

Duke, J.A. and Ayensu, E.S. (1985b) *Medicinal plants of China.* Millets. Earth360. (2010-13). http://earth360.in/web/Millets.html

Galinato, M.I., Moody, K. and Piggin, C.M. (1999) *Upland rice weeds of South and Southeast Asia.* p. 156. Los Banos, Philippines: International Rice Research Institute.

Kapoor, P.N., Netke, S.P., Bajpai, L.D. (1987) Kodo (*Paspalum scorbiculatum*) as a substitute for maize in chick diets. *Indian Journal of Animal Nutrition*, 4(2), 83–88.

Kitamura, M. and Nada, Y. (1986) Preliminary evaluation of 24 tropical grasses introduced into sub-tropical Japan. *Journal of Japanese Society of Grassland Science* 32(3), 278–280.

Nishimoto K., Ito, M. and Natori, S. (1968) The structures of arundoin, cylindrin and fernenol: triterpenoids of fernane and arborane groups of *Imperata cylindrica* var. *koenigii*. *Elsevier*, 24(2), 735–752.

Orskov, E.R. (1982) *Protein nutrition in ruminants*. pp. 735–752. London: Academic Press. doi:10.1016/0040-4020(68)88023-8. Panicum. Natural Resources Conservation Service Plants Database. USDA. Retrieved 15 May 2015.

Prawirodigdo, S. (1985) Green feeds for rabbits in West and Central Java. *The Journal of Applied Rabbit Research*, 8(4), 181–182.

Ramasamy, M., Vairavan, K. and Srinivasan, K. (1996) Production potential and economics of cereal based cropping system in red lateritic soils of Pudukkottai district. *Madras Agricultural Journal*, 83(4), 236–239.

Ruffo, C.K., Birnie, A. and Tengnas, B. (2002) *Edible wild plants of Tanzania*. 766 p. RELMA Technical Handbook Series 27. Nairobi, Kenya: Regional Land Management Unit (RELMA), Swedish International Development Cooperation Agency (Sida).

Samarakoon, S.P., Shelton, H.M. and Wilson, J.R. (1990) Voluntary feed intake by sheep and digestibility of shaded *Stenotaphrum secundatum* and *Pennisetum clandestinum* herbage. *The Journal of Agricultural Science*, 114(2), 143–150.

Scholz, Ursula. (1981) *Monograph of the genus* Oplismenus *(*Gramineae*) (PDF)*. Translated by McIntyre, Anthony, Atkins, Spencer, Twerase, Felix. Vaduz, Liechtenstein: J. Cramer. ISBN 3-7682-1292-0. Archived from the original (PDF) on 26 April 2012.

Senthivel, S., Solaiappan, U. and Subramanian, S. (1994) *Agricultural Science Digest, Karnal*, 14(3–4), 197–200.

Van Oudtshoorn, F. (1999) *Guide to grasses of southern Africa*. Pretoria, South Africa: Briza Publications.

Part Two

Application of Medicinal Plants for Specific Diseases

7 Medicinal Plants for Malaria and Parasitic Infections

Malaria and several other endemic parasitic diseases play a pivotal role in the downward spiral cycle of poverty, disease and underdevelopment that is characteristic of much of sub-Saharan Africa. These diseases constitute a major obstacle to the socio-economic progress of the region and take up to 55% of public health subventions in some countries. These infectious diseases are often referred to as *neglected tropical diseases* (NTDs). They include malaria, leishmaniasis, Chagas disease (CD) (American trypanosomiasis), human African trypanosomiasis (sleeping sickness), dengue fever, leprosy (Hansen's disease), trachoma, dracunculiasis (guinea-worm disease), fascioliasis, rabies, onchocerciasis (river blindness), Buruli ulcer (*Mycobaterium ulcerans* infection), bouba (yaws), lymphatic filariasis (elephantiasis), schistosomiasis (bilharzia), helminthiases, cysticercosis, and taeniasis. They affect about one-sixth of the world's population and are a global public health problem, predominantly occurring in less-developed countries.

Although these diseases are preventable and curable, they remain a major health and economic burden because they are found in countries with low income and without political influence, so that the patients affected with NTDs and the public health systems of less-developed countries cannot afford the financial return required by most pharmaceutical companies to develop effective treatments. Additionally, the lack of drugs for these diseases is due not only to the limited scientific knowledge about effective therapeutic targets, or the gap between basic and preclinical research, but is also a direct result of insufficient public policies and financial support for research and drug development in developing countries. The burden of these diseases on the Nigerian population can be greatly reduced by adopting a bench-to-bedside approach, which will allow for the clinical adaptation of some of the plant remedies discussed in this chapter with ethnomedical evidence and significant pre-clinical laboratory studies.

7.1 Malaria

Malaria remains one of the most persistent and pressing public health problems in Nigeria, despite decades of efforts towards global eradication and control. It is endemic in all parts of the country, except in the Mambilla highlands of Taraba state, where the high altitude and seasonally temperate climate moderate the incidence of malarial infections. Nigeria accounts for a quarter of all malaria cases in Africa (WHO, 2008). Malaria is, in many cases, a fatal disease caused by infection with any of five protozoan parasites belonging to the genus *Plasmodium*: *Plasmodium vivax*, *Plasmodium malariae*, *Plasmodium falciparum*, *Plasmodium ovalis*, or *Plasmodium knowlesi*. The mortality due to malaria has declined in recent years because of acquired immunity and the relative ease of its diagnosis by lay people. The majority of deaths are caused by infections with *P. falciparum*.

It is one of the deadliest vector-borne diseases, which affects millions worldwide, especially in sub-Saharan Africa and Southeast Asia. The disease is endemic in 104 countries, with about 3.4 billion people, mainly in the tropical parts of the world, at risk of malaria. It is estimated that 216 million malaria cases and about 445,000 deaths occurred globally in 2016, with about 80% of the cases and 90% of the deaths recorded in Africa. Nigeria suffers the world's greatest malaria burden, with approximately 30% of the total malaria burden in Africa. According to the WHO, there were an estimated 57.3 million cases of malaria and about 100,700 deaths in Nigeria in 2017. Most of the deaths (77%) were of children younger than five years old (WHO, 2018).

The endemic poverty associated with malaria burden and the magnitude of its effects on the socio-economic development of the country call for an action paradigm that links the traditional treatment approaches to modern technology, while addressing the nexus of environmental, economic, and political dimensions that sustains this deadly disease. The disease has been associated with much of the infant and premature deaths. In some untreated cases, it may leave patients with permanent neurological damage and metabolic dysfunction. Malaria is partly responsible for very low productivity in some communities and increased school absenteeism. Affected families sometimes spend up to 25% of their income on malaria treatment and prevention, a massive cost for already-impoverished households. The use of readily available and affordable medicinal plants and their constituents to combat malaria offers a broad-spectrum approach that meaningfully addresses both the proximal and fundamental causes of malaria. Furthermore, nearly all known modern malaria drugs (pharmaceuticals) were derived from medicinal plants identified through traditional medicine.

Most lay people classify all fevers together as malaria, but an expert herbalist or traditional healer easily distinguishes the various types and recommends appropriate treatment. There is often no distinction between the various forms of malarial fever, and both the tertian and quartan fevers are treated alike. The pathophysiology of malaria fever is still poorly understood. Although the fever in malaria is synchronous with the intermittent release of malarial parasites during rupture of schizonts and a minimum parasite density (pyrogenic threshold) is needed before a febrile response can occur in malaria, the manifestation of malaria can be devoid of a febrile condition but expressed through a plethora of other pathological conditions. Severe malaria is often associated with inflammatory, metabolic or haematologic complications and organ failure, and requires emergency hospital treatment. If not adequately treated, malaria can manifest in severe anaemia, haemoglobinuria, and acute respiratory distress syndrome, hypotension due to cardiovascular collapse, metabolic acidosis, hypoglycaemia, acute renal failure, and psychotic conditions, due to cerebral malaria.

7.1.1 Signs and Symptoms of Malaria

Malaria characteristically begins with a fever. Infection may lead to several different patterns of symptoms in the host, namely asymptomatic state, uncomplicated disease, or severe disease. Severe malaria occurs mostly in young children where it is a major cause of death. Since the minimum period between being bitten by an infected mosquito and developing symptoms of malaria is eight days, it means that a febrile illness that develops within a week of arrival in a malarial area is likely not to be due to malaria. Cold, chill with shivery, shaky, and sweaty episodes may develop. Other symptoms include headache, jaundice (yellowness of the eyes), a sick feeling, joint pains, aching muscles, and vomiting. These signs and symptoms may not be present in all cases and the most reliable confirmation of a malarial diagnosis is by laboratory tests.

The disease is thought to result from the sequestration of parasites in the small blood vessels of the brain and the deregulation of key immune system elements. The cellular and molecular regulatory mechanisms underlying the pathogenesis of disease are, however, not fully understood. What is known is that the genetic determinants of the host play an important role in the severity of the disease and the outcome of infection (Marquet, 2017).

7.1.2 Malaria Control in Nigeria

Control of malaria is undertaken through three main approaches: a) the use of insecticides (including treated bed nets) and insect-repelling herbs, or other methods for the elimination of the disease-carrying *Anopheles* mosquitoes; b) the administration of preventive therapy, with such drugs like primaquine, which is the only drug capable of destroying the parasitic forms found in hepatocytes, and is currently the most effective agent for the prevention of the transmission of the disease; and c) the use of drugs that treat the acute attack of the disease. The antimalarial drugs are usually classified, according to their mode of action, into two major groups. The first group consists of compounds related to the alkaloid quinine, from the bark of the *Cinchona* tree, and their synthetic derivatives aminoquinolines (amodiaquine, primaquine, chloroquine, and mefloquine), in addition to acridines (such as halofantrine). They act by interfering with the metabolism of the glucose pathway in the parasite and the ability of the parasite to digest haemoglobin, by inhibiting the production of haemozoin. In this way, the drugs interfere with *Plasmodium* nutrition, or poison it with high levels of ferriprotoporphyrin-IX, a toxic by-product of the digestion of haemoglobin. This group also includes artemisinin, from the *Artemisia annua* herb, and its synthetic derivatives, artemether and artesunate sodium, which are active against some *Plasmodium* strains resistant to other drugs, and also interacts with the haem group, generating a free radical lethal to the parasite. The second category of antimalarials includes pyrimidines (such as pyrimethamine) and biguanides (such as proguanil), that are able to inhibit the dihydrofolate reductase enzyme, thus interfering with the synthesis of folic acid, an important cofactor in the process of the synthesis of DNA and amino acids (Kobe de Oliveira *et al.*, 2015).

7.1.3 Phytotherapy of Malaria

There are two parallel tracks in the treatment of malaria in Nigeria and in most malaria-endemic poor countries. The first track consists of the use of a chemotherapeutic strategy, based on the administration of modern antimalarial drugs as described in Section 7.1.2., an approach that hinges upon species- and stage-specific treatments, guided by laboratory diagnosis and screening. This approach suits malaria, as the treatment is designed for city dwellers or travellers from the developed, malaria non-endemic world. However, limiting treatment to that which affirms diagnosis may not be rational in endemic zones. Most of the endemic malarias remain out of diagnostic reach, either by inaccessibility of care during the parasite stage, insensitivity of the diagnostic technology, or unavailability of diagnostic services. The partial and fragmented chemotherapeutic attack of malaria, guided by confirmed diagnostics, leaves most of the endemic malarias unchallenged (Baird, 2012).

The second approach consists of the use of safe and effective herbal products for the treatment of diseases that have not been confirmed by laboratory diagnosis, which enables the healer to administer a disease-elimination therapy, a single course of treatment aimed at all species and stages of malaria and related diseases, identified as major killers which are known collectively as 'malaria'. The prescription of herbal anti-malarial drugs is not usually restricted to professional herbalists; people with some knowledge of herbs often volunteer their services when malaria occurs in the community. Cases of drug-resistant malaria (both herbal and allopathic) are usually referred to the professional herbalists (Iwu, 2014).

Many plant genera are used either alone or in combination with each other for the treatment of malaria (Table 7.1). The plants used are mainly of the families Fabaceae, Rubiaceae and Apocynaceae, with a few from the families Bombaceae, Compoitae, Loganaceae, Rutaceae, Solanaceae, Meliaceae, and Gramineae. Antimalarial phytomedicines vary enormously from one community to another, in both the choice of plants and the method of preparation.

Complementary to the oral therapy is the steam treatment, whereby the patient is covered with a thick blanket or cloth and subjected to the vapors from

Table 7.1 Major Therapeutically Important Antimalarial Plants Used in Nigeria

Plant Species	Family	Plant Part Used
Adansonia digitata	Bombacaceae	Leaves, fruit
Aframomum melegueta	Zingiberaceae	Aerial part, rhizome
Ageratum conyzoides	Compositae	Whole plant
Albizzia lebbek	Fabaceae	Leaves
Alchornia cordifolia	Euphorbiaceae	Leaves, stem bark
Allium cepa	Amaryllidaceae	Bulb
Alstonia boonei	Apocynaceae	Stem bark
Anacardium occidentale	Anacardiaceae	Stem bark, leaves
Andira inermis	Fabaceae	Seed, stem bark
Andrographis paniculata	Acanthaceae	Leaves, stem
Annickia chlorantha	Annonaceae	Stem
Anogeissus leiocarpa	Combretaceae	Leaves, stem bark
Argemone mexicana	Papaveraceae	Aerial parts, latex
Artemisia afra	Compositae	Leaves, whole plant
Artemisia annua	Compositae	Leaves
Azadirachta indica	Meliaceae	All plant parts
Bambusa vulgaris	Gramineae	Leaves
Barringtonia acutangula	Lecythidaceae	Stem bark
Bidens pinosa,	Compositae	Leaves
Borassus flabellifer	Aracaceae	Fruit, leaves
Bridelia ferruginea	Euphorbiaceae	Leaves, stem bark
Cajanus cajan	Fabaceae	Leaves
Calotropic procera	Asclepiadaceae	Aerial parts, root
Canaga odorata	Annonaceae	Flower, leaves
Carica papaya	Caricaceae	All plant part
Cassia sieberiana	Fabaceae	Leaves, root bark
Catharanthus roseus	Apocynaceae	All plant parts
Ceiba pentandra,	Bombacaceae	Leaves, stem bark
Chromolaena odoata	Compositae	Aerial parts
Citrus sinensis	Rubiaceae	Leaves, root bark
Cochlospermum planchonii	Cochlospermaceae	Leaves, root
Cochlospermum tinctorium	Cochlospermaceae	Leaves, root
Combretum molle	Combretaceae	Leaves
Conyza sumatrensis	Compositae	Leaves
Cryptolepis sanguinoleta,	Apocynaceae	Seed, fruit rind
Curcuma longa	Zingiberaceae	Rhizome, whole plant
Cymbopogon citratus,	Gramineae	Aerial part, leaves
Dichroa febrifuga	Saxifragaceae	Leaves, root
Enantia chlorontha	Annonaceae	Stem bark
Erythrina senegalensis	Fabaceae	Leaves, stem bark
Eucalyptus spp	Myrtaceae	Leaves, exudate
Ficus platyphylla	Moraceae	Leaves, stem
Garcinia kola	Clusiaceae	Fruit, stem, leaves
Guiera senegalensis	Combretaceae	Leaves, root

(*Continued*)

Table 7.1 (Continued) Major Therapeutically Important Antimalarial Plants Used in Nigeria

Plant Species	Family	Plant Part Used
Harungana madagascariensis	Guttiferae	Aerial parts, stem
Hibiscus sabdariffa	Malvaceae	Calyx, whole plant
Hyptis suaveolens	Labiatae	Aerial part, root
Khaya senegalensis	Meliaceae	Stem bark, fruit
Kigelia africana	Bignoniaceae	Stem bark, fruits
Lipia multiflora	Verbenaceae	Aerial part
Mangifera indica	Anacardiaceae	Leaves, stem, root
Monodora myristica	Annonaceae	Fruit, leaves
Morinda lucida	Rubiaceae	Leaves, stem bark
Moringa oleifera	Moraceae	All plant parts
Nauclea latifolia	Rubiaceae	Stem, root
Nauclea pobeguinii	Rubiaceae	Stem, root
Newbouldia laevis	Bignoniaceae	Leaves, root
Ocimum gratissimum (other *Ocimum* spp.),	Labiatae	Aerial part
Parinari excelsa	Rosaceae	Stem
Phyllanthus amarus	Euphorbiaceae	Whole plant
Picralima nitida	Apocynaceae	All plant parts
Plumeria rubra	Apocynaceae	Leaves, flowers
Prosopis africana	Fabaceae	Stem bark, leaves
Psidium guaja	Myrtaceae	Leaves, exudate
Rauwolfia vomitoria	Apocynaceae	Stem bark, root
Reneilmia cincinnata	Zingiberaceae	Stem bark
Scoparia dulcis	Scrophulariaceae	Aerial part, stem
Securidaca longipedunculata	Polygalaceae	Leaves, whole plant
Senna occidentalis	Fabaceae	Seed, leaves
Sida acuta	Malvaceae	Whole plant
Solanum torvum	Solanaceae	Fruit, leaves
Spathodea campanulata	Bignoniacceae	Stem, root
Spondias mombin	Anacardiaceae	Stem bark, leaves
Starchytarpheta cayennensis	Verbenaceae	Whole plant
Sterospermum kunthiamum	Bignoniaceae	Stem bark, leaves
Tamarindus indica	Fabaceae	Fruit, stem bark
Tectona grandis	Verbenaceae	Leaves
Terminalia ivorensis	Combretaceae	Stem bark
Tetrapleura tetraptera	Fabaceae	Fruit, stem
Tinospora cordifolia	Menispermaceae	Stem bark, twigs
Tithonia diversifolia	Compositae	Aerial part
Trichilia emetica	Meliaceae	Leaves, root
Uvaria chamae	Annonaceae	Leaves, root
Vernonia amygdalina.	Compositae	Leaves, stem
Xylopia aethiopica	Annonaceae	Fruit
Zanthoxylum zanthoxyloides	Rutaceae	Root, stem bark
Zingiber officinale	Zingiberaceae	Rhizome, fruit

a steaming pot of various herbs (Iwu, 2016). Burns and blisters are prevented by surrounding the hot pot with short pieces of water-soaked wood; very frail patients are allowed to raise their heads above the blanket to avoid suffocation. The treatment works, and cases of suffocation or severe burns from this method of treatment have not been reported. It is noteworthy that essential oil-bearing plants are major ingredients in malarial steam therapy. *Azadirachta* is a common ingredient in such hot-pot herbs. Other ingredients in the steam therapy include the leaves of *Cymbopogon citratus* (lemongrass), *Psidium guajava* (guava), *Mangifera indica* (mango), and *Hyptis suaveolens* (pignut).

Malarial prevention is often accomplished by the use of 'health teas', prepared as weak infusions of leaves of some antimalarial plants, such as lemongrass (*C. citratus*), bitter kola (*Garcinia kola*), lime (*Citrus aurantiifolia*), bitter-leaf (*Vernonia amygdalina*), and sometimes guava leaves. This form of treatment is usually recommended for mild cases of malaria. Therapeutic soups prepared from some of the antimalarial vegetables and spices are also used as remedies for the treatment and prevention of malaria. Listed in Table 7.1 are the medicinal plants used in Nigeria for the preparation of antimalarial remedies. The compilation is extracted from the International Cooperative Biodiversity Group (ICBG) ethobotanical survey carried out in 1994–1999, literature report on similar regional-based studies by the Nigerian Natural Medicine Development Agency (NNMDA) 2000–2004, and published articles on phytotherapy of malaria in Nigeria.

One of the problems faced in malarial phytotherapy is the very poor rate of translation of results from laboratory studies on medicinal plants into therapeutic agents. Although phytochemical and phytopharmacological studies of natural remedies have yielded a rich bounty of important therapeutic agents, used either as single chemical moieties or as mixtures of chemically defined molecules in phytomedicines, the proportion of identified bioactive entities (molecules and mixtures) that showed activity in laboratory studies (*in vitro* and/or *in vivo*) and were subsequently found to be active in clinical studies is too low (1: 2,000,000). Clinical translation of laboratory studies has also been slow, due to the lack of ideal animal models and established solutions for precise targeting, cell internalization, and controlled drug solubility and release of the active compounds. The current approach is the promotion of the use of standardized herbal preparations rather than isolated pure compounds, in malarial therapy and drug development, since it mimics the method of use in many traditional medicine systems.

Several antimalarial drugs have been identified through a process called reverse pharmacology, or bedside-to-bench, which is based on the clinical validation of traditional knowledge, and relates to reversing the classical laboratory–to–clinic pathway to a clinic–to–laboratory practice. It is a trans-disciplinary approach, focused on traditional knowledge, experimental observations and clinical experience before bioassay-guided extraction and isolation of the active compounds (Willcox *et al.*, 2011).

This process yielded *Cinchona* extract which led to quinine, the structural template for the synthesis of chloroquine and its congeners, and *Artemisia annua* extract that yielded artemisinin and several active compounds currently undergoing development as antimalarials.

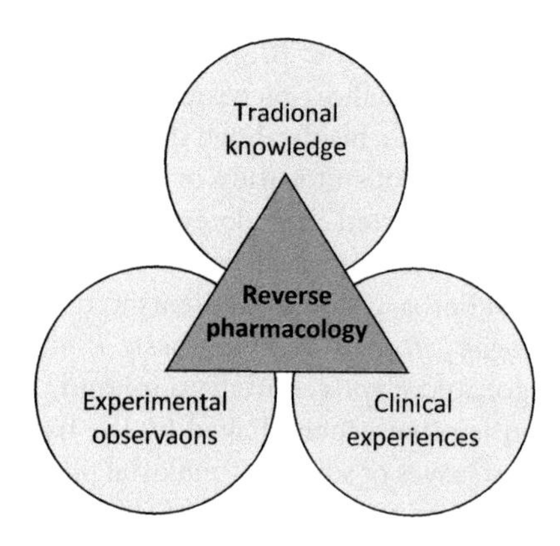

Suggestions have been advanced for combination drug therapy or the use of cocktail therapy in clinical malaria treatment, as has been successfully applied to the treatment of tuberculosis, HIV-AIDS and cancer (Teng *et al.*, 2016). Accordingly, we have included in our discussion some polyherbal preparations and profiles of four popular and commercially available antimalarial phytomedicines used in Nigeria and neighboring countries, namely Sumafoura Tiemoko Bengaly (Mali), N'Dribala (Burkina Faso), Phyto-Laria (Ghana) and PR-259 CTI from Congo (DRC). These phytomedicines, based on traditional treatment of malaria, have the added advantage of being readily available where and when modern drugs of acceptable quality may not be available.

The following pot-herb recipes or polyherbal remedies are used in Nigeria for the treatment of malaria:

Formula 1:

Four yellow pawpaw (*Carica papaya*) leaves

Thirty leaves of bitter-leaf (*Vernonia amygdalina*)

Eight bottles of water (ca. 8 L. of water)

The pulverized plant material was squeezed into water, and allowed to macerate for about 12 hours, and filtered.

Dosage: one glassful thrice daily (about 250 mL daily) for ten days (Adodo, 2004).

Formula 2:

Lemongrass leaves (*Cymbopogon citratus*)

Orange peel

Morinda lucida

Equal amounts of the plant materials were brought to the boil in water for about 40 minutes.

Dosage: One glassful thrice daily (about 250 mL daily) for seven days (Adodo, 2004).

Formula 3:

Momordica charantia

Crytolepis sanguinoleta

Vernonia amygdalina

Nauclea latifolia

Alstonia boonei

Azadirachta indica

Monodora myristica, and

Xylopia aethiopica

The proportions of the various herbs is extremely variable; ca. 1 to 5 g of each of the plant materials were macerated together in about 4 L of water or local gin overnight or boiled for about 30 minutes, and the preparation is sold as a weak alcoholic decoction. The usual dosage is about 250 mL taken daily for seven days.

Formula 4:

Azadirachta indica (leaves or stem bark)

Vernonia amygdalina (leaves)

Magnifera indica (bark)

Equal parts of the different materials were boiled together in water for about 40 mins. A cupful of the decoction was taken each day until the malaria symptoms disappeared.

Formula 5:

Enantia chlorontha (leaves)

Picralima nitida (seeds)

Physalis angulata (leaves)

Piper guineense (seeds)

The plant material was boiled in water for about one hour and one cupful was taken orally twice daily after meals for about one week.

Formula 6 (Nefang):

Mangifera indica (bark and leaf)

Psidium guajava

Carica papaya

Cymbopogon citratus

Citrus sinensis

Ocimum gratissimum (leaves).

The ratio of the herbs in the formulation varies enormously between communities that use the Nefang remedy for malaria in the Cameroon–Nigeria border towns (Tarkang *et al.*, 2014b).

Formula 7 (Akom Sirop):

Picralima nitida (seeds)

Gongronema latifolium (leaves)

Alstonia boonei (stem bark).

The ratio of the constituents varies considerably but a typical formulation consists of a ratio (by weight) of four parts of *Picralima*, four parts of *Alstonia* and one part of *Gongronema*.

Phytomedicine A: Phyto- Laria

Phyto-Laria consists of the standardized formulation of roots of *Cryptolepis sanguinolenta* in a tea-bag, developed in Ghana for the treatment of malaria. Cryptolepine, an indoloquinoline alkaloid, is the main component of Phyto-Laria. It has been shown to inhibit both chloroquinine-sensitive and chloroquine-resistant *P. falciparum* in laboratory studies and showed clinical effects that are comparable to those of a pharmaceutical antimalarial (Boye, 1989). Phyto-Laria, at a dose of one teabag taken three times a day for five days, had a cure rate of 93.5% with a mean parasitaemia clearance time of 82.3 hours and a mean fever clearance of 23.2 hours. The study found that fifty percent of the patients were cleared of their *P. falciparum* parasitaemia by 72 hours, and all by Day 7. Those with symptoms of fever, chills, nausea and vomiting were cleared rapidly, all by Day 3, but the resolution of haematological and biochemical abnormalities associated with malaria was generally slow, a feature seen in malaria post-treatment. There were two cases of recrudescence on Days 21 and 28. The laboratory findings did not suggest any toxicity (Bugyei *et al.*, 2010).

Phytomedicine B: N'Dribala

N'Dribala is prepared from the tubercles and roots of *Cochlospermum planchonii* in 70 g sachets for the treatment of malaria. It was developed in Burkina Faso but sold in northern parts of Nigeria for treatment of malaria, diabetes, and obesity. Clinical studies have confirmed the efficacy of N'Dribala in the treatment of uncomplicated falciparum malaria (Benoit-Vical *et al.*, 2003).

Phytomedicine C: Sumafoura Tiemoko Bengaly, STB)

The formulation was derived from a decoction of the standardised aerial parts of *Argemone mexicana*. It has been approved by the drug regulatory authority in Mali for the treatment of malaria. At a dosage of one glass twice a day for one week, STB achieved 80% cure in a randomized controlled clinical trial, against 95% recorded with the group treated with artemisinin-based combination therapy (Graz *et al.*, 2010).

Phytomedicine D: PR 259 CT1

PR 259 CT1 was developed in the Democratic Republic of the Congo for the treatment of malaria and it is available in eastern states of Nigeria from DRC. It consists of 80% ethanolic extract of the stem bark of *Nauclea pobeguinii*, containing 5.6% strictosamide. Clinical studies have shown that the phytomedicine is effective against *P. falciparum* malaria.

7.1.4 Major Herbs Used for Malaria Treatment in Nigeria

Many plant species, belonging to a plethora of families, are used in traditional medicine in Nigeria for the treatment of malaria (Table 7.1) (Adebayo and Krettli, 2011). Many of the species do not possess antimalarial activity *per se*, based on laboratory studies, but are useful in the clinical amelioration of the different symptoms of the disease complex called "malaria". Globally, more than 1200 medicinal plants from 160 plant families have been reported as being used in malaria treatment worldwide (Wilocox and Bodeker, 2004). In Africa, nearly as many plant species have been implicated in the treatment of malaria (Chinsembu, 2015). Several studies have been carried out to screen the rich Nigerian flora for the identification of active molecules that could be developed into active pharmaceutical ingredients but the emphasis of this Chapter is on the use of the plants themselves as therapeutic agents for the treatment of malaria and the management of its life-threatening complications.

It is noteworthy that these antimalarial medicinal plants are seldom used alone but usually in combination with other herbs. A common feature is the use of essential-oil-containing plants as adjuvant therapeutic agents. The essentials oils are likely to interact *in vivo* with the main antimalarial agent either synergistically, to enhance the potency to a level higher than the activity expected of the individual herb or drug, or antagonistically, if the potency of the combination is lower than the activity of the individual herb or drug (Bell, 2005). In addition, the combination may not only potentiate the antimalarial activity of phytomedicines but could also enhance the toxicity of each component molecule of the essential oil and the main antimalarial agent, or alter their respective bioavailability, such that their therapeutic range is exceeded (Van Zyl *et al.*, 2010).

7.1.4.1 *Azadirachta indica* (Meliaceae)

The neem tree is one of the plants most widely used for the treatment of malaria. The plant is considered to be one of the world's most useful plants, since it provides medicine, shade, fuel, fodder, and an extremely important insecticide, azadirachtin. Its dual use, both in malaria therapy and in the prevention of malaria due to its insecticidal properties, earned it the name "malaria tree". Dogon yaro (as it is popularly known) is such a popular malaria remedy that it is uncommon to see an intact tree near homes, without the stem bark scraped for medicinal use. The extracts from neem have been acclaimed as being effective in malarial treatment where the allopathic antimalarial drugs were found to be ineffective, either because of the presence of drug-resistant *Plasmodium* strains or because of acquired tolerance from repeated sub-therapeutic dosing with standard drugs. The dried flowers are used as a tonic and stomachic, and applied externally for the treatment of chronic eczema and as an insect repellant. A weak infusion of the bark is drunk as an antipyretic and as a bitter. There is a limited use of the aqueous decoction of the root as an anthelmintic agent. The seed, which yields an oil of quality indistinguishable from that of the Indian "margosa oil" (technically, the oil from *Melia azedarach*), is used extensively as an ingredient in the preparation of remedies for haemorrhoids, jaundice, and peptic ulcers. In the Edo and Delta States of Nigeria, the fruit juice is said to be useful in the treatment of boils, syphilitic sores, skin diseases, glandular swellings, jaundice, and peptic ulcers. The fruit is also favored as a laxative and for the treatment of urinary diseases, leprosy, and intestinal worm infestation. The stems and roots yield bitter but effective chewing sticks.

In traditional medical practice, the leaves, stem bark, and roots are used for the treatment of malaria in the form of an aqueous decoction. The selected part is usually reduced to small-sized particles and placed in a suitable container with water and set aside to macerate for a period ranging from one day to several weeks. Portions of the extract are often dispensed as soon as an acceptable brew is achieved, and the marc replenished with fresh water. The average dosage is ca. 100 mL, taken two or three times a day for two to three days. The drug can also be prepared by macerating the coarsely powdered plant part for a few hours in cold water and using the hand to press out fluid from the marc. In urban areas, the drug is prepared either as a decoction with hot water or as an alcoholic extract.

All neem plant parts have been shown to contain some bitter principles, believed to be responsible for the antimalarial activity associated with this plant. It is composed of the meliacins nimbin, nimbidin, nimbinin, desacetylnimbin, and structurally related compounds. The seed oil contains about 45% of these bitter substances, the stem bark yields 0.04%, while the fruit pulp and leaves contain about 25%. Perhaps the most studied constituent of the seed is azadirachtin, which was isolated together with meladanin and was found to inhibit the feeding response of the desert locust. The seeds also contain tiglic acid (5-methyl-2-butanoic acid), which is believed to be responsible for the distinctive odor of the oil. The gum exudate, which is a good emulsifying agent, is a unique gum by having D-glucosamine, an amino sugar, as one of its constituents. Other constituents of the gum include simple sugars (such as fructose, rhamnose, xylose, and mannose), uronic acid and proteinaceous materials (e.g. amino acids and dipeptides).

Flavonoids, such as kaempferol and quercetin, have been isolated from the flowers. The chloroform-soluble fraction of the ethanolic extract of the leaves has been found to contain an isoprenylated flavone, nimbaflavone (8,3′-diis oprenyl-5,7-dihydroxy-4′-methoxyflavone). Rutin and quercetin-3-rhamnoside are the main constituents of the polar fraction of the alcoholic extract of the leaves.

The antimalarial activity has been estimated to be equivalent to half the therapeutic dose of chloroquine sulphate on a dry weight basis. It had earlier been reported that the leaf extract of *A. indica* showed no antiparasitic activity when tested (*in vitro*) against *Plasmodium berghei* drawn from infected albino rats. Etkin (1981) also found that preincubation of plasmodium-carrying blood with aqueous leaf extract of the herb leads to lack of infection in mice. Furthermore, a field evaluation of patients, using extracts of neem for self-medication for the treatment of malaria, showed that the drug does indeed have antimalarial activity (Iwu, 1983).

The antimalarial activity of neem may be considered as having been established, whereas what is unclear is the mechanism of its action. It does appear that the activity of this herb may be due not to a direct antiparasitic activity *per se*, but by a possible drug-catalysed parasite-host interaction. Etkin and others have studied the significance of the oxidation-reduction, or "redox" status of red blood cells on parasites and host cell biochemistry during malarial infection. They observed that increased levels of red cell oxidation attendant on plasmodial infection is a consistent feature of malaria. The suggestion has been put forward that, whereas plasmodial parasites might be responsible for generating oxidants, excessive oxidation may in the long run be detrimental to continued and successful malarial infection (Etkin, 1975; Etkin and Eaton, 1975; Eaton *et al.*, 1976).

An approximate redox balance must therefore be maintained to ensure red cell integrity and proper metabolic functioning, which is also essential for the development of the *Plasmodium* parasite. In effect, under drug-induced physiological conditions, where excessive oxidation occurs and cannot be compensated for, a range of damaging effects will ensue, resulting in the destruction of both red blood cells (haemolysis) and the malarial parasites.

It has been shown that an aqueous extract of *A. indica* leaf significantly increased the generation of methaemoglobin (by the oxidation of haemoglobin), and the conversion of glutathione (GSH) to its oxidized counterpart (GSSG) in *in vitro* studies, using normal red cell haemolysates in dilute haemoglobin suspensions. Similar oxidant effects on haemoglobin and glutathione have been observed *in vivo*. The antimalarial activity of *Azadirachta* is therefore believed to be probably due to redox perturbation in the form of the imposition of substantial oxidant stress during treatment for malaria. The aqueous leaf extract inhibits NADPH-cytochrome c (P-450) reductase activity in rats with a significant increase in microsomal protein. The aniline hydroxylase activity and the phenobarbitone metabolism are also enhanced by the oral administration of 400 mg/kg bodyweight of the extract (Iwu *et al.*, 1986).

Neem has been associated with liver toxicity. It has been suggested that it is likely that it is the nonpolar extractives that cause the hepatotoxicity, whereas the therapeutically useful water-soluble polar forms that may be devoid of any serious side effects (Iwu, 2014). Liver biopsy of experimentally induced margosa-oil-poisoned mice demonstrated pronounced fatty infiltration of the liver and proximal renal tubules as well as cerebral oedema and mitochondrial damage.

7.1.4.2 *Alstonia boonei* (Apocynaceae)

Species of the plant genus *Alstonia* are highly prized in African ethnomedicine for the treatment of malaria, especially in situations where the allopathic antimalarial drugs are found to be ineffective, either because of the presence of the drug-resistant malarial strains or because of acquired tolerance from repeated dosing with synthetic antimalarials. For this purpose, *Alstonia* stem bark or leaves are administered as strong decoctions or 'teas', and sometimes as an ingredient in malaria "steam therapy". *Alstonia* was listed in the British Pharmacopoeia of 1914 as an antimalarial drug. It is also described in the African Pharmacopoeia for the same purpose.

The biological activities of *Alstonia* are due to the presence of several indole alkaloids, including echitamine, echitamidine, akuammidine, picraline, quebrachidine and its esters, vincamajine, alstonine, and akuammiline. Other constituents are the triterpenes β-amyrine and lupeol found in the bark, and ursolic acid in the leaves.

Although *Alstonia boonei* has enjoyed a folk reputation in three continents as a remedy for malaria, there is apparently only scanty laboratory evidence to confirm its *in vivo* efficacy. *In vitro* antiplasmodial activity of *A. boonei* alkaloids against both drug-sensitive and -resistant strains of *P. falciparum* and *in vivo* activity against *Plasmodium berghei* in mice have been reported (Awe and Okpe, 1990). An alkaloid-rich extract obtained from the New Guinea species was found to significantly inhibit (*in vitro*) the growth of W-2 and D-6 clones of *P. falciparum*. An *in vivo* antimalarial evaluation of extracts of *Alstonia scholaris* showed no significant activity. An evaluation of *Alstonia congensis* in Nigeria showed that the methanol extract of the species suppressed early infection chloroquine-sensitive *P. berghei* in mice but had no effect when infection was already established. A combination therapy with *Khaya ivorensis* has been found to be effective in malarial prophylaxis. The herbal mixture, when given to mice subjected to the fourteen-day repeated-dose toxicity test (sub-acute toxicity test), did not cause any serious toxicity, such as weight loss, liver or kidney morphological modifications, significant alterations in locomotor activity, or any other sign of illness. The herb has been successfully formulated as a solid dosage form (Majekodunmi *et al.*, 2008; Chime *et al.*, 2013).

7.1.4.3 *Argemone mexicana* (Papaveraceae)

Argemone mexicana is a plant widely distributed in Africa, Asia, and Latin America. In Nigeria it is found growing mainly in the northern part of the country, where it is known as "Kankamarkata. Bika" in Hausa. It is the main constituent of the Malian proprietary herbal antimalarial drug called "Sumafoura Tiemoko Bengaly" (section 7.1.3.).

The plant contains several alkaloids that have significant antimalarial activities in low doses. Berberine, allocryptolepine, and protopine, for example, showed *in vitro* antimalarial activity with half-maximal inhibitory concentration values, IC_{50} (μg/mL), against W2 (chloroquine resistant) of 0.32, 0.32, and 1.46 respectively. The alkaloid 6-acetonyl-dihydro-chelerythrine has been isolated from whole plant extracts of *A. mexicana* and was found to have significant anti-HIV activity. The alkaloids berberine, protopine, protopine hydrochloride, sanguinarine, and dihydrosanguinarine have been isolated from the seeds.

The *in-vitro* efficacy of the aerial parts of *A. mexicana* against *P. falciparum* has been reported, with IC_{50} values against the chloroquine-resistant K1 strain of *P. falciparum* of 5.89 and 1.00 g/mL for the aqueous decoction and methanol extracts, respectively (Dialo *et al.*, 2006). A prospective dose-response clinical trial of *A. mexicana* decoction (AMD), following a local recipe, found a clear dose response and a good safety profile (Willcox *et al.*, 2007).

It has been suggested that AMD could be used as an alternative first-line treatment against malaria, instead of the aretemisinin combination therapy (ACT) to delay the possibility of drug resistance to ACT. In a comparative study, treatment of malaria with AMD compared favorably with artesunate-amodiaquine ACT. A total of 301 patients, with presumed uncomplicated malaria (median age 5 years), were randomly assigned to receive AMD or ACT as first-line treatment. Both treatments were well tolerated. Over 28 days, second-line treatment was not required for 89% (95% confidence interval (CI) 84.1–93.2%) of patients on AMD, versus 95% (95% CI 88.8–98.3) on ACT. Deterioration to severe malaria was 1.9% in both groups in children aged ≤5 years (there were no cases in patients aged >5 years) and 0% had coma/convulsions. AMD is a government-approved antimalarial drug in Mali and could be used as a first-line complement to standard modern drugs in high-transmission areas, in order to reduce the pressure for development of resistance to ACT, in the management of malaria. In view of the low rate of severe malaria and high tolerability, AMD may also constitute a first-line treatment when access to other antimalarials is delayed (Graz *et al.*, 2010).

The validation of AMD as an effective traditional medicine for the treatment of uncomplicated falciparum malaria with a clinical efficacy that was found to be comparable to artesunate–amodiaquine, gave impetus for the use of a reverse pharmacology approach, instead of the classical drug development process aimed at single chemical entities. Allocryptopine, protopine and berberine were isolated through bio-guided fractionation, and had their identity confirmed by spectroscopic analysis. The three alkaloids showed antiparasitic activity *in vitro*, of which allocryptopine and protopine were selective towards *P. falciparum*. Furthermore, the amount of the three active alkaloids in the decoction was determined by quantitative NMR, and preliminary *in vivo* assays were conducted. These alkaloids can therefore be considered as phytochemical markers for quality control and standardization of the phytomedicine made from *A. mexicana* (Simoes-Pires *et al.*, 2014).

7.1.4.4 *Cochlospermum planchonii* (Bixaceae)

The leaves and rhizomes of the two *Cochlospermum* species found in West Africa, *C. planchonii* and *Cochlospermum tinctorium*, have been reported to be commonly and similarly used by traditional healers in Nigeria to treat malaria, fevers, jaundice, diabetes, diarrhoea, stomach disorders, typhoid fever and urinary tract infections (Benoit-Vical *et al.*, 2003; Isah *et al.*, 2013; Nafiu *et al.*, 2011). The use of these plants in the treatment of malaria extends up to the Sahel region of North Africa (Togola *et al.*, 2005; Yakubu *et al.*, 2010). The antimalarial activity of preparations from dried rhizomes (decoction) has been evaluated against the chloroquine-sensitive *Plasmodium* strain 3D7 using the histidine-rich protein II (HRP2) drug susceptibility assay. Two major apocarotenoids were isolated from rhizomes of *C. planchonii* and unambiguously identified as dihydrocochloxanthine and cochloxanthine by spectroscopic methods. Comparative high-performance liquid chromatography (HPLC) analyses of thirty-nine samples from markets and from collections from the natural habitats of both species showed a high variability in the concentrations of cochloxanthines and related carotenoids, which were proven to be characteristic for rhizomes of both species and were generally absent in leaves. Furthermore, concentrations of total phenolics and antioxidant activities (DPPH [2,2-diphenyl-1-picrylhydrazyl] and FRAP [ferric reducing ability of plasma]), as well as the haemolytic activity of various extracts, were evaluated.

Extracts from both species and pure cochloxanthine offered pronounced antioxidant activities and weak haemolytic activity, whereas, in contrast, dihydrocochloxanthine had a strong haemolytic effect at the highest concentration analysed. However, cochloxanthine, as well as dihydrocochloxanthine, showed erythroprotective effects against the haemolytic activity of the reference saponin. Moderate antiplasmodial activity between 16 and 63 μg/mL was observed with all tested extracts, and lower IC_{50} values were obtained with pure dihydrocochloxanthine (IC_{50} = 6.9 μg/mL), cochloxanthine (IC_{50} = 6.8 μg/mL), and the dichloromethane fraction (IC_{50} = 2.4 μg/mL) and the ethyl acetate fraction (IC_{50} = 11.5 μg/mL) derived from a methanolic extract of *C. planchonii*.

Studies have shown that there is a major variability in carotenoid concentration and antiplasmodial activity within both *C. planchonii* and *C. tinctorium*. The high haemolytic activity of dihydrocochloxanthine (at 100 μg/mL) should be considered to be a selection criterion for choosing species phenotypes for use in treatment (Lamien-Meda *et al.*, 2015).

7.1.4.5 *Cryptolepis sanguinolenta* (Apocynaceae)

The main medicinal use of the root of *Cryptolepis* is in the treatment of fevers. Extracts of the plant have also been employed for the treatment of urogenital infections, especially *Candida* infections of non-venereal origin. In southern Nigeria, a weak decoction of the root is administered as a tonic and also for the treatment of anti-inflammatory conditions. The plant appears to be used as an ingredient in several folk remedies for a multiplicity of diseases, but the major uses are in the preparation of remedies for the treatment of malaria, hypertension, microbial infections, and anti-inflammatory conditions. The principal constituent is the quindoline alkaloid cryptolepine, which occurs at a yield of 0.52% (w/w) in the roots, 0.48% in the stem, and 1.03% in the leaves. The compound co-occurs with the related bases and their derivatives.

Clinical studies have shown that extracts of *Cryptolepis* achieved antimalarial cures in patients, with the concomitant elimination of parasitaemia from the blood (Tempesta 2010). The extract also showed antipyretic activity in the patients evaluated. Several modifications have been made on the structure of the naturally occurring antiplasmodial compound cryptolepine, which has resulted in remarkable improvements in the antimalarial activity (Larvado *et al.*, 2008). The plant seems to possess broad-spectrum antiprotozoal activity since its major constituent, cryptolepine and its derivatives, containing alkyldiamine side-chains, also possess potent inhibitory activity against *Trypanosoma brucei* (Wright *et al.*, 1996). These derivatives are also potent inhibitors of the trypanosome papain-like cysteine protease cruzain, which could, at least in part, explain their antitrypanosomal activity. Such compounds, with good antiprotozoal activity and selectivity, provide an encouraging starting point for the rational design of new and effective antitrypanosomal agents (Lavrado *et al.*, 2012). Cryptolepis alkaloids and their semi-synthetic derivatives are subjects of on-going studies to evaluate their antiprotozoal, antidiabetic, analgesic, anti-inflammatory, and antituberculosis activities. It is also interesting that other structurally planar indoles, such as olivacine and ellipticine, exhibit similar antiplasmodial activity (Rocha e Silva *et al.*, 2012).

7.1.4.6 *Ocimum gratissimum* (Lamiaceae)

The whole herb (African scent leaf) is used throughout West Africa as a febrifuge and as an ingredient of many antimalarial remedies. The crushed leaves are instilled into the eye as a treatment for conjunctivitis. The oil from the leaves is regarded as highly antiseptic and has been applied for the dressing of wounds, mouth gargle, and to prevent postpartum sepsis. The leaves are used in Nigeria as a stomachic and as a catarrh remedy, and also as a general tonic and antidiarrhoeal. The oil is used in many parts of West Africa to prevent mosquito bites. The oil, mixed with alcohol, is applied as a lotion for skin infections and taken internally for bronchitis. An infusion of the leaves, the so-called "ocimum tea" or "bush tea" is dispensed as a remedy for malaria fever and as a diaphoretic. The effects of these oils on the growth of *P. berghei* have been investigated by various groups. The oils showed significant antimalarial activities in the four-day suppressive *in vivo* test in mice. At dosages of 200, 300 and 500 mg/kg bodyweight in mice per day, the essential oil of *O. gratissimum* exhibited 55.0%, 75.2% and 77.8% suppressive activity, respectively. The positive control, chloroquine (CQ, at 10 mg/kg) had a suppressive activity of 100%.

The leaves yield a very aromatic volatile oil that consists mainly of thymol and eugenol (32–65%). The plant oil also contains xanthones, terpenes, and lactones. The chemical composition of *Ocimum gratissimum* essential oil varies enormously, according to the chemotypes, the ratio of thymol, eugenol and geraniol being used as a marker. *Ocimum* leaf extract has been shown to be effective in the treatment of malaria (Kaou *et al.*, 2008). The volatile oil exhibited antimicrobial, insect repellent, and anthelmintic activity (Sofowora, 1982).

Nefang, a polyherbal product, which consists of *Ocimum gratissimum* (leaves) in combination with *Mangifera indica* (bark and leaf), *Psidium guajava*, *Carica papaya*, *Cymbopogon citratus* and *Citrus sinensis*, is extensively used in parts of Cross Rover State of Nigeria and in border towns between Nigeria and Cameroon as an effective therapy against *P. falciparum* malaria. *In vitro* antiplasmodial activities of the solvent extracts of the component plants have been analyzed on CQ-sensitive (3D7) and multidrug-resistant (D_2) *P. falciparum* strains. The interactions involving the different solvent extracts were further analyzed using a variable potency ratio drug combination approach. Half-maximal effective concentration (EC_{50}) values were determined by nonlinear regression curve-fitting of the dose-response data and used in calculating the half-maximal fractional inhibitory concentration (FIC_{50}) and combination indices (CI) for each pair. The derived EC_{50} values (3D7/Dd2, μg/mL) are Nefang: 96.96/55.08, MiB: 65.33/34.58, MiL: 82.56/40.04, Pg: 47.02/25.79, Cp: 1188/317.5, Cc: 723.3/141, Cs: 184.4/105.1, and Og: 778.5/118.9. Synergism was observed with MiB/Pg (CI = 0.351), MiL/Pg (0.358), MiB/Cs (0.366), MiL/Cs (0.482), Pg/Cs (0.483), and Cs/Og (0.414) when analyzed at equipotency ratios. Cytotoxicity testing of Nefang and the solvent extracts on two human cell lines (Hep G2 and U2OS) revealed no significant toxicity relative to their antiplasmodial activities (SI > 20). Taken together, available information confirms the antimalarial activities of Nefang and its constituent plant extracts and identified extract pairs with promising synergistic interactions as a rational and evidence-based antimalarial phytotherapeutic agent (Arrey- Tarkang *et al.*, 2014).

7.1.4.7 *Picralima nitida* (Apocynaceae)

Picralima contains a complex mixture of indole and dihyroindole alkaloids, including alstonine, akuammiline, akuammidine, akuammine, Ψ-akuammigine, akuammicine, echitamine, picraline, picratidine, and picraphylline (Ansa-Asamoah *et al.*, 1990). The major alkaloids of the fruits of *P. nitida* have been examined for *in vitro* activity against drug-resistant and -sensitive strains of *P. falciparum*. The alkaloids showed remarkable inhibitory activity against both clones of *P. falciparum*, with IC_{50} values of 0.08–0.9 μg/mL. Among the compounds tested, those belonging to the picraline-akuammine subgroup showed greatest activity, followed by those of the akuammicine type. The alkaloid echitamine was inactive (Iwu and Klayman, 1992; Iwu, 2014). The alkaloids also demonstrated strong inhibitory activity against clinical isolates of *Leishmania* and *Trypanosoma*.

Although the mechanism of action of *Picralima* is yet to be determined, the drug is presently used as a crude extract in Nigerian medicine as an effective anti-malarial agent and for the control of blood pressure. The plant is sometimes used in combination with *Gongronema latifolium* and *Alstonia boonei* for the treatment of acute malaria that is not susceptible to first-line drugs. In a study on the *Picralima–Grongronema–Alstonia* (PGA) combination product, it was found that, in the four-day suppressive test on mice, the PGA extract achieved percentage suppression of 39.0%, 41.6% and 54.68% for the 200 mg/kg, 400 mg/kg and 800 mg/kg body-weight concentrations respectively. In the malaria curative test, high percentage suppression values of 80.97%, 83.84% and 86.16% at the 200 mg/kg, 400 mg/kg and 800 mg/kg concentrations, respectively, were observed. The PGA extracts did not induce significant changes in haematological parameters ($P>0.05$), although significant elevation in the values of ALT (Alanine transaminase) and AST (Aspartate transaminase) ($P<0.05$) and creatinine concentration ($P<0.05$) were observed at 800 mg/kg, which was indicative of possible hepatotoxicity in the mice liver cells (Idowu *et al.*, 2015). The aqueous extract of *Picralima* is also employed in the treatment of malaria-induced psychosis. Alstonine found in the seeds and fruit rind of *Picralima* has shown antipsychotic-like effects, as a putative antipsychotic, which consistently differs from the effects of known drugs in various mouse models (Elisabetsky and Costa-Campos, 2006; Linck *et al.*, 2008, 2011).

Picralima appears to be well tolerated at the doses used in traditional medicine. Biological studies have shown that its use as a male aphrodisiac in folk medicine does not have any observed toxicity. It has been reported that *P. nitida* seed extract (PNE) has both aphrodisiac and contraceptive effects (Otoo *et al.*, 2015). The study established that the size and length of the combs of white leghorn day-old chicks treated with testosterone (0.5–1.5 mg/kg), cyproterone (3–30 mg/kg), or PNE (50–500 mg/kg) for seven days, as well as cyproterone (10 or 30 mg/kg) on PNE-induced comb growth, and PNE (50–500 mg/kg) on testosterone-induced comb growth, were measured in the chick comb test. The effect of PNE on the percentage change in an oviduct-chick weight ratio of Rhode Island Red day-old chicks treated with 17-β-oestradiol (0.1–0.9 μg/kg), PNE (30–300 mg/kg) or vehicle, for six days, was determined in the chick uterotrophic assay. Liver and kidney functions were normal and haematological profile tests were conducted to assess safety. Results showed that seven-day treatment with PNE and testosterone increased comb growth significantly ($P\leq0.01$–0.001), whereas cyproterone significantly decreased ($P\leq0.001$) comb growth dose-dependently. Qualitatively, testosterone and PNE treatment resulted in relatively brighter red combs. Cyproterone caused significant inhibition ($P\leq0.001$) of both testosterone- and PNE-induced comb growth. Co-administration of testosterone and PNE suppressed comb growth significantly ($P \leq 0.001$). Administration of 17-β oestradiol and PNE increased ($P \leq 0.001$) oviduct-chick weight ratio dose-dependently. No significant changes were observed in assessing liver and kidney function, lipid profile, and haematological parameters Otoo *et al.*, 2015).

7.1.4.8 *Vernonia amygdalina* (Asteraceae)

The leaves of the popular vegetable, bitter-leaf, are reputed to be effective remedies for malaria, gastrointestinal disorders, and diabetes. Its use in the treatment of malaria is not based on the folk belief that bitter substances are good malaria remedies. Bitter-leaf has been shown to possess significant antimalarial activity both *in vitro* and *in vivo* in laboratory studies. As an antimalarial agent, it is most valuable in its ability to reverse *Plasmodium* drug-resistance in CQ-resistant strains (Iwalokun, 2008). The administration of *V. amygdalina* and *C. papaya* extracts provides synergistic effects in amelioration of *Plasmodium* infection in mice (Okpe *et al.*, 2016). Bitter-leaf contains saponins, cardiac glycosides, flavonoids, and sesquiterpene lactones (Toubiana, 1969) The major constituents include the saponin vernonin, the sesquiterpenes vernoleptin and vernodalin, and the ubiquitous flavonoid kaempferol.

The *in vivo* transmission-blocking effects of ethanolIC and aqueous extracts of *V. amygdalina* and isolated compounds have been assessed against gametocytes and sporogonic stages of *P. berghei* and on field isolates of *P. falciparum*. In the study, aqueous and ethanolic leaf extracts were tested *in vivo* for activity against sexual and asexual blood stage *P. berghei* parasites. The transmission-blocking effects of the extracts were estimated by assessing *P. berghei* oocyst prevalence and density in *Anopheles stephensi* mosquitoes and the activity targeting early sporogonic stages (ESS), namely gametes, zygotes and ookinetes, was assessed *in vitro* using *P. berghei* CTRPp.GFP strain. A bioassay-guided fractionation was used to characterize *V. amygdalina* fractions and molecules for anti-ESS activity. Fractions active against ESS of the murine parasite were tested for *ex vivo* transmission-blocking activity on *P. falciparum* field isolates. The aqueous extract reduced the

P. berghei macrogametocyte density in mice by about 50% whereas the ethanolic extract reduced *P. berghei* oocyst prevalence and density by 27 and 90%, respectively, in *Anopheles* mosquitoes. The ethanolic extract inhibited ESS development *in vitro* almost completely (>90%) at 50 μg/mL. The molecules and fractions from *Vernonia* displayed considerable cytotoxicity on the two tested cell-lines, which is consistent with the use of this herb as a cancer chemopreventive agent.

It is evident from many studies that *Vernonia amygdalina* leaves contain molecules affecting multiple stages of *Plasmodium*, supporting its potential for drug discovery. It has been suggested that chemical modification of the identified antimalarial molecules, in particular vernodalol, could generate a library of sesquiterpene lactones that can be developed as pharmaceutical and phytotherapeutic agents (Abay *et al.*, 2015). The development of a multistage phytomedicine, designed as a malaria-preventive treatment to complement existing malaria control tools, appears a challenging but feasible goal.

V. amygdalina is consumed throughout West and Central Africa as a vegetable and is generally considered to be nontoxic, but excessive consumption of the leaves can be purgative. In a chronic toxicity study, rats were fed with up to 75% (w/w body weight) powdered leaves mixed with grower mash for 65 days and at the end of the study no significant toxicity or adverse reaction was observed (Ibrahim *et al.*, 2001).

7.1.4.9 *Andrographis paniculata* (Acanthaceae)

Andrographis paniculata leaves and whole plant are used in traditional medicine as a bitter tonic for the treatment of various febrile conditions, including malaria. It is known as the "King of Bitters" and many traditional healers strongly recommend its use with bitter kola (*Garcinia kola*) for the treatment of febrile conditions of unspecific aetiology because of the folk belief that it has broad-spectrum antimicrobial activity. Its use in malaria therapy is supported by several laboratory findings. In a four-day suppressive test against *P. berghei* NK 65 strain in

Mastomys natalensis, the Natal multimammate mouse, a crude ethanolic extract and fractions reduced the level of parasitaemia in a dose-dependent manner. Four diterpenes, isolated from the *n*-butanol fraction, also suppressed the parasitaemia level, but not in a dose-dependent manner. Chemoprophylactic activity of neoandrographolide was tested, using different protocols. Fifteen days of therapy with neoandrographolide before infection suppressed the parasitaemia load (Misra *et al.*, 1992).

The anti-malarial effect of the *Andrographis* extract (AE) has been compared to those of CQ and artemisinin, as well as their combination products. The test substances consisted of AE, CQ, artemisinin, the combination of AE and chloroquine, and the combination of AE and artemisinin. Parasite density was determined by counting the number of *P. falciparum*-infected erythrocytes in 5,000 erythrocytes of the culture. The single drug (CQ only or artemisinin only) and either in combination with AE at dose 0.5 μg/mL had a killing effect against the parasite, measured by the appearance of "crisis forms" on the infected erythrocytes. This killing effect was dose dependent and reached its optimum effect at 200 μg/mL. Treatment with the single AE extract with a 0.5 μg/mL dose increased the density of the parasite, however; with every 1μg increase in dose of AE, the killing effect also increased. The reduction of the parasite density was also seen by increasing the AE dose in the group of combination of AE+CQ as well as the group of combination of AE+artemisinin. Statistically, there was no difference in the anti-malaria efficacy among the five test drugs at the effective dose tested. The correlation between the reduction of the parasite with the increasing dose in all groups was statistically significant ($P<0.001$). AE in a single dose or in a combination had an anti-falciparum malaria effect comparable to that achieved by the clinically approved antimalarial drugs (Zein *et al.*, 2013). Similar results were obtained in animal studies with *P. berghei*-infected mice, which showed that the combination therapy model of AE+CQ were able to increase the antimalarial effectiveness of CQ alone by 85.61% inhibition of parasite growth and having a lower risk of resistance, by using a low dose of CQ (0.15 mg/kg mouse body weight) (Hafid *et al.*, 2015).

A. paniculata contains four major active diterpenoids, namely andrographolide, 14-deoxy-11, 12-didehydroandrographolide, neoandrographolide, and 14-deoxyandrographolide, which exhibit differences in types and/or degrees of their pharmacological activity. Their bioavailability and the pharmacokinetics are of immense clinical importance in the management of malaria and other therapeutic applications of this phytomedicine.

A simple liquid chromatography (LC) method has been reported which was successfully validated to determine the pharmacokinetic parameters of all four major active diterpenoids in human plasma after multiple oral doses of the *A. paniculata* product. In the assay, the LC tandem-mass spectrometry was performed in the negative mode, and the multiple reaction monitoring mode was used for the quantitation. The method showed a marked linearity over a wide concentration range of 2.50–500 ng/mL for the most active compound and over the range of 1.00–500 ng/mL for the other diterpenoids, with a correlation coefficient R2 > 0.995. The lower limit of quantification was found to be 2.50 ng/mL, while those of the other diterpenoids were 1.00 ng/mL (Pholphana *et al.*, 2015).

In addition to its antimalarial activity, *Andrographis* has been shown to be useful in the prevention of malaria by its larvicidal and ascaricidal activities. The larvicidal and acaricidal mortalities were observed after 24 h exposure to extracts like petroleum ether, ethyl acetate, methanol and aqueous extracts. Of those, the methanolic extract showed maximum activity against *Anopheles aegypti, C. quinquefasciatus*, and *R. microplus* (LC_{50} = 93.00, 83.06 and 105.84 ppm; LC_{90} = 171.81, 171.76 and 198.73 ppm; χ^2 = 11.038, 1.075 and 7.867), respectively, $P < 0.05$ (Mathivanan *et al.*, 2018). The plant is also used in traditional medicine as an effective remedy for diabetes, inflammatory conditions, high blood pressure, cancers, gastrointestinal disorders, and bronchitis (Okhuarobo *et al.*, 2015).

7.1.4.10 Nauclea spp. (Rubiaceae)

The genus *Nauclea* in Africa comprises seven species. Two of these species are used in Nigeria for the treatment of malaria: *Nauclea latifolia* and *Nauclea pobeguinii*. The two species are used interchangeably in traditional medicine. A third species, *Nauclea diderrichii*, is used occasionally for similar purposes. The genus *Nauclea* has been extensively studied for their chemical constituents, with indoloquinolizidine alkaloids as the major class of compounds reported in every *Nauclea* species analysed, with numerous structures being identified. A considerable number of pharmacological studies has also been conducted to confirm their ethnomedical uses, such as in malaria therapy and painkilling activity. It has, however, been noted that bioactive compounds responsible for the activity of the extracts have rarely been identified and, therefore, there is a clear need for further evaluations as well as for toxicity experiments (Haudecoeur *et al.*, 2018).

N. latifolia, also known as the African peach or fig, is native to the savannah forest (in the middle belt geographical zone of the continent). It has a rough bark and white sweet-scented flowers. The plant is used for the treatment of many diseases (Iwu, 2014). The plant contains several indole alkaloids, as exemplified by the compounds strictoside or isovincoside.

Its antimalarial properties have been established by animal studies on experimental malarial models (Gamaniel *et al.*, 1997). The aqueous extract of *N. latifolia* root bark (50–200 mg/kg p.o.) has also been shown to significantly ($P<0.05$) attenuate writhing episodes induced by acetic acid and to increase the threshold for pain perception in the hot-plate test in mice, in a dose-dependent manner. The extract also markedly decreased both the acute and delayed phases of formalin-induced pain in rats and also caused a significant reduction in both yeast-induced pyrexia and egg-albumin-induced oedema in rats. These effects were produced in a dose-dependent manner (Abbah *et al.*, 2010). It is noteworthy that the popular analgesic tramadol has been isolated from the root bark of *N. latifolia*. This finding is a rare example of a common synthetic drug that occurs at considerable concentrations in nature (Boumendiel *et al.*, 2013).

A hot-water leaf extract of *N. latifolia* has been assessed for its effect on antioxidant status, lipid peroxidation values, and parasite levels in hepatic and brain tissue of experimental mice (BALB/c) infected with *P. berghei* malaria (Onyesom *et al.*, 2015). In the study, it was shown that *P. berghei* malaria infection induced oxidative stress in both liver and brain tissues as evidenced by the significant ($P<0.05$) decrease in antioxidants superoxide dismutase, reduced glutathione and catalase. These reductions perhaps caused a compromise in membrane integrity as indicated by the significant increase in the lipid-peroxidation product malondialdehyde. Malaria parasites were also identified in these tissues. However, treatment with *N. latifolia* extract eliminated the parasites in tissues and protected them from oxidative damage even better than CQ treatment did, the anti-malarial potency of which also cleared tissue parasites. The measurement of protection by *N. latifolia* against damage was strengthened by the nonsignificant microstructural changes. The bioactive phytochemical(s) in *N. latifolia* extract should be identified and the mechanism(s) of the antimalarial tendency should be further investigated.

Polyherbal formulations of *N. latifolia* root with breadfruit (*Artocarpus altilis*) stem bark, *Murraya koenigii* leaf and *Enantia chlorantha* stem bark have been used as decoctions for treatment of malaria and fevers, and combinations with standard drugs have been investigated for antiplasmodial activities, using *P. berghei*-infected mice. The respective prophylactic and curative ED_{50} values of 189.4 and 174.5 mg/kg, respectively, for *N. latifolia*, and the chemosuppressive ED_{50} value of 227.2 mg/kg for *A. altilis* showed that they were the best antimalarial herbal drugs in the mixture. A 1.6-fold increase of the survival time achieved by the negative control was elicited by *M. koenigii*, thereby confirming its curative activity. Pyrimethamine with an ED_{50} of 0.5 ± 0.1 mg/kg for the prophylactic, and CQ with ED_{50} values of 2.2 ± 0.1 and 2.2 ± 0.0 mg/kg for the chemosuppressive and curative tests, respectively, were significantly ($P<0.05$) more active. Co-administrations of *N. latifolia* with the standard drugs significantly reduced their prophylactic, chemosuppressive and curative actions, by possibly increasing the parasite's resistance. Binary combinations of *N. latifolia* or *M. koenigii* with any of the other plants significantly increased the prophylactic and suppressive activities of the individual plants, respectively. Also, combining *E. chlorantha* with *A. altilis* or *N. latifolia* enhanced their respective prophylactic or curative activities, making these combinations most beneficial against malarial infections. Combinations of three or four extracts gave varied activities. Hence, the results justified the ethnomedicine-based combinations of such plants in antimalarial herbal remedies and showed the importance of the three *in vivo* models in establishing antimalarial activity (Adebajo *et al.*, 2014).

The antimalarial activity of a standardised extract of *N. pobeguinii* has been established by clinical trials. A phase IIA clinical trial was conducted to assess the efficacy of an 80% ethanolic quantified extract (containing 5.6% strictosamide as the putative active constituent) from *N. pobeguinii* stem bark, denoted as PR 259 CT1, in a small group of adult patients diagnosed with uncomplicated falciparum malaria. Results obtained from a phase I clinical trial on healthy male volunteers indicated that the oral administration of two 500-mg capsules three times daily (every eight hours) for seven days was well tolerated and showed only mild and self-resolving adverse effects. The phase IIA study was an open cohort study on eleven appraisable adult patients suffering from proven *P. falciparum* malaria. The study was specifically designed to assess the efficacy of PR 259 CT1 administered with a dose regimen of two 500-mg capsules three times daily for three days, followed by outpatient treatment of one 500-mg capsule three times daily for the next four days, in order to prove that this therapeutic dose, which was calculated from animal studies, was effective at treating adult malaria patients and consequently useful for a future Phase IIB clinical trial. Although the number of patients in the study was rather limited, the statistical analysis nevertheless suggested the efficacy and tolerability of PR 259 CT1, which indicated that this herbal medicinal product might be considered as a putative candidate for a large-scale clinical trial (Mesia *et al.*, 2012a).

Based on the results of the positive results of the Phase I and Phase IIA clinical trials with the herbal medicinal product PR 259 CT1, consisting of an 80% ethanolic extract of the stem bark of *N. pobeguinii* containing 5.6% strictosamide, a Phase IIB study has been conducted as a single-blind prospective trial in 65 patients with proven *P. falciparum* malaria to evaluate the effectiveness and safety of this herbal drug (Mesia *et al.*, 2012b). The study was carried out simultaneously using an artesunate-amodiaquine combination as a positive control; this combination is the standard first-line treatment for uncomplicated malaria recommended by the WHO. With regard to PR 259 CT1, patients were treated with a drug regimen of two 500-mg capsules three times daily for three days in the in-patient clinic, followed by out-patient treatment of one 500-mg capsule three times daily over the next four days; the positive control group received two tablets containing 100 mg artesunate and 270 mg amodiaquine (fixed-dose) once daily for three consecutive days. Antimalarial responses were evaluated using the 14-day test method. The study showed a significant decreased parasitaemia in patients treated with PR 259 CT1 and artesunate-amodiaquine with adequate clinical parasitological responses (APCR) at day 14 of 87.9 and 96.9%, respectively. The former product was better tolerated than the latter, since more side effects were observed for the artesunate-amodiaquine combination. PR 259 CT1 has been suggested as a promising candidate for the development of a herbal medicine for the treatment of uncomplicated falciparum malaria (Mesia *et al.*, 2012b).

Another potential clinical application of *N. pobeguiinii* is in the treatment of infections caused by multi-drug-resistant bacteria. Multi-drug resistance of Gram-negative bacteria constitutes a major obstacle in the antibacterial fight worldwide.

The discovery of new and effective antimicrobials and/or resistance modulators is necessary to combat the spread of resistance or to reverse the multi-drug resistance. The antibacterial and antibiotic-resistance modifying activities of methanol extracts of *N. pobeguiinii* bark (NPB) and leaves (NPL) against 29 Gram-negative bacteria including multi-drug-resistant (MDR) phenotypes has been investigated (Seukep *et al.*, 2016). The plant extract and six isolated compounds, identified as 3-acetoxy-11-oxo-urs-12-ene, *p*-coumaric acid, citric acid trimethyl ester, resveratrol, resveratrol β-D-glucopyranoside and strictosamide, were tested. The broth microdilution method was used to determine the minimal inhibitory concentrations (MIC) and minimal bactericidal concentrations (MBC) of crude extracts and the individual compounds, as well as the antibiotic-resistance modifying effects of the extract and compounds. MIC determinations indicated values ranging from 32 to 1024 μg/mL for NPB and NPL on 89.7% and 69.0% of the tested bacterial strains, respectively. Resveratrol was active towards all tested bacteria, whilst the other isolated compounds displayed weak and selective inhibitory effects. An MIC value of 16 μg/mL was obtained with resveratrol against *Klebsiella pneumoniae* KP55 strain. Synergistic effects of the combination of NPB with chloramphenicol (CHL) or kanamycin (KAN), as well as that of resveratrol with streptomycin (STR) and ciprofloxacin (CIP), were observed. The study provided significant evidence on the possible use of *N. pobeguinii* and resveratrol in the control of Gram-negative bacterial infections, including MDR phenotypes. It also indicates that NPB and resveratrol can be used as naturally occurring antibiotic-resistance modulators to tackle MDR bacteria (Seukep *et al.*, 2016).

7.1.4.11 *Ageratum conyzoides* (Asteraceae)

A. conyzoides has a folk reputation as a safe and effective antimalarial remedy (Odugbemi *et al.*, 2007). In Nigerian traditional medicine, extracts of the plant have been used in the treatment of microbial infections, fevers, wounds and burns, arthrosis, headache, inflammation and dyspnoea, pain, asthma, and muscular spasms (Iwu, 2014). The antimalarial activity of *Ageratum* has been established by laboratory studies on mice infected with *P. berghei* (Ukwe *et al.*, 2010a).

In order to determine the potential of *A. conzoides* as a possible resistance reversal agent in malarial chemotherapy, the aqueous leaf extract was evaluated for suppressive and curative activity in combination with CQ and artesunate, respectively, against *P. berghei* infection in mice. In a study that used malaria (*P. berghei*)-infected albino mice of both sexes, aqueous extracts of *A. conyzoides* in combination with CQ and artesunate were tested for antimalarial activity using the four-day suppression assay and the curative Rane Test. The study indicated that the aqueous extract of *A. conyzoides* had the ability to potentiate the effect of CQ and artesunate (Ukwe *et al.*, 2010b).

7.1.4.12 *Artemisia annua* (Asteraceae, Compositae)

A. annua or wormwood is an annual herbaceous, aromatic, bitter-tasting plant that is a globally acclaimed antimalarial herb (Dalrymple, 2012). The plant is used extensively in traditional Chinese medicine for the treatment of fevers, malaria, and jaundice. An endoperoxide, artemisinin has been shown to be responsible for the antimalarial activity of the herb. The pharmacognosist, Youyou Tu won the 2015 Nobel Prize in Medicine for her work, in which she led a team that evaluated more than 380 extracts from about 200 Chinese plants in the quest for novel antimalarial agents, using a rodent malaria model (Tu, 2011). Although the plant was well known (as qinghaosu) in Chinese herbal medicine, it was only in 1971 that scientists isolated a substance from *A. annua* that showed 100% activity against parasitaemia in *P. berghei*-infected mice, which was also found to be effective on monkeys infected with *Plasmodium cynomolgi*. In Nigeria, *A. annua* is an introduced crop but the Nigerian cultivars appear to have higher concentrations of artemisinin than the Asian cultivars. The indigenous *A. affra* is devoid of artemisinin. but it also possesses antimalarial activity.

There are many unresolved issues regarding the antimalarial profile of *Artemisia*, such as the low concentration of artemisinin in *A. annua* teas, which is the traditional method of use in malaria treatment, the low bioavailability of artemisinin when the traditional formulation is administered, and the high levels of recrudescence. In order to resolve these problems, artemisinin was subjected to intensive modifications. It was determined that artemisinin, when modified to artesunate or artemether, improved bioavailability and was more effective when used in combination with other antimalarial drugs, mainly mefloquine, which became known as Artemisinin Combination Therapy (ACT). Although the combination therapy has been very effective in malaria chemotherapy, incidences of malaria infections that are resistant to the ACTs have been reported in many parts of the world.

Fortunately, studies have shown that the use of formulations of the whole plant can overcome parasite resistance and are actually more resilient to evolution of parasite resistance, with parasites taking longer to evolve resistance, thus increasing the effective life span of the therapy.

Dried leaf *A. annua* (DLA) has been used as a last-resort treatment of patients with severe malaria, who were responding to neither ACT nor to intravenous (i.v.) artesunate (Bati-Daddy *et al.*, 2017). In the study, it was observed that, of many patients treated with ACTs and i.v. artesunate during a 6-month study period, 18 did not respond and were subsequently treated successfully with DLA. The patients were given a dose of 0.5 g DLA *per os*, twice daily for five days. Of the 18 ACT-resistant severe malaria cases compassionately treated with DLA, all fully recovered. Two were paediatric cases (Bati-Daddy *et al.*, 2017).

7.2 Other Neglected Tropical Diseases

Other members of this class of diseases, that are in close competition with malaria in terms of their morbidity, mortality and relationship to poverty, are lumped into a group of diseases that is called neglected tropical diseases (NTDs), or poverty-related diseases. They are mainly diseases caused by protozoa and helminths, with an estimated annual combined number of deaths of over one million. They also share with malaria the inadequate attention being paid to them by the major pharmaceutical companies. Although the twentieth century witnessed a strong commitment to ethics, and human-interest-driven research by the pharmaceutical industry, in recent years, these qualities seem to be gradually lost and the commitment to science has been replaced by commercial marketing, where research and development efforts are directed to the invention of the so-called blockbuster drugs, with annual sales exceeding $1 billion. In this scenario, the logic of pharmaceutical economics dictates that the therapeutic class to be studied depends entirely on the drugs that guarantees the largest immediate returns to the pharmaceutical industry and the neglect of the group of drugs that addresses the needs of the poor, located mainly in tropical countries.

These parasitic diseases pose an enormous health, social and economic impact in almost all tropical countries. The global burden of these diseases is exacerbated by the lack of licensed vaccines and access to the few drugs approved for the prevention and treatment of these diseases. The effect of the few available drugs are increasingly diminished by the occurrence of parasite drug resistance. There is a global need, therefore, for new drug discovery efforts targeted specifically against these diseases. In Nigeria, two approaches have been adopted to develop new therapies for such parasitic diseases. The first approach is to repurpose existing drugs that have been studied and approved for other conditions and that are of interest to the pharmaceutical companies, and/or drugs that have been successful in veterinary medicine. Examples include Clindamycin, doxycycline, Co-trimoxazole (trimethoprim/sulfamethoxazole), and Ivermectin for malaria, Elfornithine (also known as difluromethylornithine, DFMO) and Nifurtimox for African sleeping sickness, and Amphotericin B, Miltefosine, and Paromomycin for visceral leishmaniasis (Andrews *et al.*, 2014). The second approach is the evaluation of natural products, especially those from plants, used in traditional medicine with evidence of safe use historically as phytotherapy for the treatment of such diseases. The advantage of this approach is that the treatment option is implemented within the context that addresses other associated problems with NTDs, such as noncommunicable diseases, access to clean water and sanitation in disease-endemic countries, and eradicating poverty through trade in such herbs.

The diseases under consideration include the following: leishmaniasis, schistosomiasis, Chagas disease (American trypanosomiasis), human African trypanosomiasis (sleeping sickness), lymphatic filariasis (elephantiasis), helminthiasis (different types), leprosy (Hansen's Disease), trachoma, buruli ulcer, and tuberculosis.

7.2.1 Leishmaniasis

Leishmaniasis is a major global public health problem and is endemic in some parts of Nigeria. Infection with various species and strains of *Leishmania* causes a wide spectrum of diseases in humans, with many different clinical presentations. The manifestation and severity of the disease are often determined by the immunological health of the infected individual and the type of *Leishmania* species involved. The pentavalent antimonials (SbV), sodium stibogluconate and N-methylglucamine antimonate are the most-prescribed drugs for the treatment of most forms of leishmaniasis, while amphotericin and pentamidine are used as the secondary chemotherapeutic agents. Treatment with these agents is not consistently effective (particularly for the most virulent *Leishmania* disease forms. Treatment with these drugs is associated with serious toxic side effects, including cardiac and/or renal failure. Although there are several compounds under various stages of development, no drug has yet been discovered which is definitively effective in achieving a complete (or radical) cure of the infections. Several clinical reports have noted the persistence of the parasite after treatment, and 'clinical cure' has been demonstrated in both experimental animals and humans. What is often achieved in most chemotherapies of leishmaniasis is a suppression of the symptoms of the disease without elimination of the parasite.

A number of plants with activity against leishmania have been reported and their active components identified (Table 7.2). Extracts of *Picralima nitida*, discussed earlier for their antimalarial activity, have been shown to possess significant antileishmanial activity. The hydroalcholic extract of *O. gratissimum* showed good leishmanicidal activity against *Leishmania*. At a concentration of 100 µg/mL, it showed 91.5% inhibition of the parasite *in vitro*. Along with leishmanicidal activity, haemolytic activity of the extract was also observed. At a concentration of 1000 µg/mL, the extract showed 25% lysis of the red blood-cells, whereas no lysis was seen at concentrations of 500 and 100 µg/mL. At the end of 120 min incubation at 1000 µg/mL, there was an increase in cell lysis to 75% but no lysis was seen at concentrations of 500 and 100 µg/mL (Iwu, 2014). The sequiterpenes isolated from the rhizomes of *Renealmia cincinnata* have also been shown to be effective for the treatment of leishmaniosis (Tchuendema *et al.*, 1999).

Table 7.2 **Medicinal Plants With Antileishmanial Activity**

Plant	Family	Constituent(s)
Berberiso ristot	Berberidaceae	Berberine
Cola attiensis	Sterculiaceae	Aromatic polysulphur compounds
Dracaena mannii	Agavaceae	Saponins
Dorstenia multiradiata	Moraceae	Anthocyanidins
Desmodium gangeticum	Fabaceae	Alkylamines
Diospyros montana	Ebenaceae	Diospyrin
Gongronema latifolia	Asclepiadaceae	Lignans
Hedera helix	Araliaceae	Saponins
Jacaranda coeaia	Bignoniaceae	Jacaranone, quinol
Nyctanthes arbortristis	Verbenaceae	Iridoid glycosides
Ocimum gratissimum	Labiatae	Diterpenes, essential oil
Peganum harmala	Zygophyllaceae	Harmaline, tryptophan derivatives
Phytolacca americana	Phytolaccaceae	Saponins, proteins
Picralima nitida	Apocynaceae	Indole alkaloids
Picrorhiza kurroa	Scrophulariaceae	Picroliv, other iridoid glycosides
Plumbago zeylanica	Plumabaginaceae	Plumbagin
Polyalthia macropoda	Annonaceae	Labdane
Ricinus communis	Euphorbiaceae	Proteins
Tabebuia rosea	Bignoniaceae	Lapachol and related quinones.

7.2.2 Human African Trypanosomiasis (Sleeping Sickness)

Although both types of trypanosomiasis disease are found in Africa, it is the human African trypanosomiasis or 'sleeping sickness' that is the more serious disease on the continent. The other type of trypanosomiasis, known as Chagas disease, caused by the protozoan *Trypanosoma cruzi*, is of less importance in Africa and is rarely diagnosed in Nigeria. The parasite that causes Chagas disease is usually transmitted to humans through an injury caused by the bite of bugs of the Triatominae subfamily, known as "kissing bugs." The trypanosomes that cause sleeping sickness, on the other hand, is transmitted by the bite of the tsetse fly (*Glossina* spp., family Glossinidae). It causes serious diseases of both humans and livestock in Nigeria. The human African trypanosomiasis is a potentially fatal disease of humans, caused by two distinct subspecies of *T. brucei*, namely *T. b. gambiense* (West and Central Africa) and *T. b. rhodesiense* (East and South Africa). Other subspecies of trypanosomes are also endemic in Africa but are limited to infection of animals (*T. b. brucei*, *Trypanosoma congolense*, and *Trypanosoma evansi*) (Kobe de Oliveira *et al.*, 2015).

Human African trypanosomiasis is a disease complex which manifests differently in patients, depending on the subspecies of the parasite, the immunity of the patient and the stage of the infection. It is estimated that about 60 million people are at risk of contracting sleeping sickness, with *T. b. gambiense* accounting for more than 90% of the infections. It is one of the very few infectious diseases with a mortality rate of 100%, if untreated. In infected individuals, the parasite multiplies in the lymph and blood, causing headaches, fever, malaise, weakness, weight loss, arthralgia, and eventually vomiting, and skin lesions. In the latter stages, the parasite crosses the blood–brain barrier, migrates to the CNS, and the cerebral spinal fluid, and causes severe neurological and psychiatric disorders, leading to death (WHO, 2010; Kobe-de-Oliveira *et al.*, 2015). The prevalent subspecies in Nigeria and other parts of West Africa, *T. b. gambiense* causes a chronic infection, characterized by low parasitaemia and a longer time for the onset of CNS involvement. The classical sleeping sickness symptoms manifest over many years, leading to serious neurological symptoms and progression to coma and death. In contrast, infection by the East African subspecies, *T. b. rhodesiense,* causes a rapidly fatal disease with high parasitaemia, with death occurring within weeks to a few months. It develops quickly, invading the CNS, and leads to multiple organ involvement, including endocrine and gastrointestinal problems, and significant cardiac symptoms.

There are only a few pharmaceutical agents that are approved and available for the treatment of sleeping sickness. The commonly used ones include Suramin, Pentamidine, Eflornithine and Melarsoprol. All of them are poorly tolerated by patients and cause serious side effects, such as renal toxicity, neurological complications, hypoglycaemia, infections, hypertension, diarrhoea, neutropenia, and encephalopathy. In many parts of Nigeria, polyherbal preparations are used for the treatment of the disease. The sequential application of the remedies in many cases suggest a clear understanding of the need to address both the symptomatic manifestation of the disease and the causative parasites. Four of the most frequently used species in the preparations of the remedies are *Picralima nitida* (seeds), *Nauclea latifolia* (stem bark) *Ocimum gratissimum* (leaves), and *Uvaria chamae* (stem or root bark). Others include *Uvariopsis congolana*, *Ageratum conyzoides* (Asteraceae), *Saussurea costus*, *Eupatorium cannabinum*, *Inula montbretiana*, *Liriodendron tulipifera* (Magnoliaceae), *Porella densiflora* (Porellaceae), *Curcuma aromatica* (Zingiberaceae), *Pellia endiviifolia* (Pelliaceae), *Argemone mexicana* (Papaveraceae), *Lycoris traubii* (Amaryllidaceae), *Polyalthia suaveolens* (Annonaceae) and *Khaya senegalensis*.

7.2.3 Elephantiasis (Lymphatic Filariasis)

Elephantiasis, known in medical terms as lymphatic filariasis, is a painful and disfiguring disease, caused by three species of thread-like nematode worms, called filariae, namely *Wuchereria bancrofti*, *Brugia malayi*, and *Brugia timori*. The worms form "nests" in the human lymphatic system, an essential component of the immune system, that maintains the fluid balance between blood plasma and body tissues. The infection is usually acquired during childhood but remains dormant for a long time. The clinical manifestations include lymphoedema of the limbs and genitals (causing hydrocele, chylocele, and swelling of the scrotum and penis) which causes significant pain and discomfort, large-scale lost productivity, and discrimination.

There is a Global Program to Eliminate Lymphatic Filariasis, which aims to produce doses of a multidrug scheme in a mass drug administration program yearly for at least five years (http://www.filariasis.org) in order to interrupt transmission and reduce morbidity. The existing antifilarial therapy is inadequate in many respects; they are either marginally effective or possess intolerable side effects. The drugs currently used include the microfilaricidal diethylcarbamazine (DEC), DEC plus albendazole, or ivermectin plus albendazole. None of these is effective in killing the adult worms (macrofilaricidal activity), which can live in the host for several years and lead to the reappearance of microfilaria after several months of treatment.

Most of the medicinal plants used in the treatment of lymphatic filariasis in Nigeria are aimed at the adult worm, which makes their use to be complementary to, rather than a substitute for the prescription drugs. Plants containing essential oils, flavonoids and indole alkaloids seem to dominate in the preparation of remedies used for the management of elephantiasis. The spice *Trachyspermum ammi*, as well as *Vitex negundo* (roots), *Aegle marmelos* (leaves), *Azadirachta indica* (seeds), *Alstonia boonei* (stem bark), *Rauwolfia vomitoria*, *Picralima nitida* (seeds and fruit rind) and *Ricinus communis* (aerial parts) have been used in the preparation of remedies for lymphatic filariasis.

Xylocarpus granatum fruit decoctions have also been used for the treatment of elephantiasis. The extract has been found to possess *in vitro* and *in vivo* activity against the *Br. malayi* parasite. The active constituents were identified as gedunin and photogedunin, which have been found to possess significant *in vitro* adulticidal and microfilaricidal activity, with low IC_{50} (0.239 and 0.213 mg/mL for adults, 2.03 and 2.23 mg/mL for microfilaria, respectively). Bioactivity-guided structural modifications are on-going in various laboratories to develop more active chemical analogs (Misra *et al.*, 2011). Another plant that is widely used either alone or in combination with other herbs for the treatment of filariasis is *Bauhinia racemosa*, a small tree that is widely distributed throughout the tropics. Laboratory studies have shown that the *n*-butanol fraction from the ethanolic extract of the leaf possesses significant *in vitro* anti-filarial activity. Three galactolipids isolated from the fraction have been shown to be the active molecules against adult worms and microfilaria (Sashidhara *et al.*, 2012).

Studies have also shown that some flavonoids and flavonoid-containing extracts possess anthelmintic and nematicidal activity. For example, six pure common compounds, namely flavone, chrysin, rutin, naringenin, naringin, and hesperetin, were subjected to *in vitro* and *in vivo* assays to ascertain their potential to treat filariasis disease. All the tested flavonoids showed *in vitro* anti-filarial activity without strong cytotoxicity towards the host cell line. The widely occurring citrus flavonoid, naringenin, was found to be the most active (LC_{100} 7.8 mg/mL). In addition, naringenin showed a high Selectivity Index against the motility of adult parasites, which is interesting, since the existing anti-filarial therapy is not able to kill adult worms (Lakshmia *et al.*, 2010).

7.2.4 Buruli Ulcer

Buruli ulcer (BU) is an infectious disease caused by the bacterium *Mycobacterium ulcerans*. It is the third most common mycobacterial infection in the world, after tuberculosis and leprosy. It has increasingly been recognized as an important emerging disease in many parts of the tropics. The disease usually affects the skin but can also affect the bone. Buruli infection often leads to ulcers on the arms or legs, which can also destroy skin or soft tissue. The early stage of the infection is characterized by a painless nodule or area of swelling, which later can turn into an ulcer. The ulcer may be larger inside than at the surface of the skin and can be surrounded by swelling. Bone ulcers are extremely rare. More than 99% of the Buruli ulcer burden occurs in West Africa and most affected people live in remote areas with traditional medicine as the primary or only option available. Some of the major plants used for the treatment of Buruli include *Mangifera indica* (leaves), *Azadirachta indica* (seed and leaves), *Moringa oleifera* (seed oil), *Ricinus communis (*seed oil and leaves*)*, *Vernonia amygdalina* (leaves and twigs) *Cyperus cyperoides*, *Solanum rugosum*, *Alchornea cordifolia* (roots), *Ocimum gratissimum* (leaves) *Carica papaya* (seeds and leaves), *Chromolaena odorata* (aerial parts), *Bridellia ferruginea* (aerial parts), *Zanthoxylum zanthoxyloides* (stem bark), *Bryophyllum pinnatum* (leaves), and *Nicotiana tabacum* (leaves). Most of these plants have proved useful in hospital treatment of Buruli ulcer. An effective polyherbal preparation, composed of leaves of *Mangifera indica*, pulverized roots of *Garcinia kola*, seeds and leaves of *Carica papaya*, *Aloe vera* and seed oil from *Ricinis communis*. has been used successfully in herbal homes in Southeastern Nigeria, with impressive results.

A team of investigators from Noguchi Memorial Institute for Medical Research, College of Health Sciences, University of Ghana, University of Yaoundé and the Institute of Medical Research and Medicinal Plants Studies (IMPM),

Yaoundé, Cameroon, has conducted *in vitro* studies on some of the plants used in traditional treatment of BU against clinical *M. ulcerans* and the team found that seven species, namely *Ricinus communis, Cyperus cyperoides* (cited as *Mariscus alternifolius*), *Nicotiana tabacum, Mangifera indica, Solanum rugosum, Carica papaya*, and *Moringa oleifera*, demonstrated efficacy in hospitalized BU patients (Tsouh-Fokou *et al.*, 2015). Four isolated and characterized compounds were also reported to have moderate bioactivity *in vitro* against *M. ulcerans*. A phytomedical approach is, therefore, a feasible option in the development of effective treatments aganst Buruli ulcer.

7.2.5 Tuberculosis

Tuberculosis (TB) is a major global public health problem which has become of greater concern to healthcare providers because of the high incidence of TB as an opportunistic infection among those with HIV infection. The disease is caused by *Mycobacterium tuberculosis* bacilli, mycobacteria related to the actinomycete group (the mycobacterium TB complex), which includes human and bovine types, *Mycobacterium africanum, Mycobacterium microtti* and *Mycobacterium. canetti*. These aerobic bacilli invade the lungs in 80 to 90% of the cases, and some cases also colonize other extra-pulmonary organs, mainly lymph nodes, bones, and joints, and the genitourinary system (Tiemersma *et al.*, 2011). The clinical symptoms include prolonged cough for more than three weeks, initially without fever, followed by sputum (with or without blood), chest pain, weakness, weight loss, fever, and night sweats. Tuberculosis is called 'the white plague', due to the high mortality rate. It is responsible for more than 2 million deaths worldwide annually. Despite the availability of antibiotics and the Bacillus Calmette–Guérin (BCG) vaccine, *M. tuberculosis* is the agent of infectious origin that causes most deaths in the world. According to the World Health Organization, almost 9,600,000 people had TB in 2014, 5,400,000 of whom were men, 3,200,000 women and 1 million children. Of these numbers, 6,000,000 were new cases of TB, that is 63% of the total estimated figures for that year. This means that almost 37% of patients with TB were undiagnosed and/or unreported (WHO, 2015). Reduction of the global TB burden is one of UN's sustainable development goals (SDG), by which all countries are expected to reduce the number of death from TB by 90% by the year 2030, and to reduce the number of new cases by 80%, simultaneously. The countries are also obligated to plan in a manner where no family will be compromised because of this disease.

TB is clearly a poverty-associated disease, since it is well established that over-crowded habitation promotes the spread of the infection. Another key challenge of TB control is the upsurge in drug-resistant strains (Center for Disease Control 2005). The problem is further complicated by the poor patient compliance in adhering to the treatment regime with the effective antimicrobial agents in current use, which often requires prolonged duration of treatment. In addition, treatment of TB and HIV- AIDS is difficult due to the adverse drug interactions involved (Chan and Iseman, 2002). In clinical settings, there are challenges posed by multi drug-resistant tuberculosis (MDR TB) and extensively drug-resistant tuberculosis (XDR TB), which are resistant to the most affordable, efficacious and readily available TB drugs. MDR TB is defined as cases infected with tubercle bacilli that are resistant to the major essential drugs, rifampicin (RIF) and isoniazid (INH), while XDR (extensively drug-resistant) refers to cases which are MDR and resistant to fluoroquinolone and two injectable drugs (amikacin and capreomycin) (Shariffi-Rad *et al.*, 2017).

Medicinal plants have been used for centuries by traditional medical practitioners (TMPs) for the treatment of TB, in the form of decoctions, exudates, oils, macerations, tinctures, and infusions from various plant parts such as leaves, roots, stem bark, stem, flowers, and fruits (Gupta *et al.*, 2010). Studies have shown that traditional medical practitioners have good knowledge of TB and have remedies that can adequately address the problem, and some of their traditional remedies and their metabolites have inhibitory or bactericidal activities *in vitro* against *M. tuberculosis* at low concentrations (Newman and Cragg, 2007; Ibekwe *et al.*, 2014). Efforts are now directed at the development of standardized whole herbs and plant extracts or fully characterized phytomedicines as anti-tuberculosis drugs.

Phytomedical approaches are useful either as direct anti-tuberculosis agents or as adjuvants to supplement the classical drugs in order to synergistically improve their activity, overcome resistance, minimize some of their side-effects, and even control the adverse drug–drug interaction observed in the co-administration of drugs for HIV and TB therapy. In Nigeria, the frequently used plants for the treatment of tuberculosis and tuberculosis-similar infections include *Picralima nitida, Anogeissus leocarpus, Garcinia kola* (kolaviron), *Combretum molle, Khaya senegalensis, Ficus sur, Pavetta crassipes, Waltheria indica, Crotolaria lachnosema, Anogiessus leocarpus, Calliandra portoricensis, Cassia sieberiana, Erythrina senegalensis, Abrus precatorius, Callistemon citrinus, Securidaca longepedunculata, Toddalia asiatica, Lantana camara, Vernonia amygdalina, Allium sativum, Aloe vera, Azadirachta indica, Bidens pilosa, Carica papaya, Catharanthus roseus, Centella asiatica, Cinnamomum zeylanium, Mangifera indica, Zingiber officinale, Ocimum gratissimum, Maytenus senegalensis, Pterocarpus osun, Pentaclethra macrophylla, Tapinanthus sessifolia, Abrus precatorius, Cussonia arborea. Erythrina abyssinica, Tetrapleura tetraptera, Moringa oleifera*, and *Andrographis sanguinolenta*. The term "tuberculosis-similar infections" was used to underline the fact that, in most cases, therapy was initiated without a conclusive laboratory-based diagnosis of tuberculosis, especially among AIDS patients. Although there were good clinical outcome reports in most cases, the lack of good pre-treatment record of diagnosis reduced the usefulness of the ethnomedical reports of TB cure.

An evaluation of plants based on ethnomedical leads has shown that 69% of the extracts of such plants exhibited activity against the animal TB strain BCG, while 64% of the extracts were active against *M. tuberculosis* (Ibekwe *et al.*, 2014). The activities varied from weak, ≤ 2500 μg/mL to strong (33 μg/mL) activities of the extracts against *M. tuberculosis* . varied from weak, ≤ 2500 μg/mL to highly active, 128 μg/mL. There was 77% agreement in results obtained using BCG or *M. tuberculosis* as test organisms. The results show clear evidence for the efficacy of the majority of indigenous Nigerian herbal recipes used in the ethnomedical management of tuberculosis and related ailments. BCG may be effectively used, to a great extent, as the organism for screening for potential anti-*M. tuberculosis* agents. A set of prioritisation criteria for the selection of plants for initial further studies for the purpose of anti-tuberculosis drug discovery research has been proposed, which identified *Ficus sur, Pavetta crassipes, Combretum molle, Waltheria indica, Crotolaria lachnosema, Anogiessus leocarpus, Calliandra portoricensis, Cassia sieberiana, Abrus precatorius* and *Cussonia arborea* as potential candidates for further study (Ibekwe *et al.*, 2014).

It is noteworthy that traditional medical practitioners in Uganda, like their Nigerian counterparts, were reported as being able to accurately diagnose and successfully treat TB. *Zanthoxylum leprieurii, Piptadeniastrum africanum, Albizia coriaria,* and *Rubia cordifolia* were reported as the plants most mentioned in terms of anti-TB effectiveness by TMPs in that country (Bunalema *et al.*, 2014). The investigators also found that the TMPs had knowledge of how TB is transmitted, and they admitted that it is closely associated with HIV. As in Nigeria, decoctions of multiple plant species were commonly used, and the plant parts frequently used were leaves followed by the stem bark and root bark. In the continent of Africa taken together, the following plants were most frequently reported as being useful in the treatment of TB: *Erythrina abyssinica, Allium sativum, Ficus platyphylla, Bidens pilosa, Asparagus africanus, Carissa edulis, Pavetta crassipes, Combretum molle, Waltheria indica, Crotolaria lachnosema, Anogiessus leocarpus, Calliandra portoricensis,* and *Allium cepa* (Shariffi-Rad *et al.*, 2017). Some of these plants have shown impressive activity in laboratory studies against TB. For example, *E. abyssinica* crude methanol extract showed antimicrobial activity towards the sensitive strain H37Rv and the rifampicin-resistant strain TMC-331, with MIC values of 0.39 mg/mL and 2.35 mg/mL, respectively (Bunalema, 2010). *A. sativum* extract inhibited both non-MDR and MDR *M. tuberculosis* isolates, with MIC values ranging from 1000 to 3000 μg/mL (Hannan *et al.*, 2011). The leaf ethanol extract of *B. pilosa* exhibited activity against *M. tuberculosis* at 100 μg/mL (Gautam *et al.*, 2012). *C. edulis* showed antibacterial activity against slow- (*M. tuberculosis, Mycobacterium kansasii*) and fast- (*Mycobacterium fortuitum* and *Mycobacterium smegmatis*) growing mycobacteria (Mariita, 2006). Anti-TB biological activities have also been reported for the following plants: *Toddalia asiatica, Lantana camara, Vernonia amygdalina, Allium sativum, Aloe vera, Azadirachta indica, Bidens pilosa, Carica papaya, Catharanthus roseus, Centella asiatica, Cinnamomum zeylanium, Mangifera indica, Zingiber officinale, Maytenus senegalensis* and *Eucalyptus* spp. (Mariita *et al.*, 2010; Green *et al.*, 2010; Shariffi-Rad *et al.*, 2017). These examples are illustrative of the enormous potential in the use of ethnomedically identified plants for the treatment of TB in Nigerian communities. Although the crude extracts of the above-mentioned plants have significant antimycobacterial activity, only a few have had their compounds isolated and their minimum inhibitory concentrations (MIC) against *Mycobacterium tuberculosis* recorded. These shortcomings should not prevent their continued use to combat such an endemic deadly disease.

References

Abay, S.M., Lucantoni, L., Dahiya, N., Dori, G., Dembo, E.G., Esposito, F., Lupidi, G., Ogboi, S., Ouédraogo, R.K., Sinisi, A., TaglialatelaScafati, O., Yerbanga, R.S., Bramucci, M., Quassinti, L., Ouédraogo, J-B., Christophides, G. and Annette Habluetzel, A. (2015) Plasmodium transmission blocking activities of *Vernonia amygdalina* extracts and isolated compounds. *Malaria Journal*, 14, 288–307.

Abbah, J., Amos, S., Chindo, B., Ngazal, I., Vongtau, H.O., Adzu, B., Farida, T., Odutola, A.A., Wambebe, C. and Gamaniel, K.S. (2010) Pharmacological evidence favouring the use of *Nauclea latifolia* in malaria ethnopharmacy: effects against nociception, inflammation, and pyrexia in rats and mice. *Journal of Ethnopharmacology*, 127, 85–90.

Adebajo, A.C., Odediran, S.A., Aliyu, F.A., Nwafor, P.A., Nwoko, N.T. and Umana, U.S. (2014) *In vivo* antiplasmodial potentials of the combinations of four Nigerian antimalarial plants. *Molecules*, 19(9), 13136–13146.

Adebayo, J.O. and Krettli, A.U. (2011) Potential antimalarials from Nigerian plants: a review. *Journal of Ethnopharmacology*, 133(2), 289–302.

Adodo, A. (2004) *Herbs for healing – receiving God's healing through nature*. Revised ed. Ewu-Esan, Nigeria: Pax Herbal Clinic and Research Laboratories. Benedictine Monastery, pp. 122.

Andrews, K.T., Fisher, G. and Skinner-Adams, T.S. (2014) Drug repurposing and human parasitic protozoan diseases. *International Journal for Parasitology: Drugs and Drug Resistance*, 4, 95–111.

Ansa-Asamoah, R., Kapadia, G.J., Lloyd, H.A. and Sokoloski, E.A. (1990) Picratidine, a new indole alkaloid from *Picralima nitida* seeds. *Journal of Natural Products*, 53(4), 975.

Awe, S.O. and Opeke, O.O. (1990) Effects of *Alstonia congensis* on *Plasmodium berghei* in mice. *Fitoterapia*, 61, 225–229.

Baird, J.K. (2012) Chemotherapeutics challenges in developing effective treatments for the endemic malarias. *International Journal for Parasitology: Drugs and Drug Resistance*, 2, 256–261.

Bati-Daddy, N., Kalisya, L.K., Bagire, P.G., Watt, R.L., Towler, M.J. and Weathers, P.J. (2017) *Artemisia annua* dried leaf tablets treated malaria resistant to ACT and i.v. artesunate. *Phytomedicine*, 32, 37–70.

Bell, A. (2005) Antimalarial drug synergism and antagonism: mechanistic and clinical significance. *FEMS Microbiology Letters*, 253, 171–184.

Benoit-Vical, F., Valentin, A., Da, B., Dakuyo, Z. and Mallié, M. (2003) N'Dribala (*Cochlospermum planchonii*) versus chloroquine for treatment of uncomplicated *Plasmodium falciparum* malaria. *Journal of Ethnopharmacology*, 89(1), 111–114.

Boumendjel, A., Sotoing-Taïwe, G., Ngo-Bum, E., Chabrol, T., Beney, C., Sinniger, V., Haudecoeur, R., Marcourt, L., Challal, S., Ferreira-Queiroz, E., Souard, F., Le-Borgne, M., Lomberget, T., Depaulis, A., Lavaud, C., Robins, R., Wolfender, J.-L., Bonaz, B. and De-Waard, M. (2013) Occurrence of the synthetic analgesic tramadol in an African medicinal plant. *Angewandte Chemie - International Edition*, 52(45), 11780–11784.

Boye, G.L. (1989) Studies on antimalarial action of *Cryptolepis sanguinolenta* extract. *Proceedings of the International Symposium on East-West Medicine*, pp. 243–251. October 10–11, Seoul, Korea.

Bugyei, K.A., Boye, G.E. and Addy, M.E. (2010) Clinical efficacy of a tea-bag formulation of cryptolepis sanguinolenta root in the treatment of acute uncomplicated falciparum malaria. *Ghana Medical Journal*, 44(1), 3–9.

Bunalema, L. (2010) *Anti-mycobacterial activity and acute toxicity of Erythrina abyssinica, Cryptolepis sanguinolenta and Solanum incanum*. Kampala, Uganda: Makerere University.

Bunalema, L., Obakiro, S., Tabuti, J.R.S. and Waako, P. (2014) Knowledge on plants used traditionally in the treatment of tuberculosis in Uganda. *Journal of Ethnopharmacology*, 151, 999–1004.

Centre for Disease Control. (2005) Worldwide emergence of *Mycobacterium tuberculosis* with extensive resistance to second-line drugs. *Morbidity and Mortality Weekly Report*, 55, 250–253.

Chan, E.D. and Iseman, M.D. (2002) Current medical treatment for tuberculosis. *British Medical Journal*, 325, 1282–1286.

Chime, S.A., Ugwuoke, E.C., Onyishi, I.V., Brown, S.A. and Onunkwo, G.C. (2013) Formulation and evaluation of *Alstonia boonei* stem bark powder tablets. *Indian Journal of Pharmaceutical Sciences*, 75(2), 226–230.

Chinsembu, K.C. (2015) Plants as antimalarial agents in Sub-Saharan Africa (review). *Acta Tropica*, 152, 32–48.

Dalrymple, D.G. (2012) *Artemisia annua, artemisinin, ACTs and malaria control in Africa*. Washington, DC: Tradition, Science and Public Policy, Politics and Prose Bookstore, 253 pp.

Diallo, D., Graz, B., Falquet, J., Traore, A.K., Giani, S., Mounkoro, P.P., Berthé, A., Sacko, M. and Diakité, C. (2006) Malaria treatment in remote areas of Mali: use of modern and traditional medicines, patient outcome. *Transactions of the Royal Society of Tropical Medicine and Hygiene*, 100, 515–520.

Eaton, J.W., Eckman, J.R., Berger, E. and Jacob, H.S. (1976) Suppression of malaria infection by oxidant-sensitive host erythrocytes. *Nature*, 264(5588), 758–760.

Elisabetsky, E. and Costa-Campos, L. (2006) The alkaloid alstonine: a review of its pharmacological properties. *Evidence-Based Complementary and Alternative Medicine*, 3, 39.

Elkin, N.L. (1975) The human red cell, glucose-6-phosphate dehydrogenase deficiency and malaria. *Ph.D. dissertation*, Washington University, St. Louis, MO.

Elkin, N.L. and Eaton, J.W. (1975) Malaria induced oxidant sensitivity. In: G.J. Brewer (ed.), *Proceedings of 3rd international conference on red cell metabolism and function*, p. 219. New York: Alan R. Liss.

Etkin, N.L. (1981) A hausa herbal pharmacopoeia: biomedical evaluation of commonly used plant medicines. *Journal of Ethnopharmacology*, 4(1), 75–98.

Gamaniel, K., Wambebe, C., Amupitan, J., Hussaini, I.M., Amos, S., Awodogan, A., Dunah, A.W., Ekuta, J.E., Akeju, M.O., Usman, H. and Enwerem, N. (1997) Active column fractions of *Nauclea latifolia* on *Plasmodium berghei* and rabbit ileum. *Journal of Pharmaceutical Research and Development*, 2, 44–47.

Gautam, H.A., Sharma, R. and Rana, A.C. (2012) Review on herbal plants useful in tuberculosis. *International Research Journal of Pharmacy*, 3(7), 64–67.

Graz, B., Willcox, M.L., Diakite, C., Falqueta, J., Dackuo, F., Sidibeb, O., Gianie, S. and Diallo, D. (2010) *Argemone mexicana* decoction versus artesunate-amodiaquine for the management of malaria in Mali: policy and public-health implications. *Transactions of the Royal Society of Tropical Medicine and Hygiene*, 104, 33–41.

Green, E., Samie, A., Obi, C.L., Bessong, P.O. and Ndip, R.N. (2010) Inhibitory properties of selected South African medicinal plants against *Mycobacterium tuberculosis*. *Journal of Ethnopharmacology*, 130(1), 151–157.

Gupta, R., Thakur, B., Singh, P., Singh, H.B., Sharma, V.D., Katoch, V.M. and Chauhan, S.V. (2010) Anti-tuberculosis activity of selected medicinal plants against multidrug resistant *Mycobacterium tuberculosis* isolates. *Indian Journal of Medical Research*, 131, 809–813.

Hafid, A.F., Retnowati, D. and Widyawaruyanti, A. (2015) The combination therapy model of *Andrographis paniculata* extract and chloroquine on *Plasmodium berghei* infected mice. *Asian Journal of Pharmaceutical and Clinical Research*, 8(2), 205–208.

Hannan, A., Ikram Ullah, M., Usman, M., Hussain, S., Absar, M. and Javed, K. (2011) Antimycobacterial activity of garlic (*Allium sativum*) against multi-drug resistant and nonmulti-drug resistant *Mycobacterium tuberculosis*. *Pakistan Journal of Pharmaceutical Sciences*, 24, 81–85.

Haudecoeur, R., Peuchmaur, M., Pérès, B., Rome, M., Taïwe, G.S., Boumendjel, A. and Boucherle, B. (2018) Traditional uses, phytochemistry and pharmacological properties of African Nauclea species: a review. *Journal of Ethnopharmacology*, 212, 106–136.

Ibekwe, N.N., Nvaua, J.B., Oladosu, P.O., Usman, A.M., Ibrahim, K., Boshoff, H.I., Dowd, C.S., Orisadipe, A.T., Aiyelaagbe, O., Adesomoju, A.A., Barry III, C.I., Okogun, J.I. and in collaboration with 73 Visited Herbalists. (2014) Some Nigerian anti-tuberculosis ethnomedicines: a preliminary efficacy assessment. *Journal of Ethnopharmacology*, 155(1), 524–532.

Ibrahim, N.D.G., Abdurahman, E.M. and Ibrahim, G. (2001) Elemental analysis of the leaves of *Vernonia amygdalina* and its biological evaluation in rats. *Nigerian Journal of Natural Products and Medicine*, 5, 13–16.

Idowu, E.T., Ajaegbu, H.C.N., Omotayo, A.I., Aina, O.O. and Otubanjo, O.A. (2015) *In vivo* anti-plasmodial activities and toxic impacts of lime extract of a combination of *Picralima nitida*, *Alstonia boonei* and *Gongronema latifolium* in mice infected with chloroquine-sensitive *Plasmodium berghei*. *African Health Sciences*, 15(4), 1262–1270.

Isah, Y., Ndukwe, I.G. and Ayo, R.G. (2013) Phytochemical and antimicrobial analyses of stem-leaf of *Cochlospermum planchonii*. *Journal of Medicinal Plant and Herbal Therapy Research*, 1, 13–17.

Iwalokun, B.A. (2008) Enhanced antimalarial effects of chloroquine by aqueous *Vernonia amygdalina* leaf extract in mice infected with chloroquine resistant and sensitive *Plasmodium berghei* strains. *African Health Sciences*, 8(1), 25–35.

Iwu, M. and Klayman, D.L. (1992) Evaluation of the *in vitro* antimalarial activity of *Picralima nitida* extracts. *Journal of Ethnopharmacology*, 36, 133–135.

Iwu, M.M. (1983) *Traditional Igbo medicine*. Nsukka: Institute of African Studies, University of Nigeria.

Iwu, M.M., Obidoa, O. and Anazodo, M. (1986) Biochemical mechanism of the antimalarial activity of *Azadirachta indica* leaf extract. *Pharmacological Research Communications*, 18(1), 81–91.

Iwu, M.M. (2014) *Handbook of African medicinal plants*. Second ed. Boca Raton, FL: CRC Press/Taylor and Francis Group, pp. 476.

Iwu, M.M. (2016) *Food as medicine – functional food plants of Africa*. Boca Raton, FL: CRC Press/Taylor and Francis Group, pp. 384.

Kaou, A.M., Mahiou-Leddet, V., Hutter, S., Aınouddine, S., Hassani, S., Yahaya, I., Azas, N. and Ollivier, E. (2008) Antimalarial activity of crude extracts from nine African medicinal plants. *Journal of Ethnopharmacology*, 116(1), 74–83.

Kobe de Oliveira, S., Chiaradia-Delatorre, L.D., Mascarello, A., Veleirinho, B., Ramlov, F., Kuhnen, S., Yunes, R.A. and Maraschin, M. (2015) From bench to bedside: natural products and analogs for the treatment of neglected tropical diseases (NTDs). In: Atta-ur-Rahman (Ed.) *Studies in natural products chemistry*, Elsevier B.V., Amsterdam. Vol. 44, Chapter 2, pp. 33–90.

Lakshmi, V., Joseph, S.K., Srivastava, S., Verma, S.K. and Sahoo, M.K. (2010) Antifilarial activity in vitro and in vivo of some flavonoids tested against *Brugia malayi*. *Acta Tropica*, 116, 127–133.

Lamien-Meda, A., Kiendrebeogo, M., Compaoré, M., Meda, R.N.T., Bacher, M., Koenig, K., Pacher, T., Fuehrer, H.-P., Noedl, H., Willcox, M. and Novak, J. (2015) Quality assessment and antiplasmodial activity of West African Cochlospermum species. *Phytochemistry*, 119, 51–61.

Lavrado, J., Paulo, A., Gut, J., Rosenthal, P.J. and Moreira, R. (2008) Cryptolepine analogues containing basic aminoalkyl side-chains at C-11: synthesis, antiplasmodial activity, and cytotoxicity. *Bioorganic and Medicinal Chemistry Letters*, 18(4), 1378–1381.

Lavrado, J., Mackey, Z., Hansell, E., McKerrow, J.H., Paulo, A. and Moreira, R. (2012) Antitrypanosomal and cysteine protease inhibitory activities of alkyldiamine cryptolepine derivatives. *Bioorganic and Medicinal Chemistry Letters*, 22(19), 6256–6260.

Linck, V.M., Herrmann, A., Goerck, G., Iwu, M., Okunji, C., Leal, M.B. and Elisabetsky, E. (2008) The putative antipsychotic alstonine reverses social interaction withdrawal in mice. *Progress in Neuro-Psychopharmacology and Biological Psychiatry*, 32(6), 1449–1452.

Linck, V.M., Herrmann, A., Goerck, G., Iwu, M., Okunji, C., Leal, M.B. and Elisabetsky, E. (2011) Alstonine as an antipsychotic: effects on brain amines and metabolic changes. *Evidence-Based Complementary and Alternative Medicine*, 2011, 418597.

Majekodunmi, S.O., Adegoke, O.A. and Odeku, O.A. (2008) Formulation of the extract of the stem bark of *Alstonia boonei* as tablet dosage form. *Tropical Journal of Pharmaceutical Research*, 7, 987–994.

Mariita, M.A.R. (2006) *Efficacy of medicinal plants used by communities around Lake Victoria region and the Samburu against Mycobacteria, selected bacteria and Candida albicans*. Kenya: Kenyatta University.

Mariita, R.M., Ogol, C.K.P., Oguge, N. and Okemo, P. (2010) Antitubercular and phytochemical investigation of methanol extracts of medicinal plants used by the Samburu community in Kenya. *Tropical Journal of Pharmaceutical Research*, 9, 379–385.

Marquet, S. (2018) Overview of human genetic susceptibility to malaria: from parasitemia control to severe disease. *Infection, Genetics and Evolution*, 66: 399–409.

Mathivanan, D., Gandhi, P.R., Mary, R.R. and Suseem, S.R. (2018) Larvicidal and acaricidal efficacy of different solvent extracts of *Andrographis echioides* against blood-sucking parasites. *Physiological and Molecular Plant Pathology*, 101: 187–196.

Mesia, K., Tona, L., Mampunza, M.M., Ntamabyaliro, N., Muanda, T., Muyembe, T., Cimanga, K., Totté, J., Mets, T., Pieters, L. and Vlietinck, A. (2012a) Antimalarial efficacy of a quantified extract of *Nauclea pobeguinii* stem bark in human adult volunteers with diagnosed uncomplicated falciparum malaria part 1: a clinical phase IIA trial. *Planta Medica*, 78(3), 211–218.

Mesia, K., Tona, L., Mampunza, M.M., Ntamabyaliro, N., Muanda, T., Muyembe, T., Musuamba, T., Mets, T., Cimanga, K., Totté, J., Pieters, L. and Vlietinck, A.J. (2012b) Antimalarial efficacy of a quantified extract of *Nauclea pobeguinii* stem bark in human adult volunteers with diagnosed uncomplicated falciparum malaria. Part 2: a clinical phase IIB trial. *Planta Medica*, 78(9), 853–860.

Misra, P., Pal, N.L., Guru, P.Y., Katiyar, J.C., Srivastava, V. and Tandon, J.S. (1992) Antimalarial activity of *Andrographis paniculata* (kalmegh) against *Plasmodium berghei* NK 65 in *Mastomys natalensis*. *Pharmaceutical Biology* 30(4): 263–274.

Misra, S., Verma, M., Mishra, S.K., Srivastava, S., Lakshmi, V. and Misra-Bhattacharya, S. (2011) Gedunin and photogedunin of Xylocarpus granatum possess antifilarial activity against human lymphatic filarial parasite *Brugia malayi* in experimental rodent host. *Parasitology Research*, 109, 1351–1360.

Nafiu, M.O., Akanji, M.A. and Yakubu, M.T. (2011) Effect of aqueous extract of *Cochlospermum planchonii* rhizome on some kidney and liver function indices of albino rats. *African Journal of Traditional, Complementary and Alternative Medicines*, 8, 22–26.

Newman, D. J. and Cragg, G.M. (2007) Natural products as sources of new drugs over the last 25 years. *Journal of Natural Products*, 70, 461–477.

Odugbemi, T.O., Odunayo, R.A., Ibukun, E.A. and Fabekun, P.O. (2007) Medicinal plants useful for malaria therapy in Okeigbo, Ondo State, Southwest Nigeria. *African Journal of Traditional, Complementary and Alternative Medicines*, 4(2), 191–198.

Okhuarobo, A., Falodun, J.E., Erharuyi, O., Imieje, V., Falodun, A. and Langer, P. (2015) Harnessing the medicinal properties of *Andrographis paniculata* for diseases and beyond: a review of its phytochemistry and pharmacology. *Asian Pacific Journal of Tropical Disease*, 4(3), 213–222.

Okpe, O., Habila, N., Ikwebe, J., Upev, V.A., Okoduwa, S.I.R. and Isaac, O.T. (2016) Antimalarial potential of *Carica papaya* and *Vernonia amygdalina* in mice infected with *Plasmodium berghei*. *Journal of Tropical Medicine*, 2016, 1–5.

Onyesom, I., Osioma, E. and Okereke, P.C. (2015) *Nauclea latifolia* aqueous leaf extract eliminates hepatic and cerebral *Plasmodium berghei* parasite in experimental mice. *Asian Pacific Journal of Tropical Biomedicine*, 5(7), 546–551.

Otoo, L.F., Koffuor, G.A., Ansah, C., Mensah, K.B., Benneh, C. and Ben, I.O. (2015) Assessment of an ethanolic seed extract of *Picralima nitida* ([Stapf] Th. and H. Durand) on reproductive hormones and its safety for use. *Journal of Intercultural Ethnopharmacology*, 4(4), 293–301.

Pholphana, N., Panomvana, D., Rangkadilok, N., Suriyo, T., Ungtrakul, T., Pongpun, W., Thaeopattha, S. and Satayaviva, J. (2015) A simple and sensitive LC-MS/MS method for determination of four major active diterpenoids from *Andrographis paniculata* in human plasma and its application to a pilot study. *Planta Medica*, 82(1–2), 113–120.

Rocha e Silva, L.F., Montoia, A., Amorim, R.C.N., Melo, M.R. Henrique, M.C., Nunomura, S.M., Costa, M.R.F., Andrade Neto, V.F., Costa, D.S., Dantas, G., Lavrado, J., Moreira, R., Paulo, A., Pinto, A.C., Tadei, W.P., Zacardi, R.S., Eberlin, M.N. and Pohlit, A.M. (2012) Comparative in vitro and in vivo antimalarial activity of the indole alkaloids ellipticine, olivacine, cryptolepine and a synthetic cryptolepine analog. *Phytomedicine*, 20(1), 71–76.

Sashidhara, K.V., Singh, S.P., Misra, S., Gupta, J. and Misra-Bhattacharya, S. (2012) Galactolipids from *Bauhinia racemosa* as a new class of antifilarial agents against human lymphatic filarial parasite, *Brugia malayi*. *European Journal of Medicinal Chemistry*, 50, 230–235.

Seukep, J.A., Sandjo, L.P., Ngadjui, B.T. and Kuete, V. (2016) Antibacterial and antibiotic-resistance modifying activity of the extracts and compounds from *Nauclea pobeguinii* against Gram-negative multi-drug resistant phenotypes. *BMC Complementary and Alternative Medicine*, 16(1), 193.

Sharifi-Rad, J., Salehi, B., Stojanović-Radić, Z.Z., Fokou, P.V.T., Sharifi-Rad, M., Mahady, G.B., Sharifi-Rad, M., Masjedi, M.-R., Lawal, T.O., Ayatollahi, S.A., Masjedi, J., Sharifi-Rad, R., Setzer, W.N., Sharifi-Rad, M., Kobarfard, F., Rahman, A.-U., Choudhary, M.I., Ata, A. and Iriti, M. (2017) Medicinal plants used in the treatment of tuberculosis - ethnobotanical and ethnopharmacological approaches. *Biotechnology Advances*, https://doi.org/10.1016/j.biotechadv.2017.07.001.

Simoes-Pires, C., Hostettmann, K., Haouala, A., Cuendet, M., Falquet, J., Graz, B. and Christen, P. (2014) Reverse pharmacology for developing an anti-malarial phytomedicine. The example of *Argemone mexicana*. *International Journal for Parasitology: Drugs and Drug Resistance*, 4, 338–346.

Sofowora, A. (1982) *Medicinal plants and traditional medicine in Africa*. Chichester: John Wiley, p. 256.

Tarkang, P.A., Franzoi, K.D., Lee, E., Vivarelli, D., Freitas-Junior, L., Liuzzi, M., Nolé, T., Ayong, L.S., Agbor, G.A., Okalebo, F.A. and Guantai, A.N. (2014a) *In vitro* antiplasmodial activities and synergistic combinations of differential solvent extracts of the polyherbal product, Nefang. *BioMed Research International*, 2014: 835013.

Tarkang, P.A., Nwakiban Atchan, A.P., Kiuate, J., Okalebo, F.A., Agbor, G.A. and Guantai, A.N. (2014b) *In vitro* and *in vivo* antioxidant potential of a polyherbal antimalarial as an indicator of its therapeutic value. *European Journal of Integrative Medicine*, 6, 125–132.

Tchuendema, M.H.K., Mbaha, J.A., Tsopmo, A., Ayafor, T.F., Sterner, O., Okunji, C.O., Iwu, M.M. and Schuster, B.M. (1999) Anti-plasmodial sesquiterpenoids from the African *Reneilmia cincinnata. Phytochemistry*, 52, 1095–1099.

Tempesta, M.S. (2010) The clinical efficacy of cryptolepis sanguinolenta in the treatment of malaria. *Ghana Medical Journal*, 44(1), 1–2.

Teng, W.-C., Kiat, H.H., Suwanarusk, R. and Koh, H.-L. (2016) *Medicinal plants and malaria: applications, trends, and prospects.* Boca Raton, FL: CRC Press.

Tiemersma, E.W., Van der Werf, M.J., Borgdorff, M.W., Williams, B.G. and Nagelkerke, N.J. (2011) Natural history of tuberculosis: duration and fatality of untreated pulmonary tuberculosis in HIV negative patients: a systematic review. *PLoS One*, 6, e17601.

Toubiana, R. (1969) Structure of hydroxyvernolide, a new sesquiterpene ester from *Vernonia colorata*. Comptes Rendus des Seances de l'Academie des Sciences, Serie C: Sciences Chimiques 268(1): 82-85.

Togola, A., Diallo, D., Dembélé, S., Barsett, H. and Paulsen, B.S. (2005) Ethnopharmacological survey of different uses of seven medicinal plants from Mali, (West Africa) in the regions Doila, Kolokani and Siby. *Journal of Ethnobiology and Ethnomedicine*, 1, 7.

Tsouh-Fokou, P.V., Nyarko, A.K., Appiah-Opong, R., Tchokouaha-Yamthe, L.R., Addo, A., Asante, I.K. and Boyom, F.F. (2015) Ethnopharmacological reports on anti-Buruli ulcer medicinal plants in three West African countries. *Journal of Ethnopharmacology*, 172, 297–311.

Tu, Y. (2011) The discovery of artemisinin (qinghaosu) and gifts from Chinese medicine. *Nature Medicine*, 17(10), 1217–1220.

Ukwe, C.V., Epueke, E.A., Ekwunife, O.I., Okoye, T.C., Akudor, G.C. and Ubaka, C.M. (2010) Antimalarial activity of aqueous extract an fractions of leaves of *Ageratum conyziodes* in mice infected with *Plasmodium berghei. International Journal of Pharmacy and Pharmaceutical Sciences*, 2(1), 33–38.

Van Zyl, R.L., Seatlholo, S.T. and Viljoen, A.M. (2010) Pharmacological interactions of essential oil constituents on the in vitro growth of *Plasmodium falciparum. South African Journal of Botany*, 76, 662–667.

WHO. (2008) *World malaria report 2008*. Geneva: World Health Organization, pp. 7–15, 99–101.

WHO. (2010) *First WHO report on neglected tropical diseases: working to overcome the global impact of neglected diseases.* Geneva: World Health Organization. Available at http://whqlibdoc.who.int/publications/2010/9789241564090eng.pdf?ua=1

WHO (2015) TUBERCULOSIS REPORT 2015. WHO Library Cataloguing-in-Publication Data Global tuberculosis report 2015. WHO Press, World Health Organization, 20 Avenue Appia, 1211 Geneva 27, Switzerland.

WHO. (2018) *World malaria report 2017*. Geneva: World Health Organization. (accessed on 6.6.2018).

Willcox, M.L. and Bodeker, G. (2004) Traditional herbal medicines for malaria. *British Medical Journal*, 329(7475), 1156–1159.

Willcox, M.L., Graz, B., Falquet, J., Sidibé, O., Forster, M. and Diallo, D. (2007) *Argemone mexicana* decoction for the treatment of uncomplicated falciparum malaria. *Transactions of the Royal Society of Tropical Medicine and Hygiene*, 101, 1190–1198.

Willcox, M.L., Graz, B., Falquet, J., Diakite, C., Giani, S. and Diallo, D. (2011) A "reverse pharmacology" approach for developing an anti-malarial phytomedicine. *Malaria Journal*, 10(Suppl. 1), S8.

Wright, C.W., Phillipson, J.D., Awe, S.O., Kirby, G.C., Warhurst, D.C., Quetin-Leclercq, J. and Angenot, L. (1996) Antimalarial activity of cryptolepine and some other anhydronium bases. *Phytotherapy Research*, 10(4), 361–363.
Yakubu, M.T., Akanji, M.A. and Nafiu, M.O. (2010) Anti-diabetic activity of aqueous extract of *Cochlospermum planchonii* root in alloxan-induced diabetic rats. *Cameroon Journal of Experimental Biology*, 6, 91–100.
Zein, U., Fitri, L.E. and Saragih, A. (2013) Comparative study of antimalarial effect of sambiloto (*Andrographis paniculata*) extract, chloroquine and artemisinin and their combination against *Plasmodium falciparum in-vitro. Acta medica Indonesiana*, 45(1), 38–43.

8 Nigerian Plants with Application in the Treatment of High Blood Pressure

Blood pressure is one of the vital signs, along with respiratory rate, heart rate, oxygen saturation, and body temperature. Hypertension exists when the blood flows through the blood vessels, that is, the large arteries of the systemic circulation, with a greater force compared with normal conditions. Blood pressure is usually expressed in terms of the systolic pressure (maximum during one heartbeat) over the diastolic pressure (minimum in between two heart beats) and is measured in millimetres of mercury (mm Hg), above the surrounding atmospheric pressure (considered to be zero for convenience). Normal blood pressure is defined as when the systolic blood pressure (SBP) and diastolic blood pressure (DBP) are below 120 mm Hg and 80 mm Hg, respectively. When the SBP is 120–139 mm Hg or DBP 80–89 mm Hg, a pre-hypertension condition is diagnosed. A person with pre-hypertensive conditions is at higher risk of progression towards hypertension. The diagnostic criteria for various stages of hypertension are outlined below:

- Prehypertension: 120 to 139/80 to 89 mm Hg
- Stage 1 hypertension: 140 to 159/90 to 99 mm Hg
- Stage 2 hypertension: 160 and higher/100 and higher mm Hg

High blood pressure is considered to be a major risk factor for cardiovascular diseases (CVD) and strokes. Owing to its high prevalence and association with increased morbidity and mortality, it is a major health problem worldwide. It is estimated that more than 22% of adults, aged 18 and above, are reported with elevated blood pressure worldwide. Available reports indicate that the occurrence of hypertension among Africans and people of African descent is the highest in the world, with about 46% of adults above the age of 24 years diagnosed with the disease (WHO, 2013). High blood pressure is sometimes called a 'silent killer' because it may have no outward symptoms for years. In fact, one in five people with the condition do not know that they have it. Internally, it can quietly damage the heart, lungs, blood vessels, brain, and kidneys, if left untreated.

Hypertension is no longer considered to be a stand-alone disease but an integral component of general cardiovascular health. Given that cardiovascular disease is the principal cause of death worldwide and is increasingly common, and tackling the disease is crucial for achieving meaningful primary care, an initiative called "Global Hearts" has been established as part of the World Health Organisation (WHO) package of essential non-communicable disease interventions. It focuses on the non-communicable disease burden due to atherosclerotic cardiovascular disease and its risk factors. The initiative also provides a simple framework that enables front-line health clinics to implement the WHO's longstanding call to integrate cardiovascular disease care into primary disease prevention. Global Hearts, as envisaged, will include the building blocks for ending vertical, disease-specific approaches to cardiovascular conditions. Instead, it guides the organisation and delivery of cardiovascular disease care for the whole patient through better screening, treatment, and prevention. Although the list of non-communicable diseases causing most human illness and death is long, there are only a few risk factors for the four main conditions, namely heart disease, diabetes, cancer and lung disease. The initiative deals with them all, at least within the health sector. Its treatment protocols involve training health workers to assess and manage unhealthy diets, tobacco use and sedentary behaviour in healthy people, thereby helping to prevent not only cardiovascular disease, but also cancer and metabolic disease.

Modern healthcare approaches for the maintenance of healthy blood pressure in the human body includes changing from an unhealthy lifestyle to a more active one that includes exercise, controlled diets, and living in a non-chemically populated environment. A non-drug measure for the control of blood pressure is the Dietary Approaches to Stop Hypertension (DASH), which involves eating more fruits, vegetables, wholegrain foods, low-fat dairy, fish, poultry, and nuts, but less red meat, saturated fats, and sweets. Reducing sodium in the diet can also have a significant effect. Regular physical exercise, coupled with meditation and breathing exercises, have been found useful in managing high blood pressure. The practice of meditation can put the body into a state of deep rest, which can lower blood pressure. Yoga, tai chi, and deep breathing have also been found useful. Medicines for the treatment of hypertension follow several systemic approaches. Because of the complex nature of the causes of high blood pressure, irregularity at any of the stages in the cardiovascular system may lead to changes in blood pressure, resulting in hypertensive conditions. One such well-known system is the renin-angiotensin–aldosterone system. Angiotensin-converting enzyme (ACE) indirectly escalates the blood pressure and is known to cause constriction in blood vessels. It achieves this by converting angiotensin I to angiotensin II, which has an ability to constrict the blood vessels. ACE inhibitors block the angiotensin-converting enzyme, which results in widened and relaxed blood vessels and leads to easier blood flow through vessels, which ultimately reduces blood pressure (Atlas, 2007). Other interventions include the use of beta-blockers, diuretics, calcium channel blockers, and medications that relax the blood vessels, such as vasodilators, alpha blockers, and central agonists.

High blood pressure is recognized by most traditional medical practitioners (TMPs) as a circulatory problem and defined in many local languages as an "upsurge of blood flow" or "too much blood in the body". Some attribute the disease and its acute manifestations to evil or spiritual forces, stress or witchcraft. A report indicates that about 29% of Nigerian patients in urban areas use complementary and alternative medicines (CAM) in the management of their hypertension. It was found that, among those who used CAM, 63% took herbs and about 21% relied on consumption of garlic to control their blood pressure (Osamor and Owumi, 2010)..

A number of plants and their formulations have been in use for the treatment of hypertension by TMPs. It is important to emphasise that, apart from food plants used as components of DASH and nutraceuticals, medicinal plants used for the treatment of high blood pressure contain highly active molecules. They may act *via* the same mechanism as the pharmaceutical drugs, so that traditional or plant-based medications may interact with and affect the functions of tissues, endo-chemicals and the beneficial biome present in the human body. Accordingly, the plants discussed below are based on traditional treatment of hypertension, and include both food plants and very active drug plants. An understanding of their mechanisms of action, where available, is essential to recognize the appropriate actions, thus predicting and preventing their adverse events and possible interaction with synthetic drugs, foods and other herbs.

Listed in Table 8.1 are plants used in the treatment of hypertension in Nigeria. Some of them are not based on traditional medicine but are exotic species that have been domesticated in Nigeria and are now widely cultivated as either culinary herbs or as medicines. For example, garlic is used for the management of hypertension both in urban and rural areas. Other examples include tea leaves, coleus, and *Peganum harmala*.

These herbs are seldom prescribed alone. The method of preparation varies enormously from one part of Nigeria to another, and even among different traditional healers from the same part of the country. It is often not easy to determine the main ingredient in a given antihypertensive preparation. In Edo state alone, over 93 different formulations prepared from about 70 plant species exist for the treatment of high blood pressure (Gbolade, 2012).

Formula One:

A. 10 bulbs of garlic

B. 2 bottles of coconut water

C. 4 bottles of water

A is ground and blended in a mixture of B and C. Dosage: one "shot" (about 30 mL) thrice daily for four weeks.

Formula Two:

A. Leaves of *Nauclea latifolia,*

B. Bark and leaves of *Mangifera indica*

C. Fruits of *Citrus aurantiifolia*

The three ingredients are mixed and boiled with a sufficient volume of water for 30 minutes. Dosage: one glass (ca. 120 mL) once daily.

Formula Three:

A. *Dialum guineense*

B. Root of *Rauwolfia vomitoria*

C. Palm kernel oil (*Elaeis guineensis*)

The components are desiccated in a suitable container until charred into an ash powder. Dosage: one teaspoonful once daily. N.B. The oil palm, *Elaeis guineensis,* is listed in Table 8.1. as an antihypertensive herb because various parts of the plant are used in traditional medicine as a formulation excipient. The ash from the husk (after removing the fruits) is also employed to produce an alkaline base.

Formula Four:

A. *Talinium triangulare* leaf

B. *Gongronema latifolium* leaf

A soup is prepared with the two ingredients. Dosage: taken once per day at night after food for 4 weeks.

Formula Five:

A. *Vernonia amygdalina* leaf

B. *Allium cepa* bulb

C. *Zingiber officinale* rhizome

The three components are soaked in water for five days. Dosage: One cup taken twice daily.

Table 8.1 Major Therapeutically Important Antihypertensive Plants Used in Nigeria

Plant Species	Family	Part Used
Acalypha godseffiana	Euphorbiaceae	Leaves, seed
Aframomum melegueta	Zingiberaceae	Seeds, leaves
Allium cepa	Amaryllidaceae	Bulb
Allium sativum	Amaryllidaceae	Bulb
Annona muricata	Annonaceae	Leaves
Artocarpus altilis	Moraceae	Leaves
Ficus asperifolia	Moraceae	Leaves
Cajanus cajan	Fabaceae	Leaves
Camellia sinensis	Theaceae	Aerial part
Capsicum frutescens	Solanaceae	Fruit
Carica papaya	Caricaceae	Leaves, seed
Cassia occidentalis	Caesalpiniaceae	Leaves
Coleus forskohlii	Lamiaceae	Root
Cymbopogon citratus	Poaceae	Leaves
Dialium guineense	Fabaceae	Fruits
Elaeis guineensis	Arecaceae	Leaves, seed
Hibiscus sabdariffa	Malvaceae	Aerial parts
Hunteria umbellata	Apocynaceae	Stem bark
Gongronema latifolium	Asclepiadiaceae	Stem bark
Khaya senegalensis	Meliaceae	Leaves, stem bark
Loranthus micranthus	Loranthaceae	Leaves
Lycopersicon esculentum	Solanaceae	Fruits
Mangifera indica	Anacardiaceae	Leaves, stem
Moringa oleifera	Moringaceae	Leaves, seeds
Musanga cecropiodes	Cecropiaceae	Leaves, stem
Nigella sativa	Ranunculaceae	Seeds
Ocimum gratissimum	Lamiaceae	Leaves, stalks
Parkia biglobosa	Fabaceae	Stem bark
Peganum harmala	Nitrariaceae	Whole plant/seeds
Persea americana	Lauraceae	*Leaves, dried seeds*
Phyllanthus amarus	Euphorbiaceae	Leaves
Psidium guajava	Myrtaceae	Fruits/leaves
Pueraria lobata	Fabaceae	Roots
Raphanus sativus	Brassicaceae	Leaves
Rauwolfia vomitoria	Apocynaceae	Stem, root bark
Sesamum indicum	Pedaliaceae	Seed oil
Sida acuta	Malvaceae	Leaves
Talinum triagulare	Portulaceae	Whole plant
Tapinanthus bangwensis	Loranthaceae	Leaves
Theobroma cacao	Malvaceae	Beans
Tridax procumbens	Asteraceae	Aerial part
Vernonia amygdalina	Asteraceae	Leaves
Viscum album	Viscaceae	Leaves
Xylopia aethiopica	Annonaceae	Fruits
Zingiber officinale	Zingiberaceae	Rhizome

Formula Six:

A. *Hunteria umbellata* leaves

B. *Sida acuta* stem bark

C. *Allium sativum* bulbs

The three components are boiled with potash. Dosage: take one glass daily.

Formula Seven:

A. *Acalypha godseffiana* leaves

B. Snail

The two components are boiled together in water until the snail is lightly cooked. Dosage: one glass twice daily.

Formula Eight:

A. *Parkia biglobosa* bark

B. *Sorghum caudatum* leaves

The two components are boiled together in water with potash. Dosage: one glassful thrice daily.

Formula Nine:

A. *Phyllanthus amarus* leaves and stems

B. *Elaeis guineensis* palm kernel oil

C. *Aframomum melegueta*

The plant material is boiled in water, to make a soup. Dosage: drink the soup daily.

Formula Ten:

A. *Cajanus cajan* leaves

B. *Talinum triangulare* roots

The two plant tissues are pounded together and then boiled in water. Dosage: One cup is drunk twice daily.

Formula Eleven:

A. *Elaeis guineensis* leaves

B. *Aframomum melegueta* seeds

C. *Psidum guajava* leaves

Equal measures of palm and guava leaves are boiled together in water until a dark green extract is produced. The extract is allowed to cool and pinch of ground alligator pepper is added. Dosage: one glassful daily.

8.1 *Bryophyllum pinnatum* (Crassulaceae)

Bryophyllum is called "Never die", "Leaf of life" or "Wonder of the World" because of its many ethnomedical uses and the ease with which it is propagated directly from plantlets on the leaf. It is used in different doses for the treatment of different stages of hypertension. For this purpose, a single leaf of the plant is placed on the tongue and sucked throughout the day or a measured piece is given to the patient (depending on body size and severity of the illness) to chew. Decoction of the plant is usually prepared immediately before use for the treatment of hypertension. Studies have shown that both the aqueous and methanolic leaf extracts produced dosage-dependent significant decreases in arterial blood pressure and heart rate of anaesthetized normotensive and hypertensive rats. The hypotensive effects of the leaf extracts were more pronounced in the hypertensive than in normotensive rats (Ojewole, 2002). Although the exact mechanism of the antihypertensive action of *Bryophyllum* is not well understood, the presence of bufadienolides in *Bryophyllum* species suggest a direct effect on the heart muscles. The compound is known to be toxic to cattle and other farm livestock. *Bryophyllum* poisoning in farm animals causes anorexia, depression, ruminal atony, diarrhoea, decreased heart rate, and heartbeat rhythm abnormalities, dyspnoea, and death. It is used in most parts of Nigeria with palm oil as a dressing to heal umbilical wounds and male circumcision.

Bryophyllum has been used in clinical medicine in Europe (mainly Germany) since the 1970s as a sedative and for the treatment of premature labour (Gwehenberger *et al.*, 2004). The preparation was usually administered parenterally at a dose of 580 mg/h until contractions ceased, after which it was administered as a 50% slurry at a dose of 200 mg/h, supplemented by fenoterol at 120 mg/h, if insufficiently effective alone. Clinical outcome and inhibition of labour associated with *Bryophyllum* were similar to those of fenoterol. It has also been shown to be useful in the prevention of renal calculi formation and has been recommended as an effective therapeutic agent for the treatment of urinary calculi and related urinary disorders (Yadav *et al.*, 2016).

Bryophyllum contains the potent cytotoxic bufadienolides, bryophyllin A and B, and cardiac glycosides, known as bryotoxins. An extract of the leaves with activity against chemically induced wounds has been shown to contain bryophyllol, bryophollone and bryophollenone, bryophynol and two phenanthrene derivatives, as well as 18α-oleanane, ψ-taraxasterol, α- and β-amyrin and their acetates.

8.2 *Hibiscus sabdariffa* (Malvaceae)

Zobo or bissap, as the calyx and flowers of *H. sabdariffa* (HS) are known in Nigeria, is an ingredient in locally produced refreshing drinks, taken for the improvement of general body function, blood circulation, and hypertension, and in the manufacture of beverages, jam, and vegetable gelatin. The leaves are used as vegetables in the preparation of soups and sauces, and as remedy for fevers and liver disorders (Mohamed *et al.*, 2007). The leaf extract has been shown to possess hypoglycaemic, hypolipidaemic, and antioxidant effects, and to induce tumour cell apoptosis. It has been used as an antiobesity agent and for the treatment of benign prostate hypertrophy (BPH) and prostate cancer (Chiu *et al.*, 2015). Animal studies have consistently shown that consumption of HS extract reduces blood pressure in a dose-dependent manner. It has also been established that HS acts as a diuretic; however, in most cases, the extract did not significantly influence electrolyte levels (Herrera-Arellano *et al.*, 2004). HS extracts have a low degree of toxicity, with an LD_{50} ranging from 2,000 to over 5,000 mg/kg bodyweight/day. There is no evidence of hepatic or renal toxicity as the result of HS extract consumption, except for possible adverse hepatic effects at high doses (Hopkins *et al.*, 2013).

Hibiscus extracts have been shown to possess antihypercholesterolaemic effects, as shown in feeding experiments on hypercholestrolaemic rats (Hirunpanicha *et al.*, 2006). Similar effects were observed in human studies, as it caused lower serum cholesterol in a dose-dependent manner in a clinical study on 42 volunteers, who were observed over a period of four weeks. The volunteers ranged in age from 18 to 75 years old, with a cholesterol level of 175 to 327 mg/dL.

Subjects were randomly assigned to one of three groups: group I (one capsule of *H. sabdariffa* extract (HSE) during each meal), group II (two capsules), and group III (three capsules). It was found that serum cholesterol levels were lower after two weeks in all groups ($P<0.05$ for groups I-III), compared with baseline values, by between 7.8% and 8.2%, and by between 8.3% and 14.4% after four weeks of taking the supplement. It is important to note that the serum cholesterol level for 71% of group II volunteers was significantly lowered with a mean reduction of 12% ($P<0.05$). The authors concluded that a dosage of two capsules of HSE (with a meal) for one month can significantly lower the serum cholesterol level. The observation of lowered serum cholesterol in these subjects suggests that HSE may be effective in hypercholesterolaemic patients (Lin *et al.*, 2007).

The antihypertensive properties of HS have been validated by several clinical studies (Hopkins *et al.*, 2013; Rawat *et al.*, 2016). A meta-analysis of randomized controlled trials using HS has shown a significant and consistent effect of HS supplementation in lowering both SBP and DBP (Serban *et al.*, 2015). For example, in a randomized, double-blind, placebo-controlled clinical trial, the antihypertensive effects of HS in pre-hypertensive and mildly hypertensive adults were confirmed, with HS, in the form of a tea, showing a significant reduction in blood pressure (McKay *et al.*, 2010; Nwachukwu *et al.*, 2015a).

Another study, conducted on Nigerian patients with mild to moderate hypertension, showed that the antihypertensive effect of HS was comparable to that of Lisinopril (Nwachukwu *et al.*, 2015b). The study analysed the effects of HS aqueous extract on the three basic components of the renin–angiotensin–aldosterone system: plasma renin, serum angiotensin-converting enzyme (ACE), and plasma aldosterone (PA). Both HS and Lisinopril reduced serum ACE and PA in the patients but slightly increased plasma renin activity (Nwachukwu *et al.*, 2015a). A positive effect was also observed in the treatment of hypertension in patients with type 2 diabetes (Mozaffari-Khosravi *et al.*, 2009). The similarity in action with the standard antihypertensive drug, Lisinopril suggests that the likely mode of HS action involves modulation of the renin–angiotensin–aldosterone system. It was postulated that, since the action on aldosterone was much higher than that on ACE, this could be due to a combination of several factors, such as blockage of AT1 receptors and the inhibitory action of Mg^{2+} present in HS on aldosterone secretion (Nwachukwu *et al.*, 2015b). Other modes of action have been suggested, including direct relaxation of vascular smooth muscle, decreasing the level of acetylcholinesterase and Na^{+} in the serum, or decreasing the elevated blood lipid profile (El-Mahmoudy *et al.*, 2014).

This plant is rich in anthocyanins, polyphenols, and simple plant acids. The petals yielded 65% (dry weight) of mucilage, which, on hydrolysis, gave galactose, galacturonic acid, and rhamnose. In addition to these compounds, *Hibiscus sabdariffa* extract (HSE) contains complex polyphenolic acids (1.7% dry weight), flavonoids (1.43% dry weight) and anthocyanins (2.5% dry weight). The phytosterols campasterol, stigmasterol, ergosterol, β-sitosterol, and α-spinasterol have been reported from the seed oil. The major constituents isolated from the aerial parts include ascorbic acid, β-carotene, citric acid, malic acid, tartaric acid, glycine, and anthocyanins such as cyanidin-3-rutinoside, delphinidin, delphinidin-3-glucoxyloside (also known as ashibiscin, the major anthocyanin in *H.*

sabdariffa flowers), delphinidin-3-monoglucoside, cyanidin-3-monoglucoside, cyanidin-3-sambubioside, cyanidin-3,5-diglucoside, the flavonol glycosides hibiscetin-3-monoglucoside, gossypetin-3-glucoside, gossypetin-7-glucoside, gossypetin-8-glucoside, and sabdaritrin, as well as quercetin, protocatechuic acid, pectin, polysaccharides, mucopolysaccharides, stearic acid, and wax (Nnam and Onyeke 2003; Hirunpanich *et al.*, 2005: Maganha *et al.*, 2010).

8.3 *Nigella sativa* (Ranunculaceae)

Seeds of *N. sativa*, known in commerce as black caraway, black cumin, fennel flower, nigella, nutmeg flower, and Roman coriander, are cultivated in northern Nigeria as a spice and a medicinal herb. As a soup condiment, black seeds impact a characteristic flavour to "pepper soup" and taste like a combination of onions, black pepper, and oregano. They have a pungent, bitter taste and smell. The plant has been used for the treatment of diabetes, hypertension, cancers (leukaemia, liver, lung, kidney, prostate, breast, cervix, skin), inflammation, hepatic disorders, arthritis, kidney disorders, cardiovascular complications, and dermatological conditions (Khan *et al.*, 2011). Its reported pharmacological properties include bronchodilatory, hypotensive, antibacterial, antifungal, analgesic, anti-inflammatory and immunopotentiatory activities (Khan and Afzal, 2016).

Several *in-vitro* and animal studies have validated the antidotal and protective effects of *N. sativa* and its main constituents against natural and chemical-induced toxicities (Tavakkoli *et al.*, 2017), as well as in the treatment of cancers (Majdalawieh and Fayyad, 2016) and microbial infections (Forouzanfar *et al.*, 2014). Studies have also shown that various extracts of black seed can reduce triglycerides and low-density lipoproteins (LDL) and total cholesterol, while raising high-density lipoprotein (HDL) cholesterol (Sahebkar *et al.*, 2016a). *Nigella* has been evaluated for its antihypertensive activity in humans, in a randomized, double-blind, placebo-controlled trial, where patients with mild hypertension were treated with 100 mg or 200 mg of the seed extract. After 8 months observation, it was determined that the supplement caused a significant dose-dependent lowering of both SBP and DBP in the treatment groups (Dehkordi and Kamkhah, 2008). A meta-analysis of clinical trials found weak evidence that *N. sativa* has a short-term benefit in lowering SBP and DBP (Sahebkar *et al.*, 2016b). The probable mechanisms of action of *N. sativa* seed extract include blockage of the calcium channel and through its diuretic action (Keyhanmanesh *et al.*, 2014; Rawat *et al.*, 2016).

The seed extract has been shown by gas chromatography–mass spectrometry (GC–MS) analysis to consist of a mixture of eight fatty acids and 32 volatile terpenes. The major terpenes, thymoquinone (TQ), dithymoquinone (DTQ), *trans*-anethol, *p*-cymene, limonene, and carvone, have been identified (Nickavar *et al.*, 2003). TQ and DTQ are both cytotoxic towards various types of tumours (Worthen *et al.*, 1998). In addition, diterpenes, triterpene and terpene alkaloids have been identified in *N. sativa* seeds. The methanolic extract of the seeds contain two types of alkaloids, whereas the major principal active ingredient isolated from the volatile oil of *N. sativa* L. is TQ. Alkaloids with the isoquinoline and imidazole chromophores have also been isolated from the plant.

8.4 *Phyllanthus amarus* (Phyllanthaceae)

The stone breaker is used in traditional medicine for the treatment of various diseases, including diabetes, urinary tract infections, hypertension, cancers and wound dressings. Other applications include treatment of jaundice, dysentery, gonorrhoea, kidney and gallbladder stones, and uncontrolled menstruation (Srividya and Periwal, 1995).

Laboratory studies have shown that *P. amarus* extracts reduced BP in rabbits (Amaechina and Omogbai, 2007), and, at a dose of 200 mg/kg bodyweight, increased plasma antioxidants (GSH, GPx, SOD, and CAT) in rats (Karuna *et al.*, 2009), as well as causing inhibition of fNF-kB, TNF-a, and COX-2 in lipopolysaccharide (LPS)-treated RAW264.7 macrophages (Kiemer *et al.*, 2003).

In clinical trials, *P. amarus* decreased the hypertension in patients and increased urinary output as well as levels of Na^+ in urine (Srividya and Periwal, 1995). These actions were without any noticeable harmful side-effects in the mildly hypertensive subjects of the study (Srividya and Periwal, 1995). The mechanism for the antihypertensive activity is likely related to its diuretic, antioxidant, and anti-inflammatory activities (Kassuya *et al.*, 2005; Maity *et al.*, 2013). Phytochemical analysis of *P. amarus* showed the presence of polyphenols, including quercetin, rutin, gallic acid, phyllanthin, and hypophyllanthin (Mallaiah *et al.* 2015).

8.5 *Parkia biglobosa* (Fabaceae)

The leaves, stem bark, and seeds of *P. biglobosa* (locust bean, dawa-dawa) and related species are used extensively in traditional medicine in West Africa for treatment of many diseases, including parasitic infections, such as malaria, schistosomiasis and filariasis, circulatory system disorders, such as arterial hypertension, and disorders of the respiratory system, digestive system haemorrhages, and dermatosis (Iwu, 2016). It has also been used for the treatment of measles and chickenpox, gingivitis and as an anthelmintic. The fermented seed is the main ingredient in the preparation of the West African soup condiment and flavour agent called dawa-dawa. Studies have confirmed the antiplatelet, antioxidant and antihypertensive activities of the seed extracts. The leaves also exhibited vasorelaxant activity, which provides a probable mechanism for the antihypertensive activity of *P. biglobosa* leaves (Tokoudagba *et al.*, 2010). The activities appear to be dependent on the phenolic concentration, including mainly procyanidins. Both the leaves and the seed extract exhibit antihypertensive activity. The hydroalcoholic, ethyl acetate and butanolic extracts of *P. biglobosa* leaf, that are rich in procyanidins and monomeric flavonoids, induced endothelium-dependent nitric oxide- and endothelium-derived hyperpolarising factor-mediated relaxation in porcine coronary artery rings.

The boiled and fermented seeds contain 35% proteins, 29% lipids, and 16% carbohydrates, as well as tannins, flavonoids and coumarins, and have good organoleptic properties with a positive effect on intestinal flora (prebiotic effect). It also contains lupeol, 4-*O*-methyl-epi-gallocatechin, epigallocatechin, epi-catechin 3-*O*-gallate, and epi-gallocatechin 3-*O*-gallate. It has a rich pool of essential amino acids and magnesium, which may play a role in its antihypertensive activity. Fermentation of the seeds, apart from the possible detoxification effect, may not be necessary for the therapeutic application of the plant since proximate analysis of both the fermented and unfermented seeds revealed that fermentation confers positive benefits on the composition of the plant material. The proximate analysis revealed that the protein isolate (from fresh seeds) had the highest ash (6.0%) and protein contents (59.4%) as well as the lowest fat (5.7%) and moisture (5.1%) content when compared with the fermented and defatted samples. Similarly, the functional properties of the protein isolate were greater than those of the fermented and defatted samples, with an oil absorption capacity of 4.2% and an emulsion capacity of 82%. The magnesium and zinc contents of the protein isolate were significantly higher than in the fermented and defatted samples (Ogunyinka *et al.*, 2017).

8.6 *Ficus exasperata* (Moraceae)

The leaves of the sandpaper tree are used in traditional medicine for the treatment of a variety of conditions, including high blood pressure, diabetes mellitus, and cardiovascular dysfunctions. There is not much laboratory or clinical data on the antihypertensive activity of this plant, but it features in many herbal preparations in Nigeria for the treatment of high blood pressure. It has been shown to possess potent antioxidant activity, with strong free radical-scavenging activity and inhibition of lipid peroxidation (Akanni *et al.*, 2014).

The phenolic extract of the leaves showed *in-vitro* inhibitory effect on ACE activity and in dietary supplementation on rats fed on a high-cholesterol diet for 14 days. The study found that feeding high-cholesterol diets to rats caused a significant increase in ACE activity, with a significant decrease in ACE activity in rats fed the sandpaper leaf supplement (Oboh *et al.*, 2014). There was a significant increase in the plasma lipid profile with a concomitant increase in malondialdehyde (MDA) content in rat liver and heart tissues of rats on the high-cholesterol diet, but supplementing the diet with sandpaper leaf (at either 10% or 20% concentration) caused a significant ($P<0.05$) decrease in the plasma total cholesterol (TC), triglyceride (TG), very low density lipoprotein-cholesterol (VLDL-C), and low-density lipoprotein-cholesterol levels (LDL-C), and in MDA content in the tissues. Conversely, supplementation caused a significant ($P<0.05$) increase in plasma high-density lipoprotein-cholesterol (HDL-C) level when compared with the rats on the control diet. It has been suggested that the inhibition of ACE activity and prevention of hypercholesterolaemia by sandpaper leaf could be part of the possible mechanism underlying its antihypertensive property. Phytochemical analysis of the bioactive extract revealed the presence of quercitrin (43.7 mg/g), chlorogenic acid (42.8 mg/g) and caffeic acid (33.9 mg/g) as the major phenolics in sandpaper leaf.

8.7 *Allium sativum* (Amaryllidaceae)

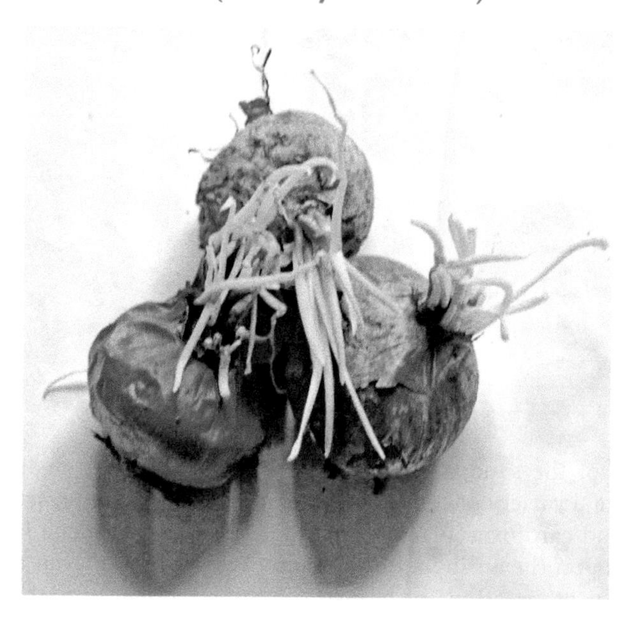

Garlic is a herbaceous plant cultivated globally for food and as medicine. It is the underground part which is valued, occurring as a bulb of fleshy-scaled leaves on the lower part of the stem. Each bulb consists of a number of bulblets or "cloves" resting on a common bulb base and covered with membranous bracts. It has a very strong and disagreeable odour that is noticeable on the breath and a strongly pungent and persistent taste – both the odour and taste linger for several days following ingestion of the plant. Garlic includes two basic varieties: hardneck and softneck. Hardneck garlics, which include *A. sativum* var. *ophioscorodon* and *A. sativum* var. *pekinense*, are characterized by a hard, woody central stalk that extends down to the basal plate at the bottom of the bulb Iwu, 2014).

Garlic possesses many pharmacological properties, including antihyperglycaemic, antithrombotic, anti-inflammatory, and antimicrobial activities. It has demonstrable antioxidant properties, which may have an indirect relationship to the use of garlic in the treatment of hypertension. It has been shown to cause a significant decrease in SBP and DBP in both a dose- and duration-dependent manner in patients (n = 210) with stage 1 essential hypertension at doses of 300 mg, 600 mg, 900 mg, 1200 mg or 1500 mg (Ashraf *et al.*, 2013). In another study, processed garlic (PG) was tested on spontaneously hypertensive rats (SHR) and hypertensive humans, for its effect on the SBP and DBP. Significant reductions in SBP and DBP were observed. The lowering of SBP was observed over a time-period of two weeks, whereas the lowering of DBP took a much longer time of eight weeks. A meta-analysis of clinical studies reported a mean decrease of 4.6 2.8 mm Hg for SBP in the group taking garlic (administered as a powder, extract or oil), compared to those taking the placebo, whereas the mean decrease in the hypertensive subgroup was 8.4 2.8 mm Hg for SBP, and 7.3 1.5 mm Hg for DBP.

A significant association between blood pressure at the start of the intervention and the level of blood pressure reduction was identified using regression analysis (Ried *et al.*, 2008). Other clinical evaluations reported consistent effects on the hypotensive effect of garlic products (Ried *et al.*, 2013).

Several bioactive constituents that are structurally related to S-allyl cysteine and allicin have been identified as being responsible for the antihypertensive activity of garlic. The activity threshold of active compounds in a therapeutically effective PG was found to be at a S-allyl-L-cysteine concentration of about 75.3 mg/100 g in PG(Han *et al.*, 2011).

8.8 *Allium cepa* (Amaryllidaceae)

Onion is a bulb-shaped vegetable which is extensively used globally for its strong flavor and numerous health benefits. *A. cepa* and other members of the genus *Allium* (excluding garlic, leek, chive, and the so-called Chinese onion) are used as condiments in various Nigerian dishes and as medicine. Several other exotic species, variously referred to as onions, have now been introduced into Nigeria in the form of dietary supplements, such as the Japanese bunching onion (*Allium fistulosum*), the tree onion (*Allium × proliferum*), and the Canada onion (*Allium canadense*). Although it is usually cultivated as annual crop and harvested in its first growing season, it thrives as a perennial crop, and older cultivars are frequently used in folk medicine for the treatment of diseases. Traditional healers often make clear distinctions in recipes between fresh onions and old onions.

In animal studies, 0.1, 0.2 and 0.4 mg/mL of the ethanolic extract of onion peel have been reported to elicit antispasmolytic effects on the isolated terminal ileum of a Wistar rat in a dose-dependent manner, with this effect being mediated *via* inhibition of calcium channels. The *in-vivo* antihypertensive effect of 200, 400 or 800 mg/kg dosages of the 70% hydroethanolic onion peel extract on hypertensive rats induced by a high-fructose diet, and subsequently treated with the extract for three weeks and aorta contractility has also been reported. The vasorelaxant effect of the extract was reported to have been mediated *via* inhibition of Ca^{2+} influx into vascular smooth muscle cells without the involvement of nitric oxide (NO), cGMP, endothelium and prostaglandins (Adeneye, 2014). Oral ingestion of an onion-olive oil maceration capsule formulation by normotensive volunteers caused a significant reduction in the arterial blood pressure as well as a significant reduction in plasma viscosity and hematocrit, indicative of a vasodilative effect (Mayer *et al.*, 2011). In another study, 24 hypertensive (WHO stage I) patients were given either four capsules of the same onion-olive oil maceration product daily or a placebo over a period of one week, with the onion-olive oil maceration product resulting in a significant decrease in both the SBP and DBP, with a more pronounced decrease recorded for SBP (Adeneye, 2014).

Onions are rich source of flavonoids, such as quercetin, rutin, and its organosulphur conjugates, that possess strong antioxidant activities and contribute to the observed pharmacological activities. Onions also contain other polyphenols, especially anthocyanin pigments. Onions exhibit enormous variation in the composition of polyphenols. For example, red onions have a considerable content of anthocyanin pigments, with at least 25 different compounds identified, representing 10% of the total flavonoid content. Yellow cultivars have the highest total flavonoid content, about 11 times higher than that in white onions, whereas shallots having the highest level of total polyphenols, about six times the amount found in Vidalia onions.

8.9 *Rauwolfia vomitoria* (Apocynaceae)

The plants belonging to the genus *Rauwolfia* (also spelled "Rauvolfia") are used in modern clinical medicine for the treatment of hypertension and schizophrenia, similar to their uses in traditional medicine. The antihypertensive activity is attributed to the presence of indole alkaloids, such as reserpine, alstonine, ajmalicine, etc., which exert their antihypertensive effect by a combination of various biochemical mechanisms, including the depletion of norepinephrine through inhibition of catecholamine storage in postganglionic adrenergic nerve endings. The hypotensive effect was often accompanied by a reduction in heart rate, contraction of the pupils, and stimulation of intestinal peristalsis. *Rauwolfia* does not cause significant alterations in cardiac output or renal blood flow. The carotid sinus reflex is inhibited but postural hypotension is rarely observed. A significant characteristic of *Rauwolfia* therapy is that the cardiovascular and CNS effects may persist even after withdrawal of the drug.

Although *Rauwolfia* is very effective in controlling most of the symptomatic manifestations of hypertension, adequate reduction in blood pressure is achieved in only about 40% of mild and 30% of medium/severe cases of hypertension (Iwu, 2014). It is, however, not quick acting and may not be suitable for those conditions that require rapid lowering of the blood pressure. Clinical studies have shown that reserpine is largely responsible for the antihypertensive properties of *Rauwolfia*. A typical study on the compound, in a 2-year double-blind evaluation of patients with essential hypertension showed a significant fall of SBP and DBP in the patients receiving the *Rauwolfia* treatment (Sheldon and Kotte, 1957). The mode of action of reserpine has also been investigated and it was found that the compound acts through control over the vasomotor centre, leading to vasodilation (Gawade and Fegade, 2012).

The most common adverse side-effect of reserpine is mental depression, which may persist for several months after the drug has been withdrawn. Treatment with *Rauwolfia* should be discontinued at first sign of despondency, early morning insomnia, or self-deprecation. Severe cases, resulting in manic depression or suicidal tendencies, are extremely rare. Other side- effects include lethargy, sedation, psychiatric depression, hypotension, nausea, vomiting, abdominal cramping, gastric ulceration, nightmares, brachycardia, angina-like symptoms, bronchospasm, skin rash, itching and withdrawal psychosis in one case, galactorrhoea, breast enlargement, and sexual dysfunction. The drug is contraindicated in patients with a past history of bronchial asthma or allergy, active peptic ulcer, ulcerative colitis, or pheochromocytoma, and in patients receiving electroconvulsive therapy. The safety and efficacy of its use in pregnancy, lactation, and in children have not been established. Chronic toxicity and interaction with other drugs are similar to that of reserpine. The plant fell out of popularity when adverse side-effects, including depression and cancer, became associated with it, becoming a major concern to clinicians (Lobay, 2015).

8.10 *Psidium guajava* (Myrtaceae)

Guava is used globally as a delicious fruit and it yields a nutrient-rich pleasant-tasting juice. The leaves and fruits of guava are used in traditional medicine for the treatment of various diseases, including diabetes, inflammations and viral infections. Many studies have established its *in-vitro* and *in-vivo* pharmacological activities against diabetes mellitus, cardiovascular diseases, cancer, and parasitic infections (Díaz-de-Cerio *et al.*, 2017). The *in-vitro* activity against lipid peroxidation in rats has been confirmed after checking the antihypertensive effect with leaves of red and white guava. The major constituents isolated from the active leaf extract include rosmarinic acid, eugenol, carvacrol, catechin, and caffeic acid (Díaz-de-Cerio *et al.*, 2017).

A comparative *in-vitro* study on the antihypertensive and antioxidant properties of fruit and leaf extracts from four varieties (giant white, small white, striped and pink) of guava (*P. guajava*) has been undertaken (Ademiluyi *et al.*, 2015). The report indicated that the extracts were assayed for their ACE-inhibitory effect, total phenol and flavonoid contents, reducing property, radical- (DPPH [2,2-diphenyl-1-picrylhydrazyl], $ABTS^{\bullet+}$ [2,2′-azino-di-(3-ethylbenzthiazoline sulfonic acid)], hydroxyl and nitric oxide) scavenging ability, Fe^{2+}-chelating ability, and inhibition of Fe^{2+}- and sodium nitroprusside (SNP)-induced lipid peroxidation reactions *in vitro*. The results showed that all the extracts significantly ($P < 0.05$) inhibited ACE activity, scavenged (DPPH, $ABTS^{\bullet+}$, nitric oxide and hydroxyl) radicals, chelated Fe^{2+} and also inhibited Fe^{2+}- and SNP-induced lipid peroxidation in rat heart *in vitro*. The pink guava variety had the highest ACE-inhibitory and antioxidant properties. In addition, rosmarinic acid, eugenol, carvacrol, catechin and caffeic acid were the dominant phenolics found in the extracts. The ACE-inhibitory effects and antioxidant properties of the guava extracts, which correlate significantly with their phenolic concentrations, could largely contribute to their antihypertensive properties, as revealed by traditional medicine (Ademiluyi *et al.*, 2015).

In another *in-vitro* study, the observation that elevated uric acid level, an index of gout resulting from the over-activity of xanthine oxidase (XO), increases the risk of developing hypertension and the fact that plant-derived inhibitors of XO and ACE, two enzymes implicated in gout and hypertension, respectively, can prevent or ameliorate both diseases, without noticeable side-effects, informed the laboratory investigation of guava leaves extract on the ability to inhibit the effect on XO and ACE. It was found that the extract effectively inhibited XO, ACE and Fe^{2+}-induced lipid peroxidation in a dose-dependent manner, as shown by the earlier studies, with half-maximal inhibitory concentrations (IC_{50}) of 38.24 ± 2.32 μg/mL, 21.06 ± 2.04 μg/mL and 27.52 ± 1.72 μg/mL against XO, ACE and Fe^{2+}-induced lipid peroxidation, respectively. The extract also strongly scavenged $DPPH^{*}$ and $ABTS^{*+}$. The flavonoids present in the active extract were in the order of quercetin > kaempferol > catechin > quercitrin > rutin > luteolin > epicatechin, while phenolic acids were in the order caffeic acid > chlorogenic acid > gallic acids (Irondi *et al.*, 2016).

In human clinical studies, a randomized, single-blind, controlled trial on the effect of *P. guajava* fruit intake on the blood pressure of patients with essential hypertension showed that a significant reduction in both mean SBP and DBP was observed as compared with the control group (Singh *et al.*, 1993).

8.11 *Theobroma cacao* (Malvaceae)

Cocoa beans yield large and often colourful pods that contain the bitter seeds used in the manufacture of cocoa, chocolate and other beverages. There are three main types of the cacao plant: Criollo, Forastero and Trinitario. Each cultivar has distinguishable characteristics such as yield, susceptibility to diseases, resilience to climate variations, taste and colour of the finished product etc. Cocoa contains a mixture of polyphenols that are collectively called cocoa-polyphenols (up to 10% (w/w) of dry seed), consisting mainly of catechins or flavan-3-ols (ca. 37%), anthocyanins (ca. 4%), and proanthocyanidins (ca. 58%). Cocoa polyphenols exhibit strong antioxidant properties, as illustrated by the very high oxygen radical absorbance capacity (ORAC) value. These polyphenols have been shown to reduce blood pressure, improve insulin sensitivity, and induce vasodilation of the peripheral and cerebral vascular system, mainly by improving nitric oxide bioavailability in endothelial cells Fisher *et al.*, 2003; Mastroiacovo *et al.*, 2015).

The effect of the long-term intake of a cocoa powder, with a high concentration of polyphenols, named CocoanOX (CCX), has been evaluated on the development of hypertension in spontaneously hypertensive rats (SHR) in order to test the suitability of cocoa as a functional food ingredient with antihypertensive activity. Rat SBP was measured weekly, from week 6 to 24 of life, by the tail cuff method (Quiñones *et al.*, 2010), The development of hypertension was attenuated in the groups treated with captopril or CCX. The antihypertensive effect was greater in the group treated with captopril. A surprising observation in the study was the fact that the group treated with the lowest dose of CCX showed greater antihypertensive response than in the other CCX groups. Both CCX and the standard cocoa preparation improved the aorta endothelial function in the SHR and the arterial blood pressure increased in the treated SHR when the corresponding antihypertensive treatment was removed (Quiñones *et al.*, 2010).

Many clinical studies have been published that validate the attenuating effect of cocoa (administered in the form of beans, powder or chocolate) on the blood pressure of hypertensive individuals (Kozuma *et al.*, 2005; Grassi *et al.*, 2008; Cienfuegos-Jovellanos *et al.*, 2009; Rawat *et al.*, 2016).

Apart from polyphenols, carbohydrates, protein, vitamins, and minerals, the alkaloid theobromine and trace amounts of caffeine and related methylxanthines have also been identified in cocoa. Cocoa contains over 300 volatile compounds, including esters, hydrocarbonslactones, monocarbonyls, pyrazines, and pyrroles. The pharmacological activities attributed to cocoa extracts and formulations include antioxidant, anti-inflammatory, cardioprotective, anticarcinogenic, and neuroprotective effects (Iwu, 2016). Considerable interest has been shown on the dietary use of cocoa for the improvement of cardiovascular health and its role in prevention of high blood pressure (Grassi *et al.*, 2010; Santos and Macedo, 2018).

8.12 *Solanum lycopersicum* (Solanaceae)

Tomatoes have been shown in numerous studies to be beneficial in maintaining good cardiovascular health. Tomato extracts have the ability to improve the metabolic profile and to reduce arterial stiffness and have been shown in numerous studies to exert a favourable effect in patients with subclinical atherosclerosis, metabolic syndrome, hypertension, peripheral vascular disease, stroke, and several other cardiovascular disorders. The antihypertensive properties of tomato are attributed largely to the presence of lycopene in the fruit. Lycopene is a lipophilic, unsaturated carotenoid, which is found in tomatoes, and, to lesser extent, in other red-colored fruits and vegetables, including oil palm, watermelon, papaya, red grapefruits, and guava. The main physiological activities associated with lycopene include antioxidant, anti-inflammatory, antihypertensive, antiplatelet, antiapoptotic, antiatherosclerotic, and protective endothelial effects. An encapsulated tomato extract containing lycopene (6%), beta-carotene (0.15%), phytoene, and phytofluene

(1%), plus vitamin E (2%), phospholipids (15%) and phytosterol (0.6%), suspended in tomato oleoresin oil, has been assessed in a double-blind placebo-controlled pilot study on 40 patients with grade 1 hypertension over a 12-week period. It was found that the tomato extract significantly attenuated SBP and DBP in the treatment group receiving *S. lycopersicum* extract for a period of 8 weeks (Engelhard *et al.*, 2006). The patients in the study were administered one capsule of the formulation per day during a treatment period of 8 weeks after a 4-week control period. In another study on 54 patients with moderate hypertension who had been treated with ACE inhibitors or calcium channel blockers, after 6 weeks of tomato extract supplementation, a decrease was observed in both SBP and DBP, which is indicative of a cause-effect relationship (Paran *et al.*, 2009).

The putative active constituent, lycopene, exhibits antihypertensive effects due its ability to inhibit ACE, and its antioxidant effect, reducing oxidative stress induced by angiotensin-II and indirectly enhancing production of nitric oxide in the endothelium (Mozos *et al.*, 2018). Angiotensin II is known to directly cause vasoconstriction and oxidative stress, as well as vascular smooth muscle cell phenotypic transformation and production of inflammatory cytokines (Ren *et al.*, 2017), and the proposed mechanism of action of lycopene is the impairment of these pathways. Lycopene can increase the antioxidant properties of vitamin C, E, polyphenols and beta-carotene in a synergistic way (Kong *et al.*, 2010; Karppi *et al.*, 2013; Mozos *et al.*, 2018). A study including 8,556 overweight or obese adult participants demonstrated an association between a lower prevalence of hypertension and both lycopene concentration and lycopene/uric acid ratio (Han and Liu, 2017). A meta-analysis of lycopene supplementation (more than 12 mg/day) has shown that the compound might significantly reduce SBP, but not DBP, in prehypertensive or hypertensive patients (Li and Xu, 2013).

The consumption of tomato extract supplement, instead of the pure lycopene supplement, is recommended because of the conflicting pharmacological outcomes and pharmacokinetics in the clinical evaluation of lycopene. Although several studies have revealed the anti-atherosclerotic effect of lycopene, some conflicting results have also been reported, challenging the vascular effects of lycopene. Several possible reasons for this obvious discrepancy have been suggested, including methodological differences in the study designs, the use of unstandardized amounts of tomato food products, different modes of delivery, combination of lycopene with other antioxidants, different processing procedures or eating behaviour, influenced by cultural and temporal patterns among different individuals. In most cases, however, beneficial effects have been observed upon supplementation with tomato-based products (Sesso *et al.*, 2003). On the other hand, dehydrated and powdered tomatoes have poor lycopene stability.

Supplementation with tomatoes, containing lycopene (red tomatoes) or not (yellow tomatoes), showed a greater antioxidant effect than lycopene alone, probably due to the synergistic effects of other naturally occurring secondary metabolites in tomatoes. Lycopene is readily bioavailable in the presence of oil, especially in monounsaturated oils, other dietary fats and processed tomato products (Mozos *et al.*, 2018).

8.13 *Camellia sinensis* (Theaceae)

The tea plant, *C. sinensis*, is used in the most popular beverage used globally, and enjoys a folk reputation as possessing many biological activities, including antioxidant, antiobesity, antimicrobial and antifungal, anti-inflammatory, immunostimulatory, anticarcinogenic, anticaries, antiarthritic, antidiabetic and insulin-enhancing, cardioprotective, nephroprotective and chemopreventive activities. Consumption of tea is beneficial in the management of hypertension because of the presence of polyphenols as the major constituents. Four commercially important varieties of *C. sinensis* are recognized. Of these, *Camellia sinensis* var. *sinensis* and *Camellia sinensis* var. *assamica* are most commonly used for tea while *Camellia sinensis* var. *pubilimba* and *Camellia sinensis* var. *dehungensis* have applications mainly in natural medicine. The species synthesises a complex mixture of catechins, flavonoids and related polyphenols. The fermentation process involved in the production of black tea leads to the transformation of catechins to theaflavins, such as theaflavin, theaflavin-3-gallate, theaflavin-3′-gallate, and theaflavin 3,3′-digallate, and thearubigin polymers. Green tea, unfermented and fresh, contains more flavonoids and catechins, the predominant constituents being catechins (particularly epicatechin, epicatechin gallate, epigallocatechin, and epigallocatechin gallate.

A laboratory study has shown that oral administration of 50 or 100 mg/kg/day of leaf saponin extract of *C. sinensis* to 7-week-old and 15-week-old spontaneously hypertensive rats for five days caused significant time-dependent reductions in the mean arterial blood pressure, with 50 mg/kg bodyweight of the extract causing a more lasting antihypertensive action. The probable mechanism of action was suggested to be *via* the ACE-inhibitory pathway (Sagesaka-Mitane *et al.*, 1996).

Studies have shown that hypertensives have elevated levels of homocysteine, which can induce endoplasmic reticulum (ER) stress in endothelial cells (Sutton *et al.*, 1997), and it has been reported that black tea extract (BT) and theaflavin-3,3′-digallate (TF3, one of main theaflavins in black tea) protect against hypertension-associated endothelial dysfunction through alleviation of ER stress. BT extract and TF3 treatment reversed the elevations of ER stress markers (ATF3, ATF6 and eIF2α) and decreased reactive oxygen species (ROS) level as well as homocysteine metabolic enzymes in aortae from angiotensin II-infused rats (Cheang *et al.*, 2015). The treatment improved endothelium-dependent relaxation in renal arteries, carotid arteries, and aortae, as well as normalising the blood pressure of hypertensive rats. These unique cardioprotective actions of BT and the improvement of vascular dysfunction provided the rationale for developing it as a supplement for hypertensive patients (Choya *et al.*, 2018). It has been shown that regular tea consumption can be protective against cardiovascular diseases through the reversal of endothelial dysfunction (Scalbert *et al.*, 2005; Hodgson, 2008). In a meta-analysis, tea consumption appeared to have dose-dependent benefits towards cardiac health by decreasing the stroke rate, with moderate improvements in LDL and total cholesterol.

References

Ademiluyi, A.O., Oboh, G., Ogunsuyi, O.B. and Oloruntoba, F.M. (2015) A comparative study on antihypertensive and antioxidant properties of phenolic extracts from fruit and leaf of some guava (*Psidium guajava* L.) varieties. *Comparative Clinical Pathology*, 25, 363–374.

Adeneye, A.A. (2014) *Herbal pharmacotherapy for hypertension management*. Saarbrucken, Germany: Lambert Academic Publishing, pp. 146.

Akanni, O.O., Owumi, S.E. and Adaramoye, O.A. (2014) *In vitro* studies to assess the antioxidative, radical scavenging and arginase inhibitory potentials of extracts from *Artocarpus altilis, Ficus exasperate* and *Kigelia Africana*. *Asian Pacific Journal of Tropical Biomedicine*, 4(Supplement 1), s492–s499.

Amaechina, F.C. and Omogbai, E.K. (2007) Hypotensive effect of aqueous extract of the leaves of *Phyllanthus amarus* Schum and Thonn (Euphorbiaceae). *Acta Poloniae Pharmaceutica*, 64(6), 547–552.

Ashraf, R., Khan, R.A., Ashraf, I. and Qureshi, A.A. (2013) Effects of *Allium sativum* (garlic) on systolic and diastolic blood pressure in patients with essential hypertension. *Pakistan Journal of Pharmaceutical Sciences*, 26(5), 859–863.

Atlas, S.A. (2007) The renin-angiotensin aldosterone system: pathophysiological role and pharmacologic inhibition. *Journal of Managed Care and Specialty Pharmacy*, 13, S9.

Cheang, W.S., Ngai, C.Y., Yen-Tam, Y., Yu-Tian, X., Tak-Wong, Y. and Zhang, Y. (2015) Black tea protects against hypertension-associated endothelial dysfunction through alleviation of endoplasmic reticulum stress. *Scientific Reports*, 5, 10340.

Chiu, C.-T., Chen, J.-H., Chou, F.-P. and Lin, H.-H. (2015) *Hibiscus sabdariffa* leaf extract inhibits human prostate cancer cell invasion via down-regulation of Akt/NF-κB/MMP-9 pathway. *Nutrients*, 7(7), 5065–5087.

Choya, K.W., Murugana, D. and Mustafaa, M.R. (2018) Natural products targeting ER stress pathway for the treatment of cardiovascular diseases. *Pharmacological Research*, 132, 119–129.

Cienfuegos-Jovellanos, E., Quinones̃-Mdel, M., Muguerza, B., Moulay, L., Miguel, M. and Aleixandre, A. (2009) Antihypertensive effect of a polyphenol-rich cocoa powder industrially processed to preserve the original flavonoids of the cocoa beans. *Journal of Agricultural and Food Chemistry*, 57, 6156–6162.

Dehkordi, F.R. and Kamkhah, A.F. (2008) Antihypertensive effect of *Nigella sativa* seed extract in patients with mild hypertension. *Fundamental & Clinical Pharmacology*, 22(4), 447–452.

Díaz-de-Cerio, E., Verardo, V., Gómez-Caravaca, A.-M., Fernández-Gutiérrez, A. and Segura-Carretero, A. (2017) Health effects of *Psidium guajava* L. leaves: an overview of the last decade. *International Journal of Molecular Sciences*, 18(4), 897.

El-Mahmoudy, A., El-Mageid, A.A. and AbdelAleem, A. (2014) Pharmacodynamic evaluation of *Hibiscus sabdariffa* extract for mechanisms underlying its antihypertensive action: pharmacological and biochemical aspects. *Journal of Physiology and Pharmacology Advances*, 4, 379–388.

Engelhard, Y.N., Gazer, B. and Paran, E. (2006) Natural antioxidants from tomato extract reduce blood pressure in patients with grade-1 hypertension: a double-blind, placebo-controlled pilot study. *American Heart Journal*, 151, 100(e6–e1).

Fisher, N.D., Hughes, M., Gerhard-Herman, M. and Hollenberg, N.K. (2003) Flavanol-rich cocoa induces nitric-oxide-dependent vasodilation in healthy humans. *Journal of Hypertension*, 21, 2281–2286.

Forouzanfar, F., Bazzaz, B.S.F. and Hosseinzadeh, H. (2014) Black cumin (*Nigella sativa*) and its constituent (thymoquinone): a review on antimicrobial effects. *Iranian Journal of Basic Medical Sciences*, 17(12), 929–938.

Gawade, B.V. and Fegade, S.A. (2012) Rauwolfia (reserpine) as a potential antihypertensive agent: a review. *International Journal of Pharmaceutical and Phytopharmacological Research*, 2(1), 46–49.

Gbolade, A. (2012) Ethnobotanical study of plants used in treating hypertension in Edo State of Nigeria. *Journal of Ethnopharmacology*, 144, 1–10.

Grassi, D., Desideri, G., Necozione, S., Lippi, C., Casale, R., Properzi, G., Blumberg, J.B. and Ferri, C. (2008) Blood pressure is reduced and insulin sensitivity increased in glucoseintolerant, hypertensive subjects after 15 days of consuming high-polyphenol dark chocolate. *Journal of Nutrition*, 138, 1671–1676.

Grassi, D., Desideri, G.-B. and Ferri, C. (2010) Blood pressure and cardiovascular risk: what about cocoa and chocolate? *Archives of Biochemistry and Biophysics*, 501(1), 112–115.

Gwehenberger, B., Rist, L., Huch, R. and von Mandach, U. Effect of Bryophyllum pinnatum versus fenoterol on uterine contractility. (2004) *European Journal of Obstetrics and Gynecology and Reproductive Biology*, 113, 164.

Han, C.Y., Ki, S.H., Kim, Y.W., Noh, K., Lee, D.Y., Kang, B., Ryu, J.H., Jeon, R., Kim, E.H., Hwang, S.J. and Kim, S.G. (2011) Ajoene, a stable garlic by-product, inhibits high fat diet-induced hepatic steatosis and oxidative injury through LKB1-dependent AMPK activation. *Antioxidants & Redox Signaling*, 14(2), 187–202.

Han, G.M. and Liu, P. (2017) Higher serum lycopene is associated with reduced prevalence of hypertension in overweight or obese adults. *European Journal of Integrative Medicine*, 13, 34–40.

Herrera-Arellano, A., Flores-Romero, S. and Chávez-Soto, M.A. (2004) Tortoriello J. Effectiveness and tolerability of a standardized extract from *Hibiscus sabdariffa* in patients with mild to moderate hypertension: a controlled and randomized clinical trial. *Phytomedicine*, 11, 375–382.

Hirunpanich, V., Utaipat, A., Morales, N.P., Bunyapraphatsara, N., Sato, H., Herunsalee, A. and Suthisisang, C. (2005) Antioxidant effects of aqueous extracts from dried calyx of Hibiscus sabdariffa Linn. (Roselle) in vitro using rat low-density lipoprotein (LDL). *Biological and Pharmaceutical Bulletin*, 28, 477.

Hirunpanicha, V., Utaipata, A., Morales, N.P., Bunyapraphatsara, N., Sato, H., Herunsale, A. and Suthisisang, C. (2006) Hypocholesterolemic and antioxidant effects of aqueous extracts from the dried calyx of Hibiscus sabdariffa L. in hypercholesterolemic rats. *Journal of Ethnopharmacology*, 103, 252.

Hodgson, J.M. (2008) Tea flavonoids and cardiovascular disease. *Asia Pacific Journal of Clinical Nutrition*, 17(Suppl. 1), 288–290.

Hopkins, A.L., Lamm, M.G., Funk, J. and Ritenbaugh, C. (2013) *Hibiscus sabdariffa* L. in the treatment of hypertension and hyperlipidemia: a comprehensive review of animal and human studies. *Fitoterapia*, 85, 84–94.

Irondi, E.A., Agboola, S.O., Oboh, G., Boligon, A.A., Athayde, A.L. and Shode, F.O. (2016) Guava leaves polyphenolics-rich extract inhibits vital enzymes implicated in gout and hypertension in vitro. *Journal of Intercultural Ethnopharmacology*, 5(2), 122–130.

Iwu, M.M. (2014) *Handbook of African medicinal plants*. Second ed. Boca Raton, FL: CRC Press/Taylor and Francis Group, pp. 476.

Iwu, M.M. (2016) *Food as medicine – functional food plants of Africa*. Boca Raton, FL: CRC Press/Taylor and Francis Group, pp. 384.

Karppi, J., Kurl, S., Ronkainen, K., Kauhanen, J. and Laukkanen, J.A. (2013) Serum carotenoids reduce progression of early atherosclerosis in the carotid artery wall among Eastern Finnish men. *PLoS ONE*, 8, e64107.

Karuna, R., Reddy, S.S., Baskar, R. and Saralakumari, D. (2009) Antioxidant potential of aqueous extract of *Phyllanthus amarus* in rats. *Indian Journal of Pharmacology*, 41(2), 64–67.

Kassuya, C.A., Leite, D.F., de Melo, L.V., Rehder, V.L. and Calixto, J.B. (2005) Anti-inflammatory properties of extracts, fractions and lignans isolated from *Phyllanthus amarus*. *Planta Medica*, 71(8), 721–726.

Keyhanmanesh, R., Gholamnezhad, Z. and Boskabady, M.H. (2014) The relaxant effect of *Nigella sativa* on smooth muscles, its possible mechanisms and clinical applications. *Iranian Journal of Basic Medical Sciences*, 17, 939.

Khan, M.A., Chen, H.-C., Tania, M. and Zhang, D.-Z. (2011) Anticancer activities of *Nigella sativa* (black cumin). *African Journal of Traditional, Complementary and Alternative Medicines*, 8(Supplement 5), 226–232.

Khan, M.K. and Afzal, M. (2016) Chemical composition of *Nigella sativa* Linn: part 2: recent advances. *Inflammopharmacology*, 24, 67–79.

Kiemer, A.K., Hartung, T., Huber, C. and Vollmar, A.M. (2003) *Phyllanthus amarus* has anti-inflammatory potential by inhibition of iNOS, COX-2, and cytokines via the NF-kappaB pathway. *Journal of Hepatology*, 38(3), 289–297.

Kong, K.W., Khoo, H.E., Prasad, K.N., Ismail, A., Tan, C.P. and Rajab, N.F. (2010) Revealing the power of the natural red pigment lycopene. *Molecules*, 15, 959–987.

Kozuma, K., Tsuchiya, S., Kohori, J., Hase, T. and Tokimitsu, I. (2005) Antihypertensive effect of green coffee bean extract on mildly hypertensive subjects. *Hypertension Research*, 28, 711–718.

Li, X. and Xu, J. (2013) Lycopene supplement and blood pressure: an updated meta-analysis of intervention trials. *Nutrients*, 5, 3696–3712.

Lin, T.-L., Lin, H.-H., Chen, C.-C., Lin, M.-C., Chou, M.-C. and Wang, C.-J. (2007) *Hibiscus sabdariffa* extract reduces serum cholesterol in men and women. *Nutrition Research*, 27, 140–145.

Lobay, D. (2015) Rauwolfia in the treatment of hypertension – review. *Integrative Medicine (Boulder)*, 14(3), 40–46.

Maganha, E.G., Halmenschlager, R. da C., Rosa, R.M., Henriques, J.A.P., Lia de Paula Ramos, A.L. and Saffi, J. (2010) Pharmacological evidences for the extracts and secondary metabolites from plants of the genus Hibiscus. *Food Chemistry*, 118, 1–10.

Maity, S., Chatterjee, S., Variyar, P.S., Sharma, A., Adhikari, S. and Mazumder, S. (2013) Evaluation of antioxidant activity and characterization of phenolic constituents of *Phyllanthus amarus* root. *Journal of Agricultural and Food Chemistry*, 61(14), 3443–3450.

Majdalawieh, A.F. and Fayyad, M.W. (2016) Recent advances on the anti-cancer properties of *Nigella sativa*, a widely used food additive. *Journal of Ayurveda and Integrative Medicine*, 7(3), 173–180.

Mallaiah, P., Sudhakara, G., Srinivasulu, N., Sasi-bhusana-rao, B., Vijayabharathi, G. and Saralakumari, D. (2015) Assessment of *in vitro* antioxidant potential and quantification of total phenols and flavonoids of aqueous extract of *Phyllanthus amarus*. *International Journal of Pharmacy and Pharmaceutical Sciences*, 7, 439–445.

Mastroiacovo, D., Kwik-Uribe, C., Grassi, D., Necozione, S., Raffaele, A., Pistacchio, L., Righetti, R., Bocale, R., Lechiara, M.C., Marini, C., Ferri, C. and Desideri, G. (2015) Cocoa flavanol consumption improves cognitive function, blood pressure control, and metabolic profile in elderly subjects: the cocoa, cognition, and aging (CoCoA) study–a randomized controlled trial. *American Journal of Clinical Nutrition*, 101(3), 538–548.

Mayer, B., Kalus, U., Grigorov, A. Pindur, G., Jung, F., Radtke, H., Bachmann, K., Mrowietz, C., Koscielny, J., Wenzel, E. and Kiesewetter, H. (2011) Effects of an onion-olive oil maceration product containing essential ingredients of the Mediterranean diet on blood pressure and blood fluidity. *Arzneimittelforschung*, 51(2), 104–111.

McKay, D.L., Chen, C.-Y.O., Saltzman, E. and Blumberg, J.B. (2010) *Hibiscus sabdariffa*, L. Tea (Tisane) lowers blood pressure in prehypertensive and mildly hypertensive adults. *Journal of Nutrition*, 140, 298–303.

Mohamed, R., Fernandez, J., Pineda, M. and Aguilar, M. (2007) Roselle (*Hibiscus sabdariffa*) seed oil is a rich source of gamma-tocopherol. *Journal of Food Science*, 72, S207–S211.

Mozaffari-Khosravi, H., Jalali-Khanabadi, B., Afkhami-Ardekani, M., Fatehi, F. and NooriShadkam, M. (2009) The effects of sour tea (*Hibiscus sabdariffa*) on hypertension in patients with type II diabetes. *Journal of Human Hypertension*, 23, 48–54.

Mozos, I., Stoian, D., Caraba, A., Malainer, C., Jarosław, O., Horbanczu´k, J.O. and Atanasov, A.G. (2018) Lycopene and vascular health. *Frontiers in Pharmacology*, 9, 521.

Nickavar, B., Mojab, F., Javidnia, K. and Amoli, M.A. (2003) Chemical composition of the fixed and volatile oils of *Nigella sativa* L. from Iran. *Zeitschrift für Naturforschung. C*, 58, 629–631.

Nnam, N.M. and Onyeke, N.G. (2003) Chemical compositions of two varieties of sorrel (*Hibiscus sabdariffa* L.), calyces and the drinks made from them. *Plant Foods of Human Nutrition*, 58, 1–7.

Nwachukwu, D., Aneke, E., Obika, L. and Nwachukwu, N. (2015a) Investigation of antihypertensive effectiveness and tolerability of *Hibiscus sabdariffa* in mild to moderate hypertensive subjects in Enugu, South-East, Nigeria. *American Journal of Phytomedicine and Clinical Therapeutics*, 3, 339–345.

Nwachukwu, D.C., Aneke, E.I., Obika, L.F. and Nwachukwu, N.Z. (2015b) Effects of aqueous extract of *Hibiscus sabdariffa* on the renin-angiotensin-aldosterone system of Nigerians with mild to moderate essential hypertension: a comparative study with Lisinopril. *Indian Journal of Pharmacology*, 47(5), 540–545.

Oboh, G., Akinyemia, A.J., Osanyinlusia, F.R., Ademiluyia, A.O., Boligonc, A.A. and Athayde, M.L. (2014) Phenolic compounds from sandpaper (*Ficus exasperata*) leaf inhibits angiotensin 1 converting enzyme in high cholesterol diet fed rats. *Journal of Ethnopharmacology*, 157(2014), 119–125.

Ogunyinka, B.I., Oyinloye, B.E., Osunsanmi, F.O., Kappo, A.P. and Opoku, A.R. (2017) Comparative study on proximate, functional, mineral, and antinutrient composition of fermented, defatted, and protein isolate of *Parkia biglobosa* seed. *Food Science and Nutrition*, 5(1), 139–147.

Ojewole, J.A. (2002) Antihypertension properties of *Bryophyllum pinnatum* (Lam) (oken) leaf extracts. *American Journal of Hypertension*, 15, 34–39.

Osamor, P.E. and Owumi, B.E. (2010) Complementary and alternative medicine in the management of hypertension in an urban Nigerian community. *BMC Complementary and Alternative Medicine*, 10, 36.

Paran, E., Novack, V., Engelhard, Y.N. and Hazan-Halevy, I. (2009) The effects of natural antioxidants from tomato extract in treated but uncontrolled hypertensive patients. *Cardiovascular Drugs and Therapy*, 23, 145–151.

Quiñones, M., Sânchez, D., Muguerza, B., Moulay, L. and Aleixandre, A. (2010) Long-term intake of CocoanOX attenuates the development of hypertension in spontaneously hypertensive rats. *Food Chemistry*, 122(4), 1013–1019.

Rawat, P., Singh, P.K. and Kumar, V. (2016) Anti-hypertensive medicinal plants and their mode of action. *Journal of Herbal Medicine*, 6, 107–118.

Ren, X.S., Tong, Y., Ling, L., Chen, D., Sun, H.J. and Zhou, H. (2017) NLRP3 gene deletion attenuates angiotensin ii-induced phenotypic transformation of vascular smooth muscle cells and vascular remodeling. *Cellular Physiology and Biochemistry*, 44, 2269–2280.

Ried, K., Frank, O.R., Stocks, N.P., Fakler, P. and Sullivan, T. (2008) Effect of garlic on blood pressure: a systematic review and meta-analysis. *BMC Cardiovascular Disorders*, 8, 13.

Ried, K., Toben, C. and Fakler, P. (2013) Effect of garlic on serum lipids: an updated meta-analysis. *Nutrition Reviews*, 71(5), 282–299.

Sagesaka-Mitane, Y., Sugiura, T., Miwa, Y., Yamaguchi, K. and Kyuki, K. (1996) Effect of tea-leaf saponin on blood pressure of spontaneously hypertensive rats. *Yakugaku Zasshi*, 116(5), 388–395.

Sahebkar, A., Beccuti, G., Simental-Mendía, L.E., Nobili, V. and Bo, S. (2016a) *Nigella sativa* (black seed) effects on plasma lipid concentrations in humans: a systematic review and meta-analysis of randomized placebo-controlled trials. *Pharmacological Research*, 106, 37–50.

Sahebkar, A., Soranna, D., Liu, X., Thomopoulos, C., Simental-Mendia, L.E., Derosa, G., Maffioli, P. and Parati, G. (2016b) A systematic review and meta-analysis of randomized controlled trials investigating the effects of supplementation with *Nigella sativa* (black seed) on blood pressure. *Journal of Hypertension*, 34(11), 2127–2135.

Santos, H.O. and Macedo, R.C.O. (2018) Cocoa-induced (*Theobroma cacao*) effects on cardiovascular system: HDL modulation pathways. *Clinical Nutrition ESPEN*, 27, 10–15.

Scalbert, A., Manach, C., Morand, C., Remesy, C. and Jimenez, L. (2005) Dietary polyphenols and the prevention of diseases. *Critical Reviews in Food Science and Nutrition*, 45(4), 287–306.

Serban, C., Sahebkar, A., Ursoniu, S., Andrica, F. and Banach, M. (2015) Effect of sour tea (*Hibiscus sabdariffa* L.) on arterial hypertension: a systematic review and meta-analysis of randomized controlled trials. *Journal of Hypertension*, 33, 1119–1127.

Sesso, H.D., Liu, S., Gaziano, J.M. and Buring, J.E. (2003) Dietary lycopene, tomato-based food products and cardiovascular disease in women. *Journal of Nutrition*, 133, 2336–2341.

Sheldon, M.B. and Kotte, J.H. (1957) Effect of *Rauwolfia serpentina* and reserpine on the blood pressure in essential hypertension; a long-term double-blind study. *Circulation*, 16(2), 200–206.

Singh, R., Rastogi, S., Singh, N., Ghosh, S., Gupta, S. and Niaz, M. (1993) Can guava fruit intake decrease blood pressure and blood lipids. *Journal of Human Hypertension*, 7, 33–38.

Srividya, N. and Periwal, S. (1995) Diuretic: hypotensive and hypoglycaemic effect of *Phyllanthus amarus*. *Indian Journal of Experimental Biology*, 33, 861–864.

Sutton-Tyrrell, K., Bostom, A., Selhub, J. and Zeigler-Johnson, C. (1997) High homocysteine levels are independently related to isolated systolic hypertension in older adults. *Circulation*, 96(6), 1745–1749.

Tavakkoli, A., Ahmadi, A., Razavi, B.M. and Hosseinzadeh, H. (2017) Black seed (*Nigella sativa*) and its constituent thymoquinone as an antidote or a protective agent against natural or chemical toxicities. *Iranian Journal of Pharmaceutical Research*, 16(Suppl), 2–23.

Tokoudagba, J.-M., Auger, C., Bréant, L., Gom, S.N., Gerbert, P., Idris-Khodja, N., Gbaguidi, F., Gbenou, J., Moudachirou, M., Lobstein, A. and Schini-Kertha, V.B. (2010) Procyanidin-rich fractions from *Parkia biglobosa* (Mimosaceae) leaves cause redox-sensitive endothelium-dependent relaxation involving NO and EDHF in porcine coronary artery. *Journal of Ethnopharmacology*, 132(1), 246–250.

WHO. (2013) A global brief on hypertension. World Health Organisation, Geneva, Switzerland. http://www.who.int/cardiovascular_diseases/publications/global_brief_hypertension/en/ (accessed 15/6/18).

Worthen, D., Ghosheh, O. and Crooks, P. (1998) The in vitro anti-tumor activity of some crude and purified components of black seed, *Nigella sativa* L. *Anticancer Research*, 18, 1527–1532.

Yadav, M., Gulkari, V.D. and Wanjari, M.M. (2016) *Bryophyllum pinnatum* leaf extracts prevent formation of renal calculi in lithiatic rats. *Ancient Science of Life*, 36(2), 90–97.

9 Nigerian Healing Plants Used for Metabolic Syndrome, Obesity, and Diabetes

Cardiovascular disease is the major cause of death globally. Metabolic syndrome is considered to be a major factor in the onset of many cardiovascular diseases. It is a condition in which at least three of the five cardiovascular risk factors exist, namely obesity, excessive visceral fat storage, dyslipidaemia, hypertension, and hyperglycaemia or type 2 diabetes. It often manifests as insulin resistance, oxidative stress, or chronic inflammation. The term metabolism is commonly used to refer specifically to the breakdown of food and its transformation into energy. It covers the whole range of biochemical processes that occur within a living organism, both anabolism (the synthesis of substances) and catabolism (the breakdown of substances). Metabolic syndrome is thought to be caused by an underlying disorder of energy utilization and storage. The detailed pathway of the syndrome is not well understood but its effect is evident.

Metabolic syndrome (MetS) is generally defined as a clustering of at least three of five of the following medical conditions:

- **A large waistline.** This is also called abdominal or central obesity or "having an apple shape." Excess fat in the stomach area is a greater risk factor for heart disease than excess fat in other parts of the body, such as on the hips.
- **A high triglyceride level** (or you are on medicine to treat high triglycerides). Triglycerides are a type of fat found in the blood.

- **A low high-density lipoprotein (HDL) cholesterol level** (or you are on medicine to treat low HDL cholesterol). HDL is sometimes called "good" cholesterol. This is because it helps to remove cholesterol from your arteries to your liver. A low HDL cholesterol level raises your risk for heart disease.

- **High blood pressure** (or you are on medicine to treat high blood pressure). Blood pressure is the force of blood pushing against the walls of your arteries as your heart pumps blood. If this pressure rises and stays high over time, it can damage your heart and lead to plaque buildup.

- **High fasting blood sugar** (or you are on medicine to treat high blood sugar). Mildly high blood sugar may be an early sign of diabetes.

Epidemiological evidence implicates Western-style dietary patterns and sedentary lifestyle as contributory factors to the development of cardiovascular diseases, dyslipidaemia, and diabetes. Recent evidence also suggests that a high-fat diet is responsible for the development of MetS both in animals (3) and in humans. It has been observed that abdominal obesity plays a pivotal role in the onset of the other conditions associated with MetS. From a clinical perspective, persons at risk of MetS includes patients being treated for dyslipidaemia, hyperglycaemia, or systemic hypertension. Several studies suggest that MetS is also associated with a two-fold increase in mortality from cardiovascular disease and greater vulnerability to other causes of morbidity. The worldwide prevalence of MetS ranges from <10% in rural areas to as much as 84%, depending on the region, urban, or rural environment, composition (sex, age, race, and ethnicity) of the population studied, and the definition of the syndrome used. In general, the International Diabetes Federation (IDF) estimates that one-quarter of the world's adult population has MetS. In Nigeria, MetS used to be a problem for the affluent or so-called big-men and big-women but now it affects all ages, and all social strata, in both villages and urban areas. Studies have shown that, in Nigeria, there is a 18% incidence of MetS among semi-urban dwellers aged 25–64 years, whereas, among rural (village) dwellers of the same age bracket, the figure is about 10% (Ulasi *et al.*, 2010).

Therapeutic interventions for the treatment of metabolic syndrome are addressed primarily for the treatment of the major risk factors, including hypertension (with the angiotensin-converting enzyme inhibitors, the angiotensin receptor blockers, and calcium channel blockers, etc.), plasma cholesterol (controlled mainly with statins), and diabetes (with anti-hyperglycaemic agents, such as metformin, glibenclamide, etc). Several plant-based phytomedicines and their active molecules have been identified as having potential for treatment of the major MetS risk factors, with bio-equivalent pharmacological profiles. In addition, in Nigeria, certain dietary components, and over 100 plant species are known to be useful in the prevention and treatment of the major risk factors or the modulation of MetS by assisting the homoeostasis mechanisms of the body.

It appears that the central causative physiological factors underlying MetS are insulin resistance, oxidative stress, and chronic inflammatory conditions.

The plants used for the treatment of MetS include those that are capable of acting on multiple targets and the physiological parameters that affect homoeostasis (Dommermuth and Ewing, 2018). MetS is a cascade of illnesses that appear to start with obesity, where insufficient physical activity and excess calorie intake are factors contributing to its development. The resultant excess energy consumption subsequently causes hypoxia (oxygen deficiency) in the adipose tissues. This induces the adipocytes (fat cells) to secrete pro-inflammatory chemokines (e.g. cyclooxygenase-2 (COX-2) and inducible nitric oxide synthases (iNOS)) that attract immune cells, macrophages and inflammatory responses, which causes the other pathological conditions. In the diabetic cases, insulin resistance leads to decreased disposal of glucose in the peripheral tissues, overproduction of glucose by the liver, defects in pancreatic B-cell function, and decreased B-cell mass. During fasting, there is increased adipocyte secretory protein production, including fibrinogen-angiopoietin-related protein, metallothionein and resistin. Resistin induces insulin resistance, that links diabetes to obesity, while metallothionein is an antioxidant metal-binding and stress-response protein.

The white adipose tissues, in addition to secreting pro-inflammatory cytokines, have an endocrine function in producing hormones, lipid metabolism regulators, vascular haemostasis controllers, or a comparable system (e.g. leptin, angiotensinogen, adipsin, acylation-stimulating protein, adiponectin, retinol-binding protein, TNF-alpha, interleukin 6, plasminogen activator inhibitor-1, or tissue factor), and this functionality is impaired in a state of metabolic dysfunction.

Some of the Nigerian plants listed here act on one or more of these risk factors or indicators of MetS but a pragmatic approach is to use preparations that address the problem at the cellular or tissue level.

9.1 Anti-Obesity Herbs

Obesity is generally defined as a state of being grossly fat or overweight. In medical terms, it is a condition in which excess body fat has accumulated to the extent that it might have a negative effect on the patient's health. Obesity is measured in terms of body mass index (BMI), which is a person's weight in kilograms divided by height in metres squared (BMI = kg/m^2). To calculate the BMI using imperial units, divide the weight in pounds by the height in inches squared and multiply the result by 703. A person is considered overweight or obese if BMI is equal to or greater than 30, which approximates to 30 pounds of excess weight. On the other hand, a "healthy weight" is defined as a BMI equal to or greater than 19 and less than 25 among all people 20 years of age or over. The World Health Organization has classified obesity according to the BMI scores: a BMI of 25 to 29.9 is defined as "pre-obese"; BMI of 30 to 34.99 is defined as "obese class I"; BMI of 35 to 39.99 is defined as "obese class II"; and BMI of 40 or greater is defined as "obese class III".

The balance between calorie intake and energy expenditure determines a person's weight. If a person eats more calories than he or she burns (metabolises), the person gains weight, and the body will store the excess energy as fat. If a person eats fewer calories than he or she metabolises, he or she will lose weight.

Therefore, the most common causes of obesity are overeating and physical inactivity. Generally, body weight is the result of genetics, disease, metabolism, epigenetics, environment, behaviour, and culture.

In many cases, obesity is of genetic origin, where either or both parents are obese. Genetics also affect hormones involved in fat regulation. For example, people born with a deficiency in leptin, a hormone produced in fat cells and in the placenta, are likely to be obese. This hormone controls weight by signaling the brain to eat less when body fat stores are too high. In a situation where the body cannot produce enough leptin or leptin cannot signal the brain to eat less, this control is lost, and obesity occurs. The possible role of leptin replacement as a treatment for obesity is the subject of ongoing studies in several laboratories. However, epidemiological studies suggest that the metabolically healthy obesity (MHO) phenotype may be a benign health condition which poses no additional risk of cardiovascular diseases. A close association has also been found between vitamin D deficiency and obesity (Cheng, 2018; Mousa *et al.*, 2018). Vitamin D deficiency may exacerbate the risk of cardiometabolic death outcomes associated with metabolic dysfunction in normal-weight and obese individuals (Al-Khalidi *et al.*, 2018).

Obesity can also arise as a result of certain diseases such as hypothyroidism, insulin resistance, polycystic ovary syndrome, Cushing's syndrome, and Prader-Willi syndrome. Traditional doctors recognize non-dietary fat-induced obesity such as the weight gain in women transiting to menopause, and appropriate remedies such as scent-leaf soup are recommended. Studies have shown that such remedies are indeed effective (Chao *et al.*, 2017). Regardless of the origin, abdominal obesity or "having an apple shape" or having excess fat in the stomach area is a greater risk factor for heart disease than excess fat in other parts of the body, such as on the hips or arms.

Nigerian plants used for the control of obesity include:

Aloe vera (aloe vera)

Camellia sinensis (green and black tea)

Capsicum annuum (red pepper, totoshi)

Cinnamomum spp. (cinnamon)

Citrullus vulgaris (sweet watermelon)

Citrus limon (lemon)

Citrus paradisi (grapefruit)

Garcinia kola (bitter kola),

Gnetum africanum (eruh)

Hibiscus sabdariffa (zobo)

Hypericum perforatum (St John's wort)

Ilex paraguariensis (yerba mate)

Irvingia gabonensis (African mango)

Momordica charantia (bitter melon)

Persea americana (avocado pear)

Phaseolus vulgaris (common bean)

Punica granatum (pomegranate)

Rosmarinus officinalis (rosemary)

Taraxacum officinale (dandelion)

The formulations and methods of preparation of these herbs vary enormously among different tribes and even individual herbalists. A common feature, however, in the recommendation among the healers is a reduction in food portion size, the restriction on intake of alcoholic beverages, and the limit on carbohydrate consumption. Most of the plants used for the management of MetS are taken as juices or ingredients in special vegetable soups.

9.2 High Blood Cholesterol

Cholesterol is a natural substance produced by the body as an essential compound for normal metabolism. Most (75%) of the cholesterol in our bloodstream is produced by the liver, and the remaining 25% comes mainly from animal-based diets. It has been established that elevated blood cholesterol levels are deleterious to health but having the right levels of cholesterol plays a vital role in the metabolic processes, such as maintaining the structural integrity of cell membranes and hormone synthesis. It is a water-insoluble waxy fatty substance that is transported in the body through lipoproteins. The U.S. Center for Disease Control reports that one-third of adults have high cholesterol levels, but the clinical significance of this finding remains a subject of much controversy. Available evidence suggests that diets relatively high in saturated fatty acids (SFAs) increase plasma total cholesterol concentrations, compared with diets relatively higher in unsaturated fatty acids. However, not all SFA had identical effects on the body's production of cholesterol. It is interesting to observe from the literature that short-chain fatty acids (6:0–10:0) and stearic acid (18:0) appear to have little or no effect on low-density lipoprotein-cholesterol (LDL-C) and high-density lipoprotein-cholesterol (HDL-C) concentrations, whereas SFAs with intermediate chain lengths, such as lauric (12:0), myristic (14:0),and palmitic (16:0) acids appear to be the most potent in increasing plasma cholesterol concentrations. It has not been possible to rationally explain the underlying mechanism by which fatty acids with 10 or fewer carbon atoms have different effects on plasma cholesterol concentrations from those with 12–16 carbon atoms but it may be related to the mode of absorption (portal vein rather than lymphatic system) and preferential hepatic oxidation of these fatty acids.

The classical nutritional advice, to limit the consumption of tropical oils rich in SFAs, such as palm oil, palm kernel oil, and coconut oil, is not supported by scientific evidence, since such oils, which are usually liquid at room temperature because they have high concentrations of short-chain SFAs, are presumed to be beneficial to the body. The most feasible approach to reducing dietary cholesterol includes replacing a meat-based diet with a vegetable-rich diet, and minimizing the intake of fatty cuts of meat by replacing with leaner cuts, trimming excess fat and skin before and after cooking, decreasing portion size of meals, substituting whole-milk or full-fat dairy products with non-fat or low-fat (<1% fat) counterparts, and avoiding foods containing animal fats.

Several Nigerian plants have the ability to lower total serum cholesterol in the body. They include the following plants:

Citrullus vulgaris (sweet watermelon)

Citrus spp. (oranges and grapefruit)

Combretum micrathum (kinkeliba)

Garcinia kola (bitter kola)

Gnetum africanum (eruh)

Ocimum sanctum (holy basil)

Psidium guajava (guava leaves)

Vernonia amygdalina (bitter-leaf)

Hepa-vital (kinkeliba leaves and bitter kola)

9.3 High Blood Triglycerides

A triglyceride is a combination of glycerol ester and three fatty acid molecules (tri-glyceride). They are the fats in the food we eat that are carried in the blood. Examples include butter, margarines, and oils in our food. High triglyceride levels (>200 mg/dL) has been implicated in increased risk of heart attack, stroke, and renal disease.

Lipoproteins are lipid and protein complexes involved in the transport of lipids in the blood to the tissues. They are classified according to their density: high density lipoprotein (HDL), low density lipoprotein (LDL), and very low-density lipoprotein (VLDL). LDL transports cholesterol from the liver to the tissues.

It is highly linked with formation of atherosclerotic plaques, following oxidation (a chemical modification), which will eventually result in the formation of fatty streaks in the arteries that can progressively increase and cause heart attacks, strokes, and peripheral artery diseases, hence the name "bad cholesterol". Oxidized LDL is also a substrate for chronic inflammation, systemic autoimmune disease, liver disease (non-alcoholic fatty liver disease), as well as chronic kidney disease.

In contrast, HDL, or good cholesterol, is beneficial to the body. It helps to transport fat molecules from the tissues and arterial walls back mainly to the liver to be excreted (reverse cholesterol transport). There is overwhelming evidence that HDL prevents or halts the progression of atherosclerosis, and, in some cases, it is known to reverse the disease.

HDL particles have antioxidant and anti-inflammatory properties. HDL also plays a key role in the nourishment of some organs and tissues, as it is transported to the adrenal gland, ovaries, and testes from which steroid hormones are formed. Studies have shown that the higher the level of total cholesterol (>200 mg/dL) and LDL (>100 mg/dL) in the blood, the greater the risk of developing a heart attack; the reverse is true, as indicated above, for HDL, in which concentrations of HDL <40 mg/dl increase the risk of heart attack. The threshold for total cholesterol and LDL depends on the cardiovascular risk score or classification of the patients.

It is therefore not enough to artificially lower total cholesterol without compensating for the HDL/LDL ratio in the body. Some of the plants used in Nigeria for the management of cholesterol and high LDL are selected based on empirical ethnomedical information without a clearly defined or discernable mechanism of action. The plants listed above for the management of high cholesterol are equally effective at reducing blood triglyceride concentrations.

9.4 Diabetes

Diabetes is a chronic metabolic disorder characterised by a high blood glucose concentration (hyperglycaemia), due to insulin deficiency and/or insulin resistance. Hyperglycaemia occurs because the liver and skeletal muscle cannot store glycogen and the tissues are unable to take up and utilise glucose. It is a serious, life-long disorder that is, as yet, incurable. Diabetes is a disease that occurs when the blood glucose concentration, also called blood sugar, is too high. Blood glucose is the main source of energy and comes from food. Insulin, a hormone made by the pancreas, helps glucose from food get into the cells to be used for energy. Sometimes the body does not make enough — or any — insulin, or it is not used well. Glucose then stays in the blood without reaching the cells, causing health problems over a long period. Although diabetes has no cure, steps can be taken to treat or manage it and stay healthy. Symptoms include polyuria (frequent urination, polydipsia (increased thirst), and polyphagia (increased hunger). There are different types of diabetes but the most common is type 2 diabetes mellitus, which results mainly from insulin resistance, meaning the cells cannot use the insulin produced. It was formerly known as non-insulin-dependent diabetes mellitus (NIDDM), or adult-onset diabetes. One is more likely to develop type 2 diabetes at age 45 or older, have a family history of diabetes, or be overweight.

Physical inactivity, race, and certain health problems such as high blood pressure also increases the chance of developing type 2 diabetes. One is also more likely to develop type 2 diabetes if you have pre-diabetes or had gestational diabetes during pregnancy.

Diabetes in most cases is symptom-less and is often not detected until it has caused some systemic damage. The disease, if left untreated, may lead to death due to acute metabolic decompensation, producing dehydration and acidosis. Chronic diabetes often results in progressive damage to a number of organs, including the kidneys, blood vessels, nervous system, and eyes.

Type 2 diabetes can be prevented by adopting a healthy diet, physical exercise, eliminating tobacco usage, and maintaining a normal body weight. When the diet/ exercise regime fails, the use of oral hypoglycaemic agents, and/or insulin replacement therapy may be necessary. In Nigeria, diabetes is well recognized by traditional healers who often recommend a combination of certain foods and herbal remedies (Iwu, 2016). The preparations often consist of a mixture of different herbs (Katerere and Eloff 2006), therefore, the plants listed below are those which, in the opinion of the authors, constitute the major ingredients for the preparation of antidiabetic medicines.

Abelmoschus esculentus (okra)

Allium cepa (onion)

Allium sativum (garlic)

Aloe barbadensis (Aloe vera), *Aloe ferox* (Cape aloe)

Andrographis paniculata (kalmegh, kalamegha)

Annona muricata (soursop)

Balanites aegyptiaca (aduwa)

Camellia sinensis (black, green tea)

Cinnamomum zeylanicum (cinnamon)

Cocos nucifera (coconut, virgin coconut oil)

Curcuma longa (turmeric)

Eugenia aromaticum (clove oil)

Galega officinalis (French lilac)

Garcinia kola (bitter kola)

Glycine max (soybean)

Gymnema sylvestre (gymnema, melyna)

Irvingia gabonensis (ogbono, dika fat)

Lagerstroemia speciosa (crepe myrtle, banaba

Magnolia officinalis (houpu)

Marsdenia latifolia (gongronema, utazi)

Momordica elegans (bitter melon)

Mucuna pruriensis (mucuna beans, velvet bean)

Musa × paradisiaca (plantain)

Ocimum gratissimum (scent leaf)

Opuntia streptacantha (prickly pear cactus)

Psidium guajava (guava leaves)

Rhodiola rosea (golden root)

Theobroma cacao (cocoa)

Trigonella foenum-graecum (fenugreek)

Vaccinium myrtillis (huckleberry, bilberry)

Vernonia amygdalina (bitter-leaf)

Vitex agnus castus (chaste tree berry)

Zingiber officinalis (ginger)

9.5 Integrative Approach to the Management of Metabolic Syndrome and Associated Co-Morbidities

The close relationships between the cardiovascular disease risk factors caused by MetS has been outlined above. While modern medicine has developed a plethora of effective pharmacological agents that can intervene with respect to specific physiological or pathological conditions of this syndrome, a more integrative approach, which addresses the underlying causes and enhances the body's cellular integrity, is better suited to address the multiple manifestations of MetS. The integrative treatment approach in cardiology focuses on bringing the "rest and digest" system into balance with the "fight or flight" system. Boosting the "rest and digest" nervous system often requires learning the traditional art of meditation, yoga, and movement. It also involves understanding the impact of food on our bodies and putting healthy foods into our system to aid in proper digestion. It also requires us to understand about environmental exposures, such as exposure to pollutants and heavy metals (Aggarwal *et al.*, 2017).

Studies have demonstrated that endoplasmic reticulum (ER) stress plays an important role in the pathogenesis of various cardiovascular diseases, such as heart failure, ischaemic heart disease, and atherosclerosis. ER is the main organelle for the synthesis, folding, and processing of secretory and transmembrane proteins. Certain pathological stimuli, including hypoxia, ischaemia, inflammation, and oxidative stress, interrupt the homoeostatic function of ER, leading to accumulation of unfolded proteins, a condition referred to as ER stress, which triggers a complex signaling network known as the unfolded protein response (UPR) (Choy *et al.*, 2018). Several natural products, targeting components of UPR and reducing ER stress, have been suggested as an innovative strategic approach to treating cardiovascular diseases. Many Nigerian plants contain compounds that possess potential cardiovascular protective properties, targeting ER stress-signalling pathways. They include *Dorstenia manii*, *Garcinia kola*, mango leaves, guava leaves, citrus fruits, *Gnetum africanum, Vernonia amygdalina,* and *Ocimum gratissimum.*

The major Nigerian healing plants outlined below (Sections 9.5.1–9.5.6) are those that address at least two of the risk factors associated with MetS. It is important to emphasize that these plants cannot be used in the absence of a healthy lifestyle for the prevention and treatment of cardiovascular diseases, modulated by the incidence of MetS.

9.5.1 Citrus Fruits (Red Orange, Grapefruit, Orange)

Studies have shown that consumption of fruit and vegetable has beneficial effects on health and are effective for the treatment of chronic diseases such as MetS (Liu *et al.*, 2004). They synthesise complex mixtures of polyphenols, that are known to protect against or mitigate the severity of chronic diseases without the undesirable side effects of modern pharmaceuticals. Citrus fruits, such as oranges, mandarins, and grapefruit, and the acid citrus fruits, like lemons, bergamots, and limes, are notably rich in such polyphenols, especially flavonoids, which are responsible for many physiological activities, including antioxidant and anti-inflammatory properties. Citrus polyphenolic compounds (e.g. flavonoids, anthocyanins, and phenolic acids) have demonstrable health benefits for the treatment of obesity, hypertension, cardiovascular diseases, and MetS (Alam *et al.*, 2014). The most important flavonoids found in citrus fruits include naringin, naringenin, nobelitin, narirutin, and hesperidin.

However, most of the medicinal effects of citrus fruits are associated with the presence of naringin and naringenin. Studies in animal models have shown that supplementation with naringin and naringenin is efficacious for the treatment of MetS and obesity (Pu *et al.*, 2012; Alam *et al.*, 2014).

9.5.2 Red Palm Oil

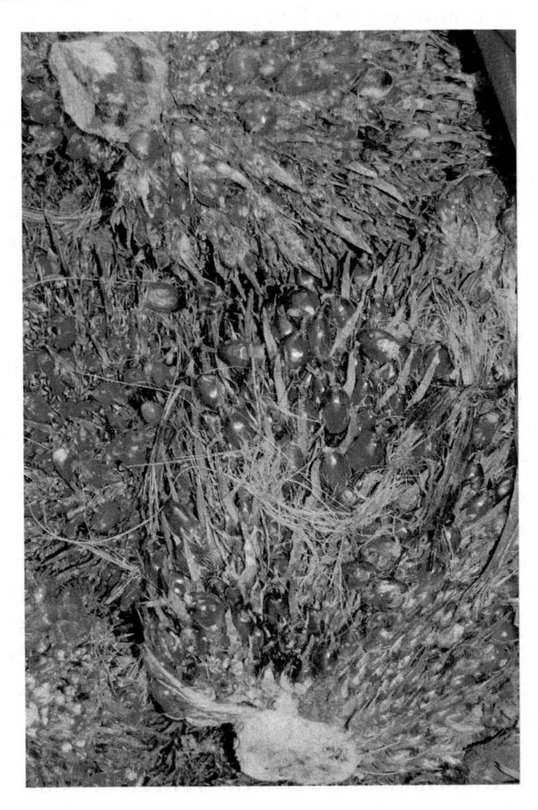

Red palm oil from the Nigerian (West African) palm tree, *Elaeis guineensis* (family Arecaceae), is one of the richest sources of carotenoids in nature. It is used as an adjuvant in many traditional remedies, often against the bias of Western-trained doctors and nutritionists, who caution against the use of palm oil because of the abundance of palmitic acid in the oil. However, palm oil has a very balanced fatty acid composition: 51% saturated fatty acids, 38% monounsaturated fatty acids, and 11% polyunsaturated fatty acids (Iwu, 2014). Red palm oil has several health benefits, due to its high content of mixed isomers of various carotenoids, which, together with vitamin E, tocotrienols, and ascorbic acid, are present in this oil (Dijkstra, 2016). Red palm oil represents a powerful network of antioxidants, which can protect tissues and cells from oxidative damage (Edem, 2002; Packer, 1992). Red palm oil is unique for its high concentration of tocotrienols, a group of vitamin E stereoisomers, with many health benefits including their ability to lower cholesterol levels and exhibit anticancer and tumour-suppressive activities (Daud *et al.*, 2012). It has been shown that a diet rich in tocotrienols, especially dietary tocotrienols from a tocotrienol-rich fraction (TRF) of palm oil, reduced the concentrations of plasma cholesterol and apolipoprotein B, platelet factor 4, and thromboxane B2, indicating its ability to protect against platelet aggregation and endothelial dysfunction (Qureshi *et al.*, 1991). Tocotrienols are 40–60 times more potent than tocopherols as antioxidants. Red palm oil has the highest concentration of tocotrienols of all edible oils, with rice bran oil as the only other edible oil that comes close, with just under one-third of the levels contained in red palm oil. Studies on rat heart showed that the isomers of -tocotrienol were more proficient in protection against oxidative stress induced by ischaemia-reperfusion than -tocopherol (Serbinova *et al.*, 1992). The tocotrienols also exhibits powerful neuroprotective activities that are not possessed by tocopherols. Tocotrienols are able to scavenge free radicals and prevent lipid peroxidation in biological membranes, and most of these functions of tocotrienols are linked to their antioxidant activities.

All parts of the oil palm tree are useful; the wood is used as frames for buildings, and the sap is fermented into palm wine, with the oil from the mesocarp and the seeds being used for cooking and for making soaps, creams, and other cosmetics. The oil palm is a unique crop that produces two oils from its fruits – palm oil from the fruit mesocarp and palm kernel oil from the seed endosperm (kernel). The most important product from the oil palm is perhaps the yellow- to red-coloured oil known as palm oil from the mesocarp, which has a very balanced fatty acid composition (51% saturated fatty acids, 38% monounsaturated fatty acids, and 11% polyunsaturated fatty acids). The major saturated fatty acids in palm oil are 38–44% palmitic acid (C16:0) and 4–5% stearic acid (C18:0). Oleic acid (C18:1) represents 39–44% and linoleic acid (C18:2) 10–12% of the unsaturated fatty acids. Palm oil is also a rich source of carotenoids and polyphenols. Advocates of the consumption of red palm oil have suggested that the oil should be considered as a functional food to ameliorate the deleterious effects of oxidation and chronic inflammation. The protein hydrosylate from palm kernel cake yields antioxidant peptides.

9.5.2.1 Virgin Red (VR) Palm Oil

Cold extraction of palm oil offers a distinctive product, that is chemically different from industrially processed palm oil. First, the absence of heating, which is used in conventional oil production, ensures that the VR palm oil retains higher amounts of tocopherols, tocotrienols, phosphatides, carotenoids and peptides than does the industrially refined red palm oil (Nagendran *et al.*, 2000). These so-called minor components not only maintain the stability and quality of palm oil but also possess significant biological properties, including antioxidant, anti-cancer, anti-inflammatory, arthrosclerosis-controlling, and cholesterol-lowering effects (Das *et al.*, 2007; Wu *et al.*, 2008; Zou *et al.*, 2012). Secondly, during extraction from the oil palm fruits, palm oil suffers some lipase-catalyzed hydrolysis that manifests itself in a relatively high content of free fatty acids and a concomitantly high partial glyceride content. While this lipase conversion is encouraged in industrial production processes, it is greatly reduced in the cold extraction process.

9.5.3 *Hibiscus sabdariffa* (Malvaceae)

Zobo (*H. sabdariffa*) has been discussed earlier in the previous chapter as an effective agent for the treatment of hypertension. It is also used in traditional medicine in Nigeria against MetS, diabetes, obesity, and liver disorders.

9.5.4 *Gnetum africanum* (Eruh, Okazi)

Eruh is used in traditional medicine to prepare special soups for the treatment of diabetes and hypertension. It has strong antioxidant properties (Iwu, 2016). Its activity is probably due to the presence of stilbenes, such as resveratrol, the latter being originally isolated from damaged grapes. Resveratrol and related stilbenes possess the following benefits: (i) calorie restriction effects, (ii) improved human adipocyte secretion profile in obesity-induced metabolic disorders, (iii) elevated basal glycerol release, (iv) reduced intracellular triglyceride (TG) concentration, (v) increased intracellular lipolysis, (vi) down-regulated extracellular matrix proteins, (vii) up-regulated processing proteins, (viii) induced protective protein secretion against cellular stress and apoptosis-regulating proteins, (ix) up-regulated adiponectin and ApoE, (x) down-regulated PAI-1 and PEDF secretion which may improve anti-inflammatory processes, (xi) increased insulin sensitivity, (xii) inhibited cyclic adenosine monophosphate-specific phosphodiesterases, and (xiii) activated 5′-adenosine monophosphate-activated kinases (Rosenow *et al.*, 2012; Xu and Si, 2012). These physiological properties may be responsible for the positive effect of stilbene- containing herbs, such as eruh, on cardiovascular health.

9.5.5 Aju Mbaise

Aju Mbaise is a poly-herbal product that originated from the Mbaise clan of Imo State in Nigeria. It is a popular traditional remedy used originally by women for the prevention of health problems after child birth, promotion of quick-healing during maternity confinement, and treatment of post-partum belly-fat, but now it is used widely for the treatment of visceral obesity, diabetes, and other symptoms of MetS. It consists of various herbs and spices which are wrapped and preserved with plantain and/or *Syncepalum dulcificum* (Etere) leaves and folded in a manner that resembles the load-wedge (Aju) used for carrying water-pots, fire-wood or other heavy objects on the head in rural villages in Africa. Mothers, whose daughters had given birth, carry *Aju Mbaise* with them when they visit for the traditional *Omugwo* to assist the young nursing mother during her maternity confinement. During the Omugwo, special soups are prepared with these herbs and spices such as *Xylopia aethiopica* (Uda), *Piper guineensis* (black pepper), and *Tetrapleura tetraptera* (Aridan). The composition of the mixture of herbs in *Aju Mbaise* varies enormously, depending on the district where it is formulated. A typical Aju Mbaise will consist of fresh leaves of the following plants: *Sphenocentrum jollyanum*, *Cnestis ferruginea*, *Psidium guajava*, *Jatropha curcas*, *Diodia sarmentosa*, and *Heterotis rotundifolia*.

The individual plant species in the mixture have various biological properties, including antioxidant, anti-inflammatory, analgesic, antidiabetic, antihyperlipidaemic, and antihypertensive activities. Ezejindu and Iro (2017) have investigated the antimicrobial properties of a decoction prepared from Aju Mbaise and showed that it possessed significant broad-spectrum antimicrobial properties. The extract was also found to be rich in minerals, namely potassium (130.30 mg/mL), calcium (12.43 mg/mL), magnesium (10.40 mg/mL), and sodium (4.60 mg/mL), with low levels of iron (2.15 mg/mL), zinc (2.23 mg/mL), phosphorus (0.80 mg/mL), copper (0.07 mg/mL), manganese (0.05 mg/mL), and chromium (0.06 mg/mL) (Ezejindu and Iro, 2017).

9.5.6 Kayan Shayi

Kayan shayi literally means "tea ingredients" or "vegetables for making tea" among the Hausa- and/or Kanuri-speaking peoples of northern Nigeria. However, the terms denote a rich and enormously varied collection of herbs and spices used for making teas for various ailments and as health-promoting decoctions. There are teas for every possible ailment and the level of patronage of the vendors is indicative of the perception of their efficacy by the communities.

There are three basic types of *kayan* or tea: those based on aromatic, fruity and flowering leaves of the tea plant (*Camellia sinensis*), with other herbs being added for flavour and mild physiological effect; a second group of *kayan* made from a selection of ingredients to "strengthen the body", which consist mainly of ginger, cinnamon and rosella as the base ingredients; and thirdly, the therapeutic teas which are blended for specific diseases or according to individual needs. *Kayan shayi*, used for the management of MetS, includes the following: *Cassia sieberiana* (cassia), *Cinnamomum* spp. (cinnamon), *Curcuma longa* (turmeric) *Azadirachta indica* (neem), *Ficus exasperata*, *Moringa oleifera* (zogala) *Zingiber officinale* (ginger), *Commiphora africana* (African myrrh), *Citrus* spp., *Eugenia app*, *Ocimum viride*, *Capsicum* spp. (sweet pepper), *Daniellia oliveri* (copal tree), *Schwenckia americana*, and *Syzygium aromaticum* (cloves).

Some of the *kayan shayi* have also featured prominently in plants used in Hausa ethnomedicine. A review of the Hausa herbal pharmacopoeia by Nina Etkin included some the plants indicated above for the treatment of a range of diseases (Etkin, 1981). In another survey on plants used for the treatment of diabetes (one of the indicators of MetS) by Hausa-Fulani people of Sokoto State in Nigeria, it was found that, among 51 correspondents on traditional medicinal plants and practices used in diabetes across the state, 54 species, belonging to 33 families, were frequently used. Of these, *Cassia sieberiana* was the most cited (19 times) and ranked first (39%) followed by *Azadirachta indica*, *Ficus exasperata* and *Schwenckia americana* ranked equal second (15%), with each being cited eight times (Shinkafi *et al.*, 2015).

9.5.7 Pepper Soups in MetS Management

The use of spices in traditional medicine serves three purposes: they are often used as therapeutic agents for their known medicinal values, they can be administered as co-drugs to enhance the "digestion" and absorption of active constituents in other herbs, and they can be used to improve the palatability of otherwise bitter-tasting medicines. Spices are known to have beneficial effects in chronic disease and are useful in the treatment of diabetes, cancer, inflammation, hypertension, and cardiovascular diseases (Kuete, 2017). They are valuable for the management of MetS largely because of their established antioxidant and antithrombotic properties (Panickar, 2013). Some of the spices used in Africa for the treatment of MetS include cinnamon, black pepper, garlic, cloves, ginger, oregano, paprika, and turmeric (Kuete, 2017). It has been suggested that the increase in the incidence of many metabolic diseases, cancers, and chronic diseases could be due, at least in part, to the abandonment of these protective traditional remedies in favour of processed imported foods that contain little or no spice (Iwu, 2016).

One of the most common methods of using the spices, apart from application as food condiments, is their use as main ingredients in the preparation of "Pepper Soups". The soup derives its name from the simplicity of its content, which consists primarily of cayenne pepper and other spices, water and salt prepared in a light sauce with little vegetable added. Pepper soup can be eaten as an appetiser/entrée in two- or three-course meals as supper or in combination with macronutrients such as boiled potatoes, pounded yam, or steamed rice (Nwose *et al.*, 2017). Pepper soups can be prepared as special health meals, as indicated above, for nursing mothers and for other therapeutic purposes such as colds, malaria, and convalescence after prolonged illness, but the simple pepper soups sold in restaurants and bars are taken only as food and for their general health benefits. The probable mechanism of the anti-MetS action of pepper soup includes the ability of the constituents to interfere with the oxidation–inflammation cascade, especially in the stabilization of the GSH-regeneration in the physiological system by γ-glutamyl cysteine ligase and GSH-synthetase enzymes. The antidiabetic, antihypertensive and anticholesterolaemic activities of some of these spices have been discussed elsewhere in this book.

References

Aggarwal, M., Aggarwal, B. and Rao, I. (2017) Integrative medicine for cardiovascular disease and prevention. *Medical Clinics of North America*, 101, 895–923.

Alam, M.A., Subhan, N., Rahman, M.M., Uddin, S.J., Reza, H.M. and Sarker, S.D. (2014) Effect of citrus flavonoids, naringin and naringenin, on metabolic syndrome and their mechanisms of action. *Advances in Nutrition*, 5, 404–417.

Al-khalidi, B., Kimball, S.M., Kuk, J.L. and Ardern, C.I. (2018) Metabolically healthy obesity, vitamin D, and all-cause and cardiometabolic mortality risk in NHANES III. *Clinical Nutrition*, 38(2): 820–828.

Chao, P.-Y., Chiang, T.-I., Chang, I.-C., Tsai, F.-L., Lee, H.-H., Hsieh, K., Chiu, Y.W., Lai, T.J., Liu, J.Y., Hsu, L.S. and Shih, Y.-C. (2017) Amelioration of estrogen-deficiency-induced obesity by *Ocimum gratissimum*. *International Journal of Medical Sciences*, 14(9), 896–901.

Cheng, J. (2018) The convergence of two epidemics: vitamin D deficiency in obese school-aged children. *Journal of Pediatric Nursing*, 38, 20–26.

Choy, K.W., Murugana, D. and Mustafa, M.R. (2018) Natural products targeting ER stress pathway for the treatment of cardiovascular diseases. *Pharmacological Research*, 132, 119–129.

Das, S., Nesaretnam, K. and Das, D.K. (2007) Tocotrienols in cardioprotection. *Vitamins and Hormones*, 76, 419–433.

Daud, Z.A.D., Kaur, D. and Khosla, P. (2012) Health and nutritional properties of palm oil and its components. In: Lai, O.-M., Tan, C.-P. and Akoh, C. C. (Eds.) *Palm Oil* - Production, Processing, Characterization, and Uses. AOCS Press, Elsevier, Amsterdam. 18: 545–560.

Dijkstra, A.J. (2016) *Palm oil. Encyclopedia of Food and Health*, Elsevier Press, Amsterdam. pp. 199–204.

Dommermuth, R. and Ewing, K. (2018) Metabolic syndrome - systems thinking in heart disease. *Primary Care: Clinics in Office Practice*, 45, 109–129.

Edem, D.O. (2002) Palm oil: biochemical, physiological, nutritional, hematological, and toxicological aspects: a review. *Plant Foods Hum. Nutr.* 57, 319–341.

Etkin, N.L. (1981) A Hausa herbal pharmacopoeia: biomedical evaluation of commonly used plant medicines. *Journal of Ethnopharmacology*, 4, 75–98.

Ezejindu, C.N. and Iro, O.K. (2017) Antibacterial activity, phytochemical properties and mineral content of "Aju Mbaise" decoction administered to nursing mothers. *Direct Research Journal of Health and Pharmacology*, 5(3), 33–38.

Iwu, M.M. (2014) *Handbook of African medicinal plants*. 2nd ed. Boca Raton, FL: CRC Press/Taylor and Francis Group, pp. 476.

Iwu, M.M. (2016) *Food as medicine – functional food plants of Africa*. Boca Raton, FL: CRC Press/Taylor and Francis Group, pp. 384.

Katerere, D.R. and Eloff, J.N. (2006) Management of diabetes in African traditional medicine. In: A. Soumyanath (ed.). *Traditional medicine for modern times: antidiabetic plants*, Chapter 11, pp. 203–220. Boca Raton, FL: CRC Press/Taylor and Francis, pp. 314.

Kuete, V. (2017) African medicinal spices and vegetables and their potential in the management of metabolic syndrome. In: V. Kuete (ed.), *Medicinal spices and vegetables from Africa*, Chapter 12, pp. 315–326. New York: Elsevier Press, pp. 694.

Liu, S., Serdula, M., Janket, S.-J., Cook, N.R., Sesso, H.D., Willett, W.C., Manson, J.E. and Buring, J.E. (2004) A prospective study of fruit and vegetable intake and the risk of type 2 diabetes in women. *Diabetes Care*, 27, 2993–2996.

Mousa, A., Naderpoor, N., de Courten, M.P.J. and Barbora de Courten, B. (2018) Vitamin D and symptoms of depression in overweight or obese adults: a cross-sectional study and randomized placebo-controlled trial. *The Journal of Steroid Biochemistry and Molecular Biology*, 177, 200–208.

Nagendran, B., Unnithan, U.R. and Choo, Y.M. (2000) Characteristics of red palm oil, a-carotene- and vitamin E-rich refined oil for food uses. *Food and Nutrition Bulletin*, 2, 189–194.

Nwose, E.U., Bwititi, P.T. and Oguoma, V.M. (2017) Metabolic syndromes and public health policies in Africa. In: V. Kuete (eds.), *Medicinal spices and vegetables from Africa*, Chapter 4, pp. 109–130. New York: Elsevier Press, pp. 694.

Panickar, K.S. (2013) Beneficial effects of herbs, spices and medicinal plants on the metabolic syndrome, brain and cognitive function. *Central Nervous System Agents in Medicinal Chemistry*, 13(1), 13–29.

Parker, S. (1996). Absorption, metabolism and transport of carotenoids. *The FASEB Journal* 10, 542–551.

Pu, P., Gao, D.M., Mohamed, S., Chen, J., Zhang, J., Zhou, X.Y., Zhou, N.J., Xie, J. and Jiang, H. (2012) Naringin ameliorates metabolic syndrome by activating AMP-activated protein kinase in mice fed a high-fat diet. *Archives of Biochemistry and Biophysics*, 518, 61–70.

Qureshi, A.A., Qureshi, N., Wright, J.J., Shen, Z., Kramer, G., Gapor, A., Chong, Y.H., DeWitt, G., Ong, A. and Peterson, D.M. (1991) Lowering of serum cholesterol in hypercholesterolemic humans by tocotrienols (palm vitee). *The American Journal of Clinical Nutrition*, 53, 1021S–1026S.

Rosenow, A., Noben, J.-P., Jocken, J., Kallendrusch, S., Fischer-Posovszky, P., Mariman, E.C.M. and Renes, J. (2012) Resveratrol induced changes of the human adipocyte secretion profile. *Journal of Proteome Research*, 11(9), 4733–4743.

Serbinova, E., Khwaja, S., Catudioc, J., Ericson, J., Torres, Z., Gapor, A., Kagan, V. and Packer, L. (1992) Palm oil vitamin E protects against ischemia/reperfusion injury in the isolated perfused Langendor V heart. *Nutrition Research*, 12, S203–S215.

Shinkafi, T.S., Bello, L., Hassan, S.W. and Ali, S. (2015) An ethnobotanical survey of antidiabetic plants used by Hausa–Fulani tribes in Sokoto, Northwest Nigeria. *Journal of Ethnopharmacology*, 172, 91–99.

Ulasi, I.I., Ijoma, C.K. and Onodugo, O.D. (2010) A community-based study of hypertension and cardio-metabolic syndrome in semi-urban and rural communities in Nigeria. *BMC Health Services Research*, 10, 71–76.

Wu, S.J., Liu, P.L. and Ng, L.T. (2008) Tocotrienol-rich fraction of palm oil exhibits anti-inflammatory property by suppressing the expression of inflammatory mediators in human monocytic cells. *Molecular Nutrition and Food Research*, 52, 921–929.

Xu, Q. and Si, L.-Y. (2012) Resveratrol role in cardiovascular and metabolic health and potential mechanisms of action. *Nutrition Research*, 32(9), 648–658.

Zou, Y., Jiang, Y., Yang, T., Hu, P. and Xu, X. (2012) Minor constituents of palm oil: characterization, processing, and application. In: Lai, O.-M., Tan, C.-P. and Akoh, C. C. (Eds.) Palm Oil - Production, Processing, Characterization, and Uses. AOCS Press, Elsevier, Amsterdam. 16: 471–526.

10 Phytotherapy of HIV-AIDS and Opportunistic Infections With Nigerian Plants

10.1 Introduction

Acquired Immunodeficiency Syndrome (AIDS), caused by infection by the Human Immunodeficiency Virus (HIV), was discovered in 1981 and it has been the cause of more than 25 million deaths worldwide. Two major types of HIV have been identified, HIV-1 and HIV-2. HIV-1 encodes three enzymes: reverse transcriptase, protease, and integrase. Combination antiviral therapy with inhibitors of reverse transcriptase and protease activity has shown the potential therapeutic efficacy of antiviral therapy for treatment of AIDS (WHO, 2010). However, the ability of HIV to rapidly evolve drug resistance, together with toxicity problems, requires the discovery and development of new classes of anti-AIDS drugs (Matsushita, 2000; Eron, 2009; Nakanjako *et al.*, 2009).

The prevalence and incidence of HIV has declined globally over the past decade (except for parts of Eastern Europe and Central Asia, where a slight increase has been observed), due to the development of highly active antiretroviral therapy (HAART), as well as the impact of preventive measures (UNAIDS, 2016 Cohen, 2012). There is also the combination antiretroviral therapy (cART), which provides better outcomes and stable viral suppression, but it is not devoid of undesirable side-effects, especially in persons undergoing long-term treatment. With the increasing reports of the emergence of multidrug resistance, it is imperative to find new drugs and novel targets to treat infected persons and furthermore to attack HIV reservoirs in the body, for example, in brain and lymph nodes, to achieve the goal of complete eradication of HIV and AIDS (Vermani and Garg, 2002; Chinsembu and Hedimbi, 2010). In this vein, the World Health Organization (WHO) suggested the need to evaluate ethnomedicines for the management of HIV/AIDS (WHO, 1989).

The United States' Food and Drug Administration (USFDA or FDA) classifies antiretroviral drugs for HIV infection into the following categories:

(1) Multi-class Combination Products,

(2) Nucleoside Reverse Transcriptase Inhibitors (NRTIs),

(3) Nonnucleoside Reverse Transcriptase Inhibitors (NNRTIs),

(4) Protease Inhibitors (PIs),

(5) Fusion Inhibitors,

(6) Entry Inhibitors—CCR5 Co-receptor Antagonists, and

(7) HIV Integrase Strand Transfer Inhibitors.

One of the positive developments in the treatment of HIV/AIDS within the past decade is the remarkable progress made in understanding the pathogenesis and aetiology of the infection and its associated diseases. Although there is no effective and safe cure for AIDS, there are now available innovative intervention measures, which are able to ameliorate the lethality of the infection and this has given much optimism to the health professions that a cure for this dreadful disease may be in sight. A major challenge faced by clinicians in the treatment of HIV/AIDS is the incidence of latent reservoirs of HIV in certain organs of the body, which may later re-infect the patient. This clinical latency occurs during the second phase of HIV infection when the virus remains in an inactive state for an indefinite period of time. This stage is marked by a gradual loss in CD41 T-cells, and this period is mainly asymptomatic and varies from three years to 20 years (Trivedi *et al.*, 2019). This is the reason why HIV is difficult to completely eradicate, as it establishes a quiescent or latent infection within the memory CD4+ T cells, which have a stem-cell-like capacity for self-renewal. Once the HIV DNA is integrated into the host chromatin, the virus can repeatedly initiate replication as long as that cell exists.

Whereas the approved antiretroviral drugs can prevent new cells from becoming infected, they cannot eliminate infection once the viral DNA has successfully integrated into the target cell. The lymph nodes harbour the virus because of limited drug penetration and limited host clearance mechanisms, and serve as a source of virus recrudescence in individuals who stop or interrupt their therapy (Giunta, *et al.*, 2011). It has been suggested that antiretroviral therapy (ART) may be needed for several decades before the viral reservoir decays to negligible levels (Salehi *et al.*, 2018). Therefore, in order to achieve the complete eradication of HIV and AIDS, new therapeutic approaches are being subjected to serious evaluation (Lucera *et al.*, 2014). For example, as part of the so-called "shock-and-kill" strategy to purge latent reservoirs of HIV, new approaches target latently infected cells, which remain a primary barrier to eradication of HIV-1. Based on a better understanding of the molecular mechanism by which HIV latency persists has led to the discovery of a number of drugs that are able to selectively reactivate latent proviruses without inducing polyclonal T cell activation. These include histone deacetylase (HDAC) inhibitors, such as vorinostat, that are able to induce HIV transcription from latently infected cells. HDAC inhibitors have been shown to increase the susceptibility of CD4+ T cells to infection by HIV in a dose- and time-dependent manner, without enhancing viral fusion with cells, but increase reverse transcription, nuclear import, and integration, and enhance viral production in a spreading-infection assay (Lucera *et al.*, 2014). Some medicinal plants that occur in Nigeria, with reported activity in the prevention of HIV penetration of host cells, include *Ancistrocladus korupensis* (source of michellamine B), *Pelargonium sidoides* and *Scutellaria baicalensis.*

The successes of these therapies notwithstanding, prevention remains by far the best option available to reduce the mortality and the spread of HIV/AIDS. It was not surprising to observe that many of the patients who have survived the infection or are able to live with AIDS, have either used herbal medicines or complemented their antiviral therapy with plant-based nutraceuticals (Im *et al.*, 2016). It is also significant to note that most of these herbal medicinal products were derived originally from recipes used in traditional medicine.

In this Chapter, plants used in Nigeria for the treatment of HIV/AIDS, based on ethnomedical information, are reviewed. Only plants with recorded evidence of biological activity and promise of possible use in the treatment of HIV/AIDS have been discussed. The objective is to summarise available information for an evidence-based medical approach to the ethnomedical use of herbal medicines for the management of HIV/AIDS (Iwu *et al.*, 2001). Evidence-based medicine has been defined as the conscientious, explicit, and judicious use of the available evidence from health-care research in the management of individual patients (Sackett *et al.*, 1996). These compounds inhibit different stages in the replication cycle of HIV, and their use, sometimes as purified mixtures, enables the drug to hit various targets of the virus in a synergistic manner. Recent evidence has demonstrated that, in treating AIDS, treatment regimens containing multiple drugs can usually amplify the therapeutic efficacies of each agent, leading to maximal therapeutic efficacy with minimal adverse effects. Therefore, our focus will be on combination therapies that comprise more than one active ingredient.

10.2 Inhibitors of HIV Reverse Transcriptase (RTase)

Many natural products have been shown to possess significant activity, at least in *in vitro* studies, at concentrations that are comparable to the synthetic inhibitors of RTases (Loya *et al.*, 1999; Jung *et al.*, 2000),. The compounds vary enormously in their chemical structures, and include alkaloids, flavonoids, lignans, coumarins, naphthoquinones, anthraquinones, polysaccharides and terpenes. The most important examples include the novel naphthoquinones michellamines A–C isolated from the rare tropical plant *A. korupensis*, calanolide A isolated from *Calophyllum lanigerum*, betulinic acid from *Syzygium claviflorum*, the dibenzylbutadiene lignans from *Anogeissus acuminata,* the sulfated polysaccharides found in many seaweeds, and putranjivain A from *Embelica officinalis* (*Phyllanthus embelica*). The protostanes garcisaterpene A and C can suppress HIV-1 RTase activity and are obtained from the ethyl acetate extract of the stems and bark of *Garcinia speciosa* (Rukachaisirikul *et al.*, 2003). Repandusinic acid has shown inhibition of HIV-1 RTase and is extracted from *Phyllanthus niruri* of the Euphorbiaceae family (Ogata *et al.*, 1992). Some of these antiretroviral natural products are found in Nigerian plants that are used in traditional medicine for the treatment of HIV/AIDS (Kapil and Sharma 1997),

The major Nigerian plants used in the preparation of ethnomedical remedies with reported antiretroviral activity include:

Abrus precatorius
Acacia nilotica
Acalypha macrostachya
Acanthospermum hispidum
Achyranthes aspera
Adansonia digitata
Alchornea cordifolia
Allium sativum
Aloe vera
Alstonia boonei
Ancistrocladus korupensis
Andrographis paniculata
Annona senegalensis
Annona squamosa
Anogeissus acuminata
Argemone mexicana
Artocarpus heterophyllus
Artemisia afra
Aspilia africana
Balanites aegyptiaca
Bidens pilosa
Boerhavia diffusa
Boerhavia erecta
Boscia senegalensis
Bridelia ferruginea
Bridelia micrantha
Cajanus cajan
Calophyllum inophyllum
Capparis decidua
Carissa bispinosa
Cassia fistula
Catharanthus roseus
Cissus quadrangularis
Citrullus colocynthis
Combretum micranthum
Combretum molle
Curcuma longa
Cynanchum paniculatum
Cyperus rotundus
Detarium microcarpum
Dioscorea bulbifera
Dioscorea dumentorum
Dioscorea hispida
Dracaena mannii
Epilobium angustifolium
Erythroxylum citrifolium
Eucalyptus citriodora
Eupatorium odorantum
Euphorbia hirta
Ficus racemosa
Ocimum basilicum
Ocimum gratissimum
Phaseolus vulgaris
Phyllanthus amarus
Phyllanthus niruri
Piper guineensis
Piper longum
Portulaca oleracea
Prunus africana
Psidium guajava
Ricinus communis
Salvadora persica
Sarcocephalus latifolius
Scoparia dulcis
Sida acuta
Sida cordata

Foeniculum vulgare
Garcinia kola
Garcinia mangostana
Garcinia speciose
Gymnema sylvestre
Hibiscus sabdariffa
Irvingia gabonensis
Jatropha curcas
Kigelia africana
Khaya senegalensis
Maytenus senegalensis
Melia azedarach
Momordica charantia
Monotes africana
Morinda citrifolia
Morinda lucida
Moringa oleifera
Myristica fragrans
Nigella sativa
Solanum incanum
Spathodea campanulata
Spondias mombin
Spondias pinnata
Syzygium aromaticum

Tabernaemontana elegans
Terminalia ivorensis
Terminalia sericea
Tetrapleura tetraptera
Tinospora cordifolia
Tribulus terrestris
Trichilia emetic
Tulbaghia alliacea
Ximenia caffra
Xylopia aethiopica
Xylopia frutescens
Zanthoxylum zanthoxyloides

10.3 Non-specific Antiviral and Immuno-Modulatory Agents

The exact mode of action of several plant-derived antiviral agents has not yet been determined but some of them have been shown to exhibit significant activity against HIV, a virus associated with AIDS (Matthee, *et al.*, 1999). This category of antiviral agents was also considered important since they may provide additional insights into the possible biochemical mechanisms of the treatment of AIDS. These compounds either interfere directly at various stages in the replication cycle of HIV (Vlietinck *et al.*, 1998) or strengthen the patients' immune system against the devastating effects of the infection (Martins *et al.*, 2002). Several plants of Nigerian origin contain compounds that have been shown both in laboratory and clinical outcome reports to have anti-HIV properties and/or are useful in ameliorating the effects of AIDS (Iwu, 1993; Iwu and Gbodossou 2000; Iwu *et al.*, 2001). Some of the traditional remedies used for the treatment of HIV/AIDS do not necessarily fit into the classical categories of antiviral chemotherapeutic agents. Given the hybrid spreading mechanisms of the virus, the clinical benefits of these drugs can only be realized from information obtained from their outcomes in human use.

Many people with HIV take herbs to support the immune system and to help it repair the damage caused by the virus (Wagner *et al.*, 1990; Mitra *et al.*, 1999; Yokozawa, *et al.*, 1999). This is one of the most important uses for herbs, but it is also an area in which it may be difficult to find enough information to make informed choices. We know that the immune system works as a result of incredibly complicated interactions between immune cells and the proteins which they use to communicate with each other. It is often difficult to predict how drugs or herbs that target one part of the immune system will impact on another part (Liu *et al.* 2015). The cell-mediated immune system includes specialized immune cells, such as CD4+ cells, CD8+ cells, and natural killer cells that work together with the immune proteins interleukin-2 (IL-2), interferon-gamma (IFN-gamma), tumour necrosis factor alpha (TNF-alpha), and many other proteins. Herbal therapies that may be useful for HIV-positive people usually enhance cell-mediated immunity (Sheng, *et al.*, 2000; Pise *et al.*, 2015). Although we say that AIDS is an immune deficiency syndrome, parts of the immune system of an HIV-positive person work very hard and may already be overstimulated by the demands of HIV infection. Some immune stimulants (or immune boosters) may actually worsen the health of HIV-positive people by stimulating the wrong parts of the immune system or by increasing the burden on the system. Immune therapies are often taken in cycles (a few days or weeks on followed by a few days or weeks off) to prevent the system from adapting to the treatment in such a way that the treatment's effects are weakened. This point is important to consider when choosing herbal therapies for immune support.

10.3.1 *Moringa oleifera*

The leaves and other parts of *M. oleifera* are used in traditional medicine for the treatment of various diseases, including diabetes, cancer, hypertension, and inflammatory disorders (Popoola and Obembe, 2013; Leone *et al.*, 2015). *Moringa* species have been studied for their antioxidant, anti-inflammatory, anticancer, antihypertensive and antihyperglycaemic activities (Abd-Rani *et al.*, 2018). The antiviral activity of *Moringa* has been extensively documented (Abd-Rani *et al.*, 2018) but the plant is used in folk medicine as a nutritional supplement to achieve effective antiretroviral therapy (ART) outcomes or as an adjunct to prescription drugs by people living with human immunodeficiency virus (PLHIV) treated with ART (Sebit *et al.*, 2000). It has been established that adherence to an antiretroviral regimen and the establishment of a good immunometabolic response are essential for a good clinical outcome in HIV/AIDS treatment (Volberding, 2000). Leaf powder of *Moringa* is recommended to combat marginal and major nutritional deficiencies. A study in DR Congo has shown that, after six months of supplementation with *Moringa* leaf powder, patients undergoing HIV treatment exhibited a significantly greater increase in BMI and albumin levels than those who did not receive the supplement (Tshingani *et al.*, 2017). In another study, it was found that co-administration of *M. oleifera* leaf powder at the traditional dose did not significantly alter the steady-state pharmacokinetics (PK) of Nevirapine in HIV-infected people on antiretroviral therapy (Monera-Penduka *et al.*, 2017). Although *Moringa* is known to inhibit cytochrome P450 3A4, 1A2 and 2D6 activities *in vitro*, this property did not affect the PK of antiretroviral drugs metabolised via the same metabolic pathways (Awodele *et al.*, 2015). The plant has also been suggested as a potential agent for the treatment of HIV-Associated Neurocognitive Disorders (HAND) (Kurapati *et al.*, 2016).

Moringa leaves contain amino acids, including aspartic acid, glutamic acid, serine, glycine, threonine, α-alanine, valine, leucine, isoleucine, histidine, lysine, cysteine, methionine, arginine, and tryptophan, as well as high concentrations of vitamin C, polyphenols, and alkaloids (Iwu, 2016). The flowers and the fruits also contain amino acids. The seeds generate an almost colourless fixed oil, known in commerce as *Beni* or *Moringa* oil. The oil consists of a 60% liquid olein fraction and 40% solid fat. The major constituents of the oil are oleic acid (65%), stearic acid (10.8%), behenic acid (8.9%), myristic acid (7.3%), palmitic acid (4.2%), and lignoacetic acid (3.0%) (Rao *et al.*, 1949). *Moringa* should be regarded as consisting of three distinct herbs: the highly nutritional leaves, the seeds and the whole aerial parts (Iwu, 2016).

Moringa extracts are generally non-toxic, even at high doses. In a toxicity study, aqueous extract of *M. oleifera* leaves did not cause any mortality in Wistar albino mice at orally administered doses of up to 6,400 mg/kg live bodyweight (Awodele *et al.*, 2012). The LD_{50}-value of the *Moringa* leaf aqueous extract was determined to be 1,585 mg/kg intraperitoneal.

10.3.2 *Nigella sativa*

Black seed oil has been used in Nigerian traditional medicine for the treatment of various diseases, including HIV infection and related opportunistic AIDS infections. There is a reported case study on the complete recovery and sero-reversion of an adult HIV patient after treatment with *N. sativa* seed concoction for a period of six months (Onifade *et al.*, 2011; Onifade *et al.*, 2013a). The patient presented to the herbal therapist with a history of chronic fever, diarrhoea, weight loss and multiple papular pruritic lesions of three months duration. Examination revealed moderate weight loss, and the laboratory tests of ELISA (Genscreen) and western blot (new blot 1 & 2) confirmed sero-positivity to HIV infection with pre-treatment viral (HIV-RNA) load and CD4 count of 27,000 copies/mL and CD4 count of 250 cells/mm^3, respectively. The patient was treated with a *N. sativa* concoction (10 mL) twice daily for six months. He was contacted daily to monitor side-effects and drug efficacy. Fever, diarrhoea and multiple pruritic lesions disappeared on days 5, 7 and 20, respectively, of black seed therapy. The investigators reported that CD4 count decreased to 160 cells/mm^3 despite significant reductions in viral load (≤1000 copies/mL) on day 30 of *N. sativa* treatment. Repeated enzyme immunoassay (EIA) and western blot tests on day 187 of *N. sativa* therapy were seronegative. The post-therapy CD4 count was found to be 650 cells/mm^3 with undetectable viral (HIV-RNA) load. Several repeats of the HIV tests remained seronegative and aviraemic, with a normal CD4 count during the 24 months without herbal therapy (Onifade *et al.*, 2013a).

In a related study by the same investigators (Onifade *et al.*, 2013b), six patients taking a herbal concoction containing black seed (called α-Zam) as an alternative therapy for HIV infection were recruited into the study and monitored for four months. All six patients were infected with HIV, as confirmed by western blot analysis in either of two Nigerian teaching hospitals (Ladoke Akintola University of Technology (LAUTECH) Teaching Hospital or Ahmadu Bello University Teaching Hospital) before commencing preliminary clinical and laboratory examinations using World Health Organization (WHO) and US Centre for Disease Control and Prevention (CDC) criteria. The patients were monitored after commencement of herbal medications to assess the efficacy (disappearance of presenting signs and symptoms associated with HIV infection), side-effects, drug toxicity, and compliance. The symptoms and signs associated with HIV infection disappeared within 20 days of commencement of herbal therapy, with significant differences in a number of parameters ($P<0.05$) between the values before treatment and at periodic intervals after herbal therapy commenced. Body weight increased from a mean ± standard error (SE) of 53 ± 2 kg to 63 ± 2 kg, viral load (HIV-RNA) decreased from 42,300 ± 1500 copies/mL to an undetectable level (≤50 copies/mL), and CD4+ count increased from an average of 227 ± 9 to 680 ± 12 mm^3/μL at four months post-therapy. The study confirmed the reported case study that a herbal remedy containing black seed was effective in the treatment of HIV infection, based on a significant improvement in both the clinical features and laboratory results of HIV infection (Onifade *et al.*, 2013b).

10.3.3 *Momordica charantia*

The bitter melon *M. charantia* and the related species *Momordica balsamina* are used in Nigerian traditional medicine for the treatment of a variety of diseases, including diabetes, fevers, and viral infections. It has been shown that extracts of both species possess antiHIV activities (Jiratchariyakul *et al.*, 2001; Bot *et al.*, 2007). The effects and molecular mechanisms of bitter melon-induced antidiabetic, antiHIV, and antitumour activities associated with over twenty active components have been reported (Jiratchariyakul *et al.*, 2001). The plant was found to inhibit HIV-1 reverse transcriptase, an effect attributable to a protein constituent, MRK29 (Jiratchariyakul *et al.*, 2001). The fruit pulp extract of *M. balsamina*, commonly used in the northern part of Nigeria for its antiviral effect on human and animal health, showed potent inhibitory activity against HIV-1 replication (Bot *et al.*, 2007).

10.3.4 *Garcinia kola*

Seeds of bitter kola are used by many HIV patients in self-medication for treatment of HIV/AIDS. It is also believed to enhance the efficacy of some of the RTase inhibitors by its antioxidant effects. Kolaviron, a mixture of C-3/C-8"-linked biflavanones, GB1, GB2, kolaflavanone, and the rare benzophenone, kolanone, found in the seeds of this West African tree, *G. kola*, was found to exhibit dose-dependent activity against certain viruses, including HIV and the Ebola virus. This food plant, used in African folk medicine as a general antidote, is an ingredient in commercial herbal formulations as an "immune tonic". Preliminary studies, using the Luminetics™ assay for T cell activation, indicated that kolaviron and one of its major components, GB-1, also showed immune-potentiating properties in whole blood cultures, from normal and HIV-infected patients, concomitant with the addition of mitogens or recall antigens. Kolaviron alone was not immunostimulatory and ATP responses were dose dependent. The compound showed no toxic effects on cells, making it a potential candidate for development as a drug for the control of HIV replication and immune reconstitution.

Nevirapine (NVP), a non-nucleoside reverse transcriptase inhibitor used in the treatment of HIV infections, has been reported to be toxic to the male reproductive system. Kolaviron has been shown to ameliorate NVP-induced testicular toxicity (Adaramoye *et al.*, 2013). It has also been reported that administration of kolaviron for 56 days attenuated the nephrotoxicity caused by nevirapine therapy (Offor *et al.*, 2017). An isoprenylated xanthone derivative, identified as 1,4,6-trihydroxy-3-methoxy-2-(3-methyl-2-butenyl)-5-(1,1-dimethyl-prop-2-enyl)xanthone, isolated from the ethanolic extract of the root bark of *Garcinia edulis*, exhibited antiHIV-1 protease (PR) activity, with an IC_{50} value of 11.3 μg/mL *in vitro*, whereas an antiHIV-1 PR activity of IC_{50} value of 2.2 μg/mL was reported for acetyl-pepstatin, which was used as a positive control (Magadula, 2010). The implication of this finding is that it is only kolaviron that may be responsible for the HIV activity of Garcinia species as isoprenylated xanthones occur in the seeds of *G. kola*.

10.3.5 *Kigelia africana*

K. africana (sausage tree) is used in different parts of Nigeria as a herbal remedy for the treatment of numerous diseases, including rheumatism, psoriasis, diarrhoea, and stomach ailments. It is also used in traditional medicine for wound healing, as an aphrodisiac, and for skin diseases. Laboratory studies have confirmed the anti-inflammatory, analgesic, antioxidant and anticancer activity of the extract of different parts of the plant. Over 150 bioactive compounds have been isolated from different part of the plant, including iridoids, naphthoquinones, flavonoids, terpenes, and phenylethanoid glycosides. They have been shown to have antioxidant, antimicrobial, and anticancer properties (Bello *et al.*, 2016).

10.3.6 *Terminalia sericea*

Terminalia species are multipurpose medicinal plants used mostly in the treatment of diarrhoea, sexually transmitted infections, skin rashes, tuberculosis, and other infections (Mongalo *et al.*, 2016). Three members of the genus, *Terminalia catappa*, *Terminalia ivorensis* and *T. sericea*, are employed as ingredients for the preparation of remedies for the treatment of AIDS and associated opportunistic infections. The antiHIV activity of the *Terminalia* species have been attributed to the presence of different types of phytochemicals in the plants (Dwevedi *et al.*, 2016). Studies on the antiHIV activity of *T. sericea* have shown that aqueous and methanolic extracts of the leaves exhibited inhibition of HIV-1 RTase RNA-dependent DNA polymerase (RDDP) function by 74.2% and 98.0%, respectively, at a concentration of 100 μg/mL, with IC_{50} values of 24.1 and 7.2 μg/mL, respectively (Bessong *et al.*, 2004). Furthermore, aqueous and methanolic leaf extracts inhibited the ribonuclease H (RNase) function of HIV-1 RTase, with IC_{50} values of 18.5 and 8.1 μg/mL, respectively, which suggests that the extracts inhibit both the RDDP and RNase H functions of HIV-1 RTase in a similar manner. Other investigators reported similar HIV-inhibitory activities of various extracts of the plant (Tshikalange, 2008; Klos *et al.*, 2009; Rege *et al.*, 2010), with the ethanolic extract of the root being the most potent, having an IC_{50} of 0.0006 μg/mL in the MAGI cell assay (Mongalo *et al.*, 2016).

References

Abd Rani, N.Z., Husain, K., and Kumolosasi, E. (2018) Moringa genus: a review of phytochemistry and pharmacology. *Frontiers in Pharmacology*, 9, 108.

Adaramoye, O.A., Akanni, O.O. and Farombi, E.O. (2013) Nevirapine induces testicular toxicity in wistar rats: reversal effect of kolaviron (biflavonoid from *Garcinia kola* seeds). *Journal of Basic and Clinical Physiology and Pharmacology*, 24(4), 313–320.

Apaydin, E.A., Maher, A.R., Shanman, R., Booth, M.S., Miles, J.N., Sorbero, M.E. and Hempel, S. (2016) A systematic review of St. John's wort for major depressive disorder. *Systematic Review*, 5(1), 148.

Awodele, O., Oreagba, I.A., Odoma, S., Texeira-Da-Silva, J.A. and Osunkalu, V.O. (2012) Toxicological evaluation of the aqueous leaf extract of *Moringa oleifera* Lam. *Journal of Ethnopharmacology*, 139, 330–336.

Awodele, O., Popoola, T., Rotimi, K., Ikumawoyi, V. and Okunowo, W. (2015) Antioxidant modulation of nevirapine induced hepatotoxicity in rats. *Interdisciplinary Toxicology*, 8(1), 8–14.

Bag, A. and Chattopadhyay, R.R. (2015) Evaluation of synergistic antibacterial and antioxidant efficacy of essential oils of spices and herbs in combination. *PLoS One*, 10(7), e0131321.

Bello, I., Shehu, M.W., Musa, M., Asmawi, M.Z. and Mahmud, R. (2016) *Kigelia africana* (Lam.) Benth. (sausage tree): phytochemistry and pharmacological review of a quintessential African traditional medicinal plant. *Journal of Ethnopharmacology*, 189, 253–276.

Bessong, P.O., Obi, C.L., Igumbor, E., Andreola, M.L. and Litvak, S. (2004) In vitro activity of three selected South African medicinal plants against immunodeficiency virus type 1 reverse transcriptase. *African Journal of Biotechnology*, 3, 555–559.

Bot, Y., Mgbojikwe, L., Nwosu, C., Abimiku, A., Dadik, J. and Damshak, D. (2007) Screening of the fruit pulp extract of *Momordica balsamina* for anti HIV property. *African Journal of Biotechnology*, 6, 47–52.

Chinsembu, K.C. and Hedimbi, M. (2010) An ethnobotanical survey of plants used to manage HIV/AIDS opportunistic infections in Katima Mulilo, Caprivi region, Namibia. *Journal of Ethnobiology and Ethnomedicine*, 6, 25.

Cohen, J. (2012) AIDS research. FDA panel recommends anti-HIV drug for prevention. *Science*, 336(6083), 792.14.

Dwevedi, A., Dwivedi, R. and Sharma, Y.K. (2016) Exploration of phytochemicals found in *Terminalia* sp. and their antiretroviral activities. *Pharmacognosy Reviews*, 10(20), 73–83.

Eron, J.J. Jr. (2009) Antiretroviral therapy: new drugs, formulations, ideas, and strategies. *Topics in HIV Medicine*, 17(5), 146–150.

Giunta, B., Ehrhart, J., Obregon, D.F., Lam, L., Le, L., Jin, J., Fernandez, F., Tan, J. and Shytle, R.D. (2011) Antiretroviral medications disrupt microglial phagocytosis of β-amyloid and increase its production by neurons: implications for HIV-associated neurocognitive disorders. *Molecular Brain*, 4(1), 23.

Green, M.H. (1988) Method of treating viral infections with amino acid analogs. *United States Patent 5,110,600*. Filed January 25, 1988

Im, K., Kim, J. and Min, H. (2016) Ginseng, the natural effectual antiviral: protective effects of Korean red ginseng against viral infection. *Journal of Ginseng Research*, 40(4), 309–314.

Iwu, M. (1993) *Handbook of African medicinal plants*. Boca Raton, FL: CRC Press, pp. 434.

Iwu, M.M. (2016) Food as medicine – functional food plants of Africa. Boca Raton, FL: CRC Press/Taylor and Francis Group, pp. 384.

Iwu, M.M. and Gbodossou, E. (2000) The role of traditional medicine. *The Lancet*, 356(1), S3.

Iwu, M.M., Okunji, C.O., Tchimene, M.K., Sokomba, E. *et al.* (2001) *Report of the international conference on traditional medicine in HIV/AIDS and malaria.* InterCEDD, BDCP Press, Silver Spring, MD.

Jiratchariyakul, W., Wiwat, C., Vongsakul, M., Somanabandhu, A., Leelamanit, W., Fujii, I., Suwannaroj, N. and Ebizuka, Y. (2001) HIV inhibitor from Thai bitter gourd. *Planta Medica*, 67, 350–353.

Jung, M., Lee, S., Kim, H. and Kim, H. (2000) Recent studies on natural products as anti-HIV agents. *Current Medicinal Chemistry*, 7, 649–661.

Kapil, S. and Sharma, S. (1997) Immunopotentiating compounds from *Tinospora cordifolia. Journal of Ethnopharmacology*, 58(2), 89–95.

Klos, M., Van de Venter, M., Milne, P.J., Traore, H.N., Meyer, D. and Oosthuizen, V. (2009) In vitro anti-HIV activity of five selected South African medicinal plant extracts. *Journal of Ethnopharmacology*, 124, 182–188.

Kurapati, K.R, Atluri, V.S., Samikkannu, T., Garcia, G. and Nair, M.P. (2016) Natural products as anti-HIV agents and role in HIV-associated neurocognitive disorders (HAND): a brief overview. *Frontiers in Microbiology*, 6, 1444.

Leone, A., Spada, A., Battezzati, A., Schiraldi, A., Aristil, J. and Bertoli, S. (2015) Cultivation, genetic, ethnopharmacology, phytochemistry and pharmacology of *Moringa oleifera* leaves: an overview. *International Journal of Molecular Sciences*, 16(6), 12791–12835.

Liu, Z.B., Yang, J.P. and Xu, L.R. (2015) Effectiveness and safety of traditional Chinese medicine in treating acquired immune deficiency syndrome: 2004–2014. *Infectious Diseases of Poverty*, 4, 59.

Loya, S., Rudi, A., Kashman, Y. and Hizi, A. (1999) Polycitone A, a novel and potent general inhibitor of retroviral reverse transcriptases and cellular DNA polymerases. *Biochemical Journal*, 344(Pt 1), 85–92.

Lucera, M.B., Tilton, C.A., Mao, H., Dobrowolski, C., Tabler, C.O., Haqqani, A.A., Karn, J. and Tilton, J.C. (2014) The histone deacetylase inhibitor vorinostat (SAHA) increases the susceptibility of uninfected CD4+ T cells to HIV by increasing the kinetics and efficiency of post entry viral events. *Journal of Virology*, 88, 10803–10812.

Magadula, J.J. (2010) A bioactive isoprenylated xanthone and other constituents of *Garcinia edulis. Fitoterapia*, 81, 420–423.

Martins, R.S., Péreira, E.S. Jr., Lima, S.M., Senna, M.I., Mesquita, R.A. and Santos, V.R. (2002) Effect of commercial ethanol propolis extract on the in vitro growth of *Candida albicans* collected from HIV-seropositive and HIV-seronegative Brazilian patients with oral candidiasis. *Journal of Oral Science*, 44(1), 41–48.

Matsushita, S. (2000) Current status and future issues in the treatment of HIV-1 infection. *International Journal of Hematology*, 72(1), 20–27.

Matthee, C., Wright, A.D. and Konig, G.M. (1999) HIV reverse transcriptase inhibitors of natural origin. *Planta Medica*, 65, 493–506.

Mitra, S.K., Gupta, M., Suryanarayana, T. and Sarma, D.N.K. (1999) Immunoprotective effect of IM-133. *International Journal of Immunophamacology*, 21, 115–120.

Monera-Penduka, T.G., Maponga, C.C., Wolfe, A.R., Wiesner, L., Morse, G.D. and Nhachi, C.F. (2017) Effect of *Moringa oleifera* Lam. leaf powder on the pharmacokinetics of nevirapine in HIV-infected adults: a one sequence cross-over study. *AIDS Research and Therapy*, 14, 12.

Mongalo, N.I., McGaw, L.J., Segapelo, T.V., Finnie, J.F. and Van Staden, J. (2016) Ethnobotany, phytochemistry, toxicology and pharmacological properties of *Terminalia sericea* Burch. ex DC. (Combretaceae) – a review. *Journal of Ethnopharmacology*, 194, 789–780.

Nakanjako, D., Colebunders, R., Coutinho, A.G. and Kamya, M.R. (2009) Strategies to optimize HIV treatment outcomes in resource-limited settings. *AIDS Reviews*, 11(4), 179–189.

Offor, U., Ajayi, S.A., Jegede, A., Kharwa, S., Naidu, E.C. and Azu, O.O. (2017) Renal histoarchitectural changes in nevirapine therapy: possible role of kolaviron and vitamin C in an experimental animal model. *African Health Science*, 17(1), 164–174.

Ogata, T., Higuchi, H., Mochida, S., Matsumoto, H., Kato, A., Endo, T., Kaji, A. and Kaji, H. (1992) HIV-1 reverse transcriptase inhibitor from *Phyllanthus niruri*. *AIDS Research and Human Retroviruses*, 8(11), 1937–1944.

Onifade, A.A., Jewel, A.P. and Okesina, A.B. (2011) Virologic and immunologic outcome of treatment of HIV infection with a herbal concoction, A-ZAM, among clients seeking herbal remedy in Nigeria. *African Journal of Traditional, Complementary, and Alternative Medicines*, 8(1), 37–44.

Onifade, A.A., Jewell, A.P. and Adedeji, W.A. (2013a) *Nigella sativa* concoction induced sustained seroreversion in HIV patient. *African Journal of Traditional, Complementary, and Alternative Medicines*, 10(5), 332–335.

Onifade, A.A., Jewell, A.P., Ajadi, T.A., Rahamon, S.K. and Ogunrin, O.O. (2013b) Effectiveness of a herbal remedy in six HIV patients in Nigeria. *Journal of Herbal Medicine*, 3(3), 99–103.

Pise, M.V., Rudra, J.A. and Upadhyay, A. (2015) Immunomodulatory potential of shatavarins produced from *Asparagus racemosus* tissue cultures. *Journal of Natural Science, Biology and Medicine*, 6(2), 415–420.

Popoola, J.O. and Obembe, O.O. (2013) Local knowledge, use pattern and geographical distribution of *Moringa oleifera* Lam. (Moringaceae) in Nigeria. *Journal of Ethnopharmacology*, 150(2), 682–691.

Rao, R., Raghunanda, I. and Geroge, M. (1949) Moringa seeds. *Indian Journal of Medical Research*, 49, 159.

Rege, A.A., Ambaye, R.Y. and Deshmuck, A.J. (2010) In vitro testing of anti-HIV activity of some medicinal plants. *Indian Journal of Natural Products and Resources*, 1, 193–199.

Rukachaisirikul, V., Pailee, P., Hiranrat, A., Tuchinda, P., Yoosook, C., J. Kasisit, J., Taylor, W.C. and Reutrakul, V. (2003) Anti-HIV-1 protostane triterpenes and digeranylbenzophenone from trunk bark and stems of *Garcinia speciose*. *Planta Med.* 69, 1141–1146.

Sackett, D. L., Rosenberg, W. M., Gray, J. A., Haynes, R. B. and Richardson, W. S. (1996). Evidence based medicine: what it is and what it isn't. *BMJ*. 13, 312(7023), 71–72.

Salehi, B., Kumar, N.V.A., Şener, B., Sharifi-Rad, M., Kılıç, M., Mahady, G.B., Vlaisavljevic, S., Iriti, M., Kobarfard, F., Setzer, W.N., Ayatollahi, S.A., Ata, A. and Sharifi-Rad, J. (2018) Medicinal plants used in the treatment of human immunodeficiency virus. *International Journal of Molecular Sciences*, 19(5), 1459.60.

Sebit, M.B., Chandiwana, S.K., Latif, A.S., Gomo, E., Acuda, S.W., Makoni, F. and Vushe, J. (2000) Quality of life evaluation in patients with HIV-I infection: the impact of traditional medicine in Zimbabwe. *Central African Journal of Medicine*, 46, 208–213.

Sheng, Y., Bryngelsson, C. and Pero, R.W. (2000) Enhanced DNA repair, immune function and reduced toxicity of C-MED-100, a novel aqueous extract from *Uncaria tomentosa*. *Journal of Ethnopharmacology*, 69(2), 115–126.

Trivedi, J., Tripathi, A., Chattopadhyay, D. and Mitra, D. (2019) Plant-derived molecules in managing HIV infection. In: Khan, M.S. A., Ahmad, I. and Chattopadhyay (Eds.) *New look to phytomedicine*, Academic Press. Elsevier Amsterdam. Chapter 11, pp. 273–298.

Tshikalange, T.E. (2008) *In vitro anti-HIV-1 properties of ethnobotanically selected South African plants used in the treatment of sexually transmitted diseases. Ph.D.* thesis. University of Pretoria, South Africa. Cited in Mongalo *et al.*, 2016 – op cit.

Tshingani, K., Donnen, P., Mukumbi, H., Duez, P. and Dramaix-Wilmet, M. (2017) Impact of *Moringa oleifera* lam. leaf powder supplementation versus nutritional counseling on the body mass index and immune response of HIV patients on antiretroviral therapy: a single-blind randomized control trial. *BMC Complementary and Alternative Medicine*, 17(1), 420.

UNAIDS (2016). Global AIDS update 2016. World Heal. Organ. 422, https://doi.org/ ISBN 978-92-9253-062-5.

Vermani, K. and Garg, S. (2002) Herbal medicines for sexually transmitted diseases and AIDS. *Journal of Ethnopharmacology*, 80(1), 49–66.

Vlietinck, A.J., De Bruyne, T., Apers, S. and Pieters, L.A. (1998) Plant-derived leading compounds for chemotherapy of human immunodeficiency virus (HIV) infection. *Planta Medica*, 64, 97–109.

Volberding, P. (2000). Consensus statement: anemia in HIV infection – current trends, treatment options, and practice strategies. *Clinical Therapeutics*, 22(9), 1004–1020.

Wagner, H. (1990) Search for plant derived natural compounds with immunostimulatory activity. *Pure and Applied Chemistry*, 62, 1217–1222.

WHO (1989) *Report of a WHO Informal Consultation on Traditional Medicine and AIDS: In Vitro Screening for Anti-HIV Activity.* World Health Organization. Geneva. Switzerland. 18 pages.

WHO. (2010) *Joint WHO/ILO policy guidelines on improving health worker access to prevention, treatment and care services for HIV and TB.* Geneva: World Health Organization. April 28.

Yokozawa, T., Wang, T.S., Chen, C.P. and Hattori, M. (1999) *Tinospora tuberculata* suppresses nitric oxide synthesis in mouse macrophages. *Biological and Pharmaceutical Bulletin*, 22(12), 1306.

11 Application of Nigerian Plants in Cancer Treatment

The word 'cancer' is used as a generic term to describe a large group of diseases characterised by the growth of abnormal cells beyond their usual boundaries, that can then invade adjoining parts of the body and/or spread to other organs. Other terms commonly used are malignant tumours and neoplasms. Cancer can affect almost any part of the body and has many anatomic and molecular subtypes that each require specific management strategies (WHO, 2018). Cancer is the fastest-growing disease worldwide. According to the World Health Organization, cancer is one of the leading causes of morbidity and mortality worldwide, with approximately 14 million new cases annually and 9.6 million cancer-related deaths estimated in 2018. More than 60% of the world's total new annual cases of cancer occur in Africa, Asia and Central and South America. These regions account for 70% of the world's cancer deaths. In 2017, estimates of 1,688,780 new cancer cases and about 600,920 cancer deaths occurred in the United States (Siegel *et al.*, 2017). For all disease sites combined, the cancer incidence rate is 20% higher in men than in women, whereas the cancer death rate is 40% higher. However, sex disparities vary by cancer type. There are different types of cancer, and cancer can occur in any part of the body, including lymphatic, digestive, urinary, and reproductive systems, skin and blood.

Cancer occurs as the outcome of the carcinogenesis process that involves three distinguishable but closely connected stages: a) initiation stage, which is the outcome of a rather rapid and irreversible outcome of assault on a normal cell, which leads to a transformed or initiated cell; b) promotion stage, in which the initiated cell changes to a preneoplastic cell; and c) the progression stage during which the preneoplastic cell becomes a neoplastic cell. Cancer, therefore, arises from one single cell. Carcinogenesis is a complex process that is not well understood, involving the division and differentiation of cells during the cell cycle, and may also involve exogenous carcinogens. The crucial initiation stage may be due to the initial uptake of a carcinogen and the subsequent stable genotoxic damage caused by its metabolic activation. Other causes of cancer initiation include oxidative stress, chronic inflammation, and hormonal imbalance. After the initiation, promotion and progression stages of cancer, the greater tissue and organ damage occur in the metastasis phase, during which the cancerous cells invade further, otherwise healthy cells, tissues and organs with the help of adhesion molecules and angiogenic factors.

Available evidence from epidemiological studies and cohorts indicate that between 30% and 50% of cancer deaths could be prevented by modifying or avoiding key risk factors, including avoiding tobacco products, reducing alcohol consumption, maintaining a healthy body weight, exercising regularly and addressing infection-related risk factors (WHO, 2018).

In Nigeria, cancer as a disease is a recent introduction to the health lexicon of traditional healers, although they have specific therapies for breast, prostate, and skin cancers, based on the similarity of the symptoms in both vernacular disease classification and modern definitions of cancers. Traditional treatment of other types of cancer is based on the outcome of ethnomedical evaluations and botanical surveys. Accordingly, this chapter discusses plants used both by Nigerian traditional healers and those identified through ethnomedical field studies.

There are also plants that have yielded isolated bioactive molecules that have shown promise as possible anticancer drugs, that will also be discussed in this chapter, but the objective is not to present a comprehensive report of Nigerian plants with possible anticancer activity; the intent is to discuss the major herbs that are in current use, based on ethnomedical information. Many therapeutic agents, currently in clinical use for cancer treatment and chemoprevention of the initiation stage and the suppression of the preneoplastic cell, have their origin in medicinal plants that grow in Nigeria. A noteworthy example is the African periwinkle, *Catharanthus roseus*, which yields two drugs (vinblastine and vincristine) that have been used in clinical oncology for almost 50 years. These alkaloids exert their anticancer properties by blocking the polymerization of tubulin molecules into microtubules, preventing the formation of the mitotic spindle which results in metaphase arrest and apoptosis. A series of semisynthetic analogues of these two important drug molecules have been developed, aimed at improving the activity and pharmacokinetics of the original compounds. Other plant-derived compounds have been developed as clinically useful chemotherapeutic agents. The traditional uses of medicinal plants for the treatment of diseases, including cancer, has been of tremendous help to drug discovery programmes in the selection of plants for investigation as a source of novel, bioactive natural products.

Several phytochemicals from many African foods, herbs and spices have been shown to act at various stages of cancer development by inhibition of the initiation and promotion of carcinogenesis, induction of tumour cell differentiation and apoptosis, and suppression of tumour angiogenesis (Sawadogo *et al.*, 2012; Mbele *et al.*, 2017). Of particular interest in this chapter are those Nigerian plants whose putative modes of action suggest biochemical pathways that may be different from those of chemotherapeutic agents in current clinical use. For example, plants that are immunomodulators, antioxidants and anti-inflammatory agents are increasingly playing positive roles in cancer therapy (Mbaveng *et al.*, 2017), and are included in this chapter. The major Nigerian plants with anticancer properties are outlined below.

11.1 Bitter leaf (*Vernonia amygdalina*)

Bitter leaf is a vegetable used for the preparation of the famed "bitter leaf soup". It is a common ingredient in traditional medicine in Nigeria and features as an important ingredient for remedies used for treating diabetes, cardiac insufficiency and gynaecological conditions (Toyang *et al.*, 2013). It contains saponins, steroidal glycosides, flavonoids, and sesquiterpene lactones (Iwu, 2016). Laboratory reports and clinical studies on *V. amygdalina* indicate the following activities: antimalarial, antidiabetic, antioxidant, antihypercholestrolaemic, prebiotic, anthelmintic, immune boosting in HIV infections, and anti-inflammatory properties, as well as affecting uterine contractility (Toyang and Verpoorte, 2013; Iwu, 2016). The organic fraction of the extracts of the plant was shown to possess cytotoxic effects towards human carcinoma cells of the nasopharynx (Kupchan *et al.*, 1969). The aqueous extract also exhibited similar cytotoxic activities (Izevbigie, 2003). Available data suggest that treatment with low concentrations of aqueous leaf extracts of *V. amygdalina* arrests the proliferative activities and induces apoptosis in oestrogen receptor-positive, oestrogen receptor-negative, and triple-negative human breast cancerous cells and in androgen-independent human prostate cancer PC-3 cells (Howard *et al.*, 2015). Also, in athymic mice, *Vernonia* extracts potentiate increased efficacies and optimise treatment outcomes when given as a co-treatment with conventional chemotherapy (Howard *et al.*, 2015).

Against prostate cancer cells, it has been established that PC-3 cells, treated with various concentrations of *V. amygdalina* extracts for 48 h, using the trypan blue test and MTT [3-(4, 5-dimethylthiazol-2-yl)-2,5-diphenyltetrazolium bromide] assay, exhibited significant growth inhibition, with an IC_{50} value of about 196.6 μg/mL. This finding was also shown in other evaluations, including cell morphology, lipid peroxidation and comet (alkaline single-cell gel electrophoresis) assays; on the other hand, apoptosis analysis showed that the *Vernonia* extract caused growth-inhibitory effects on PC-3 cells through the induction of cell growth arrest, DNA damage, apoptosis, and necrosis *in vitro* and may provide protection from oxidative stress diseases as a result of its high antioxidant content (Johnson *et al.*, 2017). The therapeutic efficacy of *V. amygdalina* (VA) leaf extracts as an anti-cancer agent against human breast cancer has been studied *in vitro* using the MTT and Comet assays (Yedjou *et al.*, 2008; Izevbigie *et al.*, 2004). Human breast adenocarcinoma (MCF-7) cells were treated with different doses of VA leaf extracts for 48 h. Data obtained from the MTT assay showed that VA significantly ($P<0.05$) reduced the viability of MCF-7 cells in a dose-dependent manner within 48 h of exposure. Data generated from the comet assay also indicated a slight dose-dependent increase in DNA damage in MCF-7 cells associated with VA treatment. It was observed that a slight increase occurred in comet tail length, tail arm and tail moment, as well as in percentages of DNA cleavage at all doses tested, showing evidence that VA induced minimal genotoxic damage in MCF-7 cells. The study suggests that VA treatment moderately ($P<0.05$) reduces cellular viability but induces only minimal DNA damage to MCF-7 cells. These findings provide evidence that VA extracts represent a DNA-damaging anti-cancer agent against breast cancer and its mechanisms of action functions, at least in part, through minimal DNA damage and moderate toxicity in tumour cells (Yedjou *et al.*, 2008). It has also been suggested that the chemopreventive properties of VA can be attributed to its abilities to scavenge free radicals, induce detoxification, inhibit stress-response proteins and interfere with the DNA-binding activities of some transcription factors (Farombi and Owoeye, 2011).

V. amygdalina aqueous leaf extracts (VA extracts) have been reported to be effective against triple-negative breast cancer (TNBC) that is unresponsive to most clinical therapies (Howard *et al.*, 2016). This is important because this cancer is the dominant biological cause of population-based racio-ethnic disparities in breast cancer mortality in the United States. VA extracts caused chemotherapeutic vulnerability of TNBC cells and stem cell-derived tumours. VA extracts arrest cell proliferation and induce apoptosis *in vitro*, inhibit growth of implanted tumours and show chemopreventive efficacy *in vivo*. The HRAS cells and MDA-MB-468 cells were subcutaneously implanted into nude mice with or without pretreatment with VA extracts before chemotherapeutic treatment with VA extracts and/or paclitaxel to evaluate their ability to inhibit tumour growth. The most significant reduction in tumour volume was observed in the MDA-MB-468 cell-induced tumours, following VA extract pre-treatment, compared with those from HRAS cell implantation. This result clearly demonstrates that VA extracts induce apoptosis, exhibit additive effects, inhibit tumour growth and display chemopreventive actions against TNBCs (Howard *et al.*, 2016).

In a study on a related species, *Vernonia condensata*, the extract (VCE) was found to possess cytotoxic properties against various cancer cells in a dose- and time-dependent manner. As with bitter leaf, it was interesting to observe that, when treated with VCE, there was no significant cytotoxicity in peripheral blood mononuclear cells (PBMCs). Flow cytometry analysis revealed that, although VCE induced cell death, arrest was not observed. VCE treatment led to disruption of mitochondrial membrane potential in a concentration-dependent manner, resulting in activation of apoptosis and culminating in cell death. Immunoblotting studies revealed that VCE activated an intrinsic pathway of apoptosis. More importantly, VCE treatment resulted in tumour regression, leading to significant extension in the life span of treated mice, without showing any detectable side-effects (Thomas *et al.*, 2016).

11.2 *Persea americana*; Avocado Pear (Family: Lauraceae)

The leaves and dried seeds are used in traditional medicine in Nigeria for treatment of various diseases, including body pain, coughs, diabetes, hypertension, and inflammatory conditions. The aqueous decoction of the leaves is used in ethnomedicine in Nigeria for the treatment of tumours and tumour-related diseases (Taiwo *et al.*, 2017). The anticancer activity of the extracts and isolated compounds from avocado leaves has been investigated, using a bioassay-guided fractionation, spectroscopy, Alamar blue fluorescence-based viability assay in cultured HeLa cells, and microscopy (Taiwo *et al.*, 2017). Four compounds, zoapatanolide A, agathisflavone, anacardicin (1,2-bis(2,6-dimethoxy-4-methoxycarbonylphenyl) ethane) and methyl gallate, isolated from the plant were evaluated for their cytotoxicity activity, and zoapatanolide A was found to be the most active with a half-maximal inhibitory concentration (IC_{50}) value of 36.2 ± 9.8 mM in the viability assay. The probable molecular basis of their observed cytotoxic effects was investigated by conducting Autodock Vina binding free energy assays of each of the isolated compounds with seven molecular targets implicated in cancer development (namely MAPK8, MAPK10, MAP3K12, MAPK3, MAPK1, MAPK7 and VEGF). Pearson correlation coefficients were obtained with experimentally determined IC_{50} values in the Alamar blue viability assay (Taiwo *et al.*, 2017). The compounds were found to be less potent than a standard anticancer compound, doxorubicin, but the results provided reasonable evidence that the plant species contains compounds with cytotoxic activity worthy of development for cancer chemotherapy.

11.3 Aloe vera gel (AVG) and *Aloe ferox*

Members of the genus *Aloe* produce a complex mixture of anthraquinone glycosides, that have many medical applications. Aloe vera is one of the most popular botanicals used in medicine, dietary supplements and cosmetics. The most important active constituents of the aloe plants are the anthraquinones like aloin, barbalion, anthranol, cinnamic acid, aloetic acid, emodin, chrysophanic acid, and resistanol, and enzymes (including cyclooxygenase and bradykinase), as well as other compounds, such as vitamins, saccharides, and amino acids. The best-known constituent, however, is aloin (10-β-D-glucopyranosyl-1,8-dihydroxy-3-(hydroxymethyl)- 9(10H)-anthracenone).

The two *Aloe* species have been used for the treatment of different types of cancer. Several studies have reported the anticancer properties of Aloe vera gel (AVG). These include studies on smoking and lung cancer, which showed that AVG can prevent pulmonary carcinogenesis and cancer of other tissues; inhibition against Ehrlich ascite carcinoma cell number in acute myeloid leukaemia and acute lymphocyte leukaemia, with concentration-dependent cytotoxicity effect; activity against two cell lines of colon cancer (DLD-1 and HT2) (Iwu, 2016) and anticancer activity *via* the *Aloe* polysaccharides by stimulating immune response through activation of macrophages.

It has been shown that *Aloe emodin* possesses selective activity against neuroectodermal tumours, without any adverse effect on normal cells. It promotes cell death through specific drug uptake by neuroectodermal tumours. In another study, it was found that *A. emodin* had an inhibitory effect on the activation of the Epstein–Barr virus (which plays a role in the emergence of cancer). The combined effect of *A. emodin* and the chemotherapeutic agent cisplatinol (doxorubicin, 5-fluorouracil) on the proliferation of an adhering variant cell line of Merkel cell carcinoma has also been demonstrated. Combinations of *A. emodin* with honey have been found to be beneficial in cancer treatment without any observable side-effects (Iwu, 2016).

11.4 *Ocimum gratissimum*

The genus *Ocimum* contains several culinary herbs used in herbal medicine for their multiple beneficial pharmacological properties, including anticancer activity (Chen *et al.*, 2011; Lee *et al.*, 2011; Reshma *et al.*, 2008). The crude extract of *O. gratissimum* (OG), and its hydrophobic and hydrophilic fractions, have been shown to differentially inhibit breast cancer cell chemotaxis and chemoinvasion *in vitro* and to retard tumour growth and temporal progression of MCF10aDCIs. com xenografts, a model of human breast comedo-ductal carcinoma *in situ* (Nangia-Makker *et al.*, 2013). The extract-induced inhibition of tumour growth was associated with decreases in basement membrane disintegration, angiogenesis, and MMp-2 and MMp-9 activities, as confirmed by *in situ* gelatin zymography and cleavage of galectin-3. There were also decreases in MMp-2 and MMp-9 activities in the conditioned media of OG-treated MCF10aT1 and MCF10aT1-eIII8 premalignant human breast cancer cells, as compared with the control (Nangia-Makker *et al.*, 2013). The MMp-2 and MMp-9 inhibitory activities of OG were verified *in vitro*, using gelatin, a synthetic fluorogenic peptide, and recombinant galectin-3 as MMp substrates. Mice fed on OG-supplemented drinking water showed no adverse effects compared with the control. These data suggest that OG is non-toxic and that the anti-cancer therapeutic activity of OG may, in part, be contributed by its MMp inhibitory activity (Nangia-Makker *et al.*, 2013). The related species, *Ocimum sanctum*, also has activity against breast cancer (Nangia-Makker *et al.*, 2007).

11.5 Green Tea

The tea plant, *C. sinensis* is not cultivated much in Nigeria, but the few tea plantations in the country produced teas of excellent quality. The polyphenols known as catechins, which are abundant in the unfermented leaves, have been shown to have antimutagenic, antigenotoxic, and anticarcinogenic activities. Green tea contains a complex mixture of polyphenols, as well as both condensed and hydrolysable tannins. Green tea polyphenols (GTPs) are thought to be primarily responsible for the putative cancer preventive effects of this beverage. GTPs have been shown to exert growth-inhibitory properties against many different types of cancers. The most abundant polyphenol constituents are gallic acid (GA), (2)-gallocatechin (GC), (1)-catechin (C), (2)-epicatechin (EC), (2)-epigallocatechin (EGC), (2)-epicatechin gallate (ECG), (2)-epigallocatechin gallate (EGCG), *p*-coumaroylquinic acid (CA), and (2)-gallocatechin-3-gallate (GCG) (Iwu, 2016).

One of the major polyphenols of green tea, epigallocatechin-3-gallate, has been shown, in several studies, to exhibit antitumour effects. Laboratory studies, as well as human epidemiological data, support the possibility that green tea has preventive effects against different types of cancer, especially liver cancer, breast cancer and prostate cancer, as well as gastric, colon and rectal cancers. For other types of cancer, evidence provided from human observation and laboratory studies supports the cancer-preventive role of green tea intake against many types of cancer. These include bladder and kidney cancers, thyroid cancer, myeloma, lymphoma and leukaemia, skin cancer, ovarian, uterine, and cervical cancers, and pancreatic cancer. The results of clinical studies have been conflicting and the conflicting results has been attributed to possible differences in bioavailability, lifestyle habits, and severity/stage of cancer. Green tea is generally well tolerated and probably plays a role in cancer prevention and/or supports conventional cancer treatments. Green tea and GTP activity can be considered to be complementary to other cancer management medications.

11.6 Monkey kola – Genus *Cola*

Monkey kola belongs to the plant genus *Cola*. They occur as three distinct types: the red variety, *Cola latertia*, yellow variety, *Cola parchycarpa*, and the white variety, *Cola lepidota*. Monkey kola occur as forest crops and are rarely cultivated in agroforestry. Their use as occasional or seasonal (July to November) snack-fruits seems to be limited to the Southeastern part of Nigeria, where it is known as "Achicha" or "Ochiricha" in Igbo and "Ndiyah" in Efik (Ene-Obong *et al.*, 2016). The fruits have a bitter flavor, unlike the kola nut, and are used in folk medicine as a tonic and aphrodisiac. The leaves of the plant are used in the preparation of a traditional medicine remedy against cancer, sexual impotence, headache and inflammation, antitussive and cardiac diseases (Iwu *et al.*, 1999). The dried seeds of monkey kola are pulverized and added to akamu (corn custard) as a remedy for breast cancer. Laboratory studies have shown that monkey kola extracts possess mild activity against cancer lines (Engel *et al.*, 2011).

11.7 *Theobroma cacao*; Cocoa (Family: Malvaceae)

Cocoa polyphenols, including its flavanols epicatechin, catechin, quercetin, naringenin, luteolin, apigenin, and procyanidins, have shown potential ability to act as effective chemopreventive agents. Their antioxidant property can exert protective effects on cell constituents and thereby limit the risk factor for cancer and other chronic diseases. Laboratory studies have also demonstrated that cocoa phenolic compounds exhibit a range of potential anti-inflammatory effects on intestinal cells, with possible contribution to their cancer chemopreventive activity. A cocoa-rich diet can prevent the early stages of colorectal cancer by reducing oxidative stress and cell proliferation by inducing apoptosis, since it has been established that patients with inflammatory bowel disease are at increased risk of developing ulcerative colitis-associated colorectal cancer. Human epidemiological studies and clinical trials, as well as laboratory investigations, have clearly demonstrated that cocoa and its main phenolic components can prevent and/or slow down the initiation-progression of different types of cancers. The available evidence suggests that cocoa polyphenols and some of its composite compounds exhibit cellular mechanisms of action that include the modulation of the redox status and multiple key elements in signal transduction pathways related to cell proliferation, differentiation, apoptosis, inflammation, angiogenesis, and metastasis (Iwu, 2016).

11.8 *Andrographis paniculata*; King of bitters (Family: Acanthaceae)

The leaves of *A. paniculata* are used as a major ingredient in the preparation of medicinal bitters and as a remedy for the treatment of many diseases, including malaria, diabetes, viral infections, hypertension, and cancers (Hossain *et al.*, 2014). Its use in cancer treatment and as a chemopreventive agent is of great interest because it is not necessarily due to strong cytotoxicity but because of its other properties, such as its antioxidant, anti-inflammatory and immune-modulatory effects. Extracts and isolates of *A. paniculata* inhibited the growth of a diverse cancer cell representing different types of human cancers (Geethangili *et al.*, 2008). The major biologically active compound isolated from the plant, andrographolide, has been found to exhibit both direct and indirect effects on cancer cells by inhibiting proliferation of cancer cells, cell-cycle arrests, or cell differentiation, enhancing the body's own immune system against cancer cells, and inducing apoptosis and necrosis of cancer cells (Vojdani and Erde, 2006). Studies have shown that the use of *Andrographis* extracts or the isolated active compound in adjunct therapy can enhance the effects of conventional chemotherapeutic agents. For example, andrographolide can inhibit human colorectal carcinoma (CRC) LoVo cell growth. However, in addition to the effect of the compound alone, andrographolide in combination with the chemotherapeutic agent, cisplatin, enhances the activity of the chemotherapeutic agent in the treatment of CRC (Lin *et al.*, 2014).

11.9 *Nigella sativa*; black seed, black cumin (Family: Ranunculaceae)

Black seed or *N. sativa* is used as a spice and medicinal herb in northern Nigeria and it is gaining increasing popularity in other parts of Nigeria. As mentioned in Chapter 8, black seed imparts a characteristic flavour to "pepper soup" as a condiment, with a characteristic taste like a combination of onions, black pepper, and oregano. Black seed has been used for the treatment of diabetes, hypertension, inflammation, hepatic disorder, arthritis, kidney disorder, cardiovascular complications and dermatological conditions (Tavakkoli *et al.*, 2017). Its reported pharmacological properties include bronchodilatory, hypotensive, antibacterial, antifungal, analgesic, anti-inflammatory, and immunopotentiating activities (Khan and Afzal, 2016).

Black seed contains a mixture of eight fatty acids and 32 volatile terpenes, including thymoquinone (TQ), dithymoquinone (DTQ), *trans*-anethol, *p*-cymene, limonene, and carvone (Nickavar *et al.*, 2003). In addition, diterpenes, triterpenes and terpene alkaloids have been identified in *N. sativa* seeds. Alkaloids with the isoquinoline and imidazole chromophores have also been isolated from the plant. For cancer therapy, black seed has been used with good outcomes for treatment of leukaemia, liver, lung, kidney, prostate, breast, cervix, and skin cancers (Khan *et al.*, 2011). TQ and DTQ are both cytotoxic for various types of cancer cells (Worthen *et al.*, 1998). For gastric cancer, it has been shown that *N. sativa* oil alone or in combination with known chemotherapy agents, such as etoposide, gave the strongest therapy when compared with other agents. Such a combination also leads to the activation of the mitochondrial pathway, which plays a significant role in the molecular mechanism of induction of apoptosis by these agents (Czajkowska *et al.*, 2017).

Many studies have demonstrated the efficacy of *N. sativa* against cancer in the blood system, kidneys, lungs, prostate, liver, and breast and on many malignant cell lines but the probable molecular mechanisms of the observed activities are still not clearly understood (Majdalawieh and Fayyad, 2016). Reported biological properties include its anti-proliferative effect, cell cycle arrest, apoptosis induction, reactive oxygen species (ROS) generation, antimetastasis/antiangiogenesis effects, Akt pathway control, modulation of multiple molecular targets, including p53, p73, STAT-3, PTEN, and PPAR-γ, and activation of caspases. However, available evidence suggests that the major anti-cancer mechanisms of *N. sativa* are its ROS-scavenging activity and the preservation of various antioxidant enzyme activities, such as glutathione peroxidase, catalase, and glutathione-S-transferase, which, in turn, modulate the molecular targets in apoptosis pathways (Mollazadeh *et al.*, 2017). It has also been suggested that *N. sativa* manifests its anticancer activity *via* the inducible nitric oxide synthase (iNOS) signalling pathway (Majdalawieh and Fayyad, 2016).

References

Chen, H.-M., Lee, M.-J., Kuo, C.-Y., Tsai, P.-L., Liu, J.-Y. and Kao, S.-H. (2011) *Ocimum gratissimum* aqueous extract induces apoptotic signalling in lung adenocarcinoma cell A549. *Evidence-Based Complementary and Alternative Medicine*, 2011, 739093.

Czajkowska, A., Gornowicz, A., Pawłowska, N., Czarnomysy, R., Nazaruk, J., Szymanowski, W., Bielawska, A. and Bielawski, K. (2017) Anticancer effect of a novel octahydropyrazino[2,1-a:5,4-a′]diisoquinoline derivative and its synergistic action with *Nigella sativa* in human gastric cancer cells. *BioMed Research International*, 2017, 9153403.

Ene-Obong, H.N., Okudu, H.O. and Asumugha, U.V. (2016) Nutrient and phytochemical composition of two varieties of Monkey kola (*Cola parchycarpa* and *Cola lepidota*): an under utilised fruit. *Food Chemistry*, 193, 154–159.

Engel, N., Oppermann, C., Falodun, A. and Kragl, U. (2011) Proliferative effects of five traditional Nigerian medicinal plant extracts of human breast and bone cancer cell lines. *Journal of Ethnopharmacology*, 137, 1003–1010.

Farombi, E.O. and Owoeye, O. (2011) Antioxidative and chemopreventive properties of *Vernonia amygdalina* and Garcinia biflavonoid. *International Journal of Environmental Research and Public Health*, 8, 2533–2555.

Geethangili, M., Rao, Y.K., Fang, S.-H. and Tzeng, Y.-M. (2008) Cytotoxic constituents from *Andrographis paniculata* induce cell cycle arrest in Jurkat cells. *Phytotherapy Research*, 22(10), 1336–1341.

Howard, C.B., Johnson, W.K., Pervin, S. and Izevbigie, E.B. (2015) Recent perspectives on the anticancer properties of aqueous extracts of Nigerian *Vernonia amygdalina*. *Botanics: Targets and Therapy*, 5, 65–76.

Howard, C.B., Mcdowell, R., Feleke, K., Deer, E., Stamps, S., Thames, E., Singh, V. and Pervin, S. (2016) Chemotherapeutic vulnerability of triple-negative breast cancer Cellcderived tumors to pretreatment with *Vernonia amygdalina* aqueous extracts. *Anticancer Research*, 36(8), 3933–3943.

Hossain, M.S., Urbi, Z., Sule, A. and Rahman, K.M.H. (2014) *Andrographis paniculata* (Burm. f.) wall. ex nees: a review of ethnobotany, phytochemistry, and pharmacology. *The Scientific World Journal*, 2014, 274905.

Iwu, M.M., Duncan, A.R. and Okunji, C.O. (1999) *New antimicrobials of plant origin.* In: J. Janick (ed.), *Perspectives on New Crops and New Uses*, ASHS Press, Alexandria, 457–462.
Iwu, M.M. (2016) *Food as medicine – functional food plants of Africa.* Boca Raton, FL: CRC Press/Taylor and Francis Group, pp. 384.
Izevbigie, E.B. (2003) Discovery of water-soluble anticancer agents (edotides) from a vegetable found in Benin City, Nigeria. *Experimental Biology and Medicine*, 228, 293–298.
Izevbigie, E.B., Bryant, J.L. and Walker, A. (2004) A novel natural inhibitor of extracellular signal-regulated kinases and human breast cancer cell growth. *Experimental Biology and Medicine*, 229, 163–169.
Johnson, W., Tchounwou, P.B. and Yedjou, C.G. (2017) Therapeutic mechanisms of *Vernonia amygdalina* Delile in the treatment of prostate cancer. *Molecules* 22(10), E1594.
Khan, M.A., Chen, H.-C., Tania, M. and Zhang, D.-Z. (2011) Anticancer activities of *Nigella sativa* (black cumin). *African Journal of Traditional, Complementary and Alternative Medicines*, 8(supplement 5), 226–232.
Khan, M.A. and Afzal, M. (2016) Chemical composition of *Nigella sativa* Linn: part 2 recent advances. *Inflammopharmacology*, 24, 67–79.
Kupchan, S.M., Hemmnigway, R.J., Karim, A. and Werner, D. (1969) Tumor inhibitors. XLVII. Vernodalin and vernomygdin. Two new cytotoxic sesquiterpene lactones from *Vernonia amygdalina* Del. *The Journal of Organic Chemistry*, 34, 3908–3911.
Lee, M.-J., Chen, H.-M., Tzang, B.-S., Lin, C.-W., Wang, C.-J., Liu, J.-Y. and Kao, S.-H. (2011) *Ocimum gratissimum* aqueous extract protects H9c2 myocardiac cells from H2O2-induced cell apoptosis through Akt signalling. *Evidence-Based Complementary and Alternative Medicine*, 2011, 578060.
Lin, H.-H., Shi, M.-D., Tseng, H.-C. and Chen, J.-H. (2014) Andrographolide sensitizes the cytotoxicity of human colorectal carcinoma cells toward cisplatin via enhancing apoptosis pathways *in vitro* and *in vivo*. *Toxicological Sciences*, 139(1), 108–120.
Majdalawieh, A.F. and Fayyad, M.W. (2016) Recent advances on the anti-cancer properties of *Nigella sativa*, a widely used food additive. *Journal of Ayurveda and Integrative Medicine*, 7(3), 173–180.
Mbaveng, A.T., Kuete, V. and Efferth, T. (2017) Potential of central, eastern and western Africa medicinal plants for cancer therapy: spotlight on resistant cells and molecular targets. *Frontiers in Pharmacology*, 8, 343.
Mbele, M., Hull, R. and Dlamini, Z. (2017) African medicinal plants and their derivatives: current efforts towards potential anti-cancer drugs. *Experimental and Molecular Pathology*, 103, 121–134.
Mollazadeh, H., Afshari, A.R. and Hosseinzadeh, H. (2017) Review on the potential therapeutic roles of *Nigella sativa* in the treatment of patients with cancer: involvement of apoptosis: - black cumin and cancer. *Journal of Pharmacopuncture*, 20(3), 158–172.
Nangia-Makker, P., Tait, L., Hogan, V., Shekhar, M.P.V., Funasaka, T. and Raz, A. (2007) Inhibition of breast tumor growth and angiogenesis by a medicinal herb: *Ocimum sanctum*. *International Journal of Cancer*, 121(4), 884–894.
Nangia-Makker, P., Raz, T., Tait, L., Shekhar, M.P.V., Li, H., Balan, V. and Raz, A. (2013) *Ocimum gratissimum* retards breast cancer growth and progression and is a natural inhibitor of matrix metalloproteases. *Cancer Biology and Therapy*, 14(5), 417–427.
Nickavar, B., Mojab, F., Javidnia, K. and Amoli, M.A. (2003) Chemical composition of the fixed and volatile oils of *Nigella sativa* L. from Iran. *Zeitschrift für Naturforschung. C*, 58, 629–631.
Reshma, K., Rao, A.V., Dinesh, M., and Vasudevan, D.M. (2008) Radioprotective effects of ocimum flavonoids on leukocyte oxidants and antioxidants in oral cancer. *Indian Journal of Clinical Biochemistry*, 23(2), 171–175.

Sawadogo, W.R., Schumacher, M., Teiten, M.-H., Dicato, M. and Diederich, M. (2012) Traditional West African pharmacopeia, plants and derived compounds for cancer therapy. *Biochemical Pharmacology*, 84, 1225–1240.

Siegel, R.L., Miller, K.D. and Jemal, A. (2017) Cancer statistics, 2017. *CA: A Cancer Journal for Clinicians*, 67, 7–30.

Taiwo, B.J., Fatokun, A.A., Olubiyi, O.O., Bamigboye-Taiwo, O.T., van Heerden, F.R. and Wright, C.W. (2017) Identification of compounds with cytotoxic activity from the leaf of the Nigerian medicinal plant, *Anacardium occidentale* L. (Anacardiaceae). *Bioorganic and Medicinal Chemistry* 25, 2327–2335.

Tavakkoli, A., Mahdian, V., Razavi, B.M. and Hosseinzadeh, H. (2017) Review on Clinical trials of black seed (*Nigella sativa*) and its active constituent, thymoquinone. *Journal of Pharmacopuncture*, 20(3), 179–193.

Thomas, E., Gopalakrishnan, V., Somasagara, R.R., Choudhary, B. and Raghavan, S.C. (2016) Extract of *Vernonia condensata*, inhibits tumor progression and improves survival of tumor allograft bearing mouse. *Scientific Reports*, 6(23255), 1–12. Available at www.nature.com/reports.

Toyang, N.J. and Verpoorte, R. (2013) A review of the medicinal potentials of plants of the genus Vernonia (Asteraceae). *Journal of Ethnopharmacology*, 146(3), 681–723.

Toyang, N.J., Wabo, H.K., Ateh, E.N., Davis, H., Tane, P., Sondengam, L.B., Bryant, J. and Verpoorte, R. (2013) Cytotoxic sesquiterpene lactones from the leaves of *Vernonia guineensis* Benth. (Asteraceae). *Journal of Ethnopharmacology*, 146(2), 552–556.

Vojdani, A. and Erde, J. (2006) Regulatory T cells, a potent immunoregulatory target for CAM researchers: modulating tumor immunity, autoimmunity and alloreactive immunity (III). *Evidence-Based Complementary and Alternative Medicine*, 3(3), 309–316.

WHO. (2018) *Cancer*. Genève: World Health Organization. Accessed October 2018. Available at www.who.int/cancer

Worthen, D., Ghosheh, O. and Crooks, P. (1998) The in vitro anti-tumor activity of some crude and purified components of black seed, *Nigella sativa* L. *Anticancer Research*, 18, 1527–1532.

Yedjou, C., Izevbigie, E. and Tchounwou, P. (2008) Preclinical assessment of *Vernonia amygdalina* leaf extracts as DNA damaging anti-cancer agent in the management of breast cancer. *International Journal of Environmental Research and Public Health*, 5(5), 337–341.

12 Control of Oxidative Stress and Chronic Inflammation with Nigerian Plants

12.1 Introduction

In normal physiological processes, when the body is under attack, it releases some biological agents as a highly regulated first-line defence against foreign challenges, such as microorganisms, toxins, allergens, or other xenobiotics. This response causes an inflammation of the surrounding tissues, which is essential to generate an adaptation (or immunity) in response to particular infectious agents. These pro-inflammatory agents are also required to remove dead or damaged tissue as well as to initiate the body's healing and repair process. The four indicative signs of inflammation in its acute phase are redness, swelling, heat, and pain. The inflammatory process involves a general accumulation of extravascular plasma proteins, including pro-inflammatory cytokines and chemokines, in addition to a variety of immune inflammatory cell types. Studies have shown that, during an inflammatory response, mediators, such as pro-inflammatory cytokines, including interleukin IL-1, tumour necrosis factor (TNF), interferon (INF)-c, IL-6, IL-12, IL-18 and the granulocyte–macrophage colony-stimulating factor, are released. The body usually regulates the pro-inflammatory response through a cascade of antagonism involving anti-inflammatory cytokines, such as IL-4, IL-10, IL-13, IFN-α, and the transforming growth factor. The nuclear factor-jB (NF-jB), a transcription factor, also plays an important role in the inflammatory response by regulating the expression of various genes encoding pro-inflammatory cytokines, adhesion molecules, chemokines, growth factors, and inducible enzymes such as cyclooxygenase-2 (COX-2) and inducible nitric oxide synthase (iNOS). Sometimes, the delicate balance between pro- and anti-inflammatory endochemicals is distorted, which becomes detrimental to the tissues and dangerous to human health. Inflammation plays an important role in various diseases, such as rheumatoid arthritis, atherosclerosis and asthma, all of which are highly prevalent conditions globally.

The balance between pro- and anti-inflammatory agents may be distorted, even if only within a small threshold, which leads to persistent inflammation caused by low levels of pro-inflammatory chemicals over a long time. The persistent presence of these pro-inflammatory agents causes the body to malfunction and causes or aggravates certain diseases, including arthritis, cardiovascular diseases, cancer, diabetes, obesity, neurologic diseases, pulmonary diseases, psychological diseases, and autoimmune diseases. Studies have shown that even prostate diseases are essentially inflammation based. Although the origins of inflammation in the prostate are multi-factorial and may include bacterial colonization, viruses, environmental and dietary components, and hormones, unchecked infection by common parasites, like *Toxoplasma gondii*, is also found to trigger chronic inflammation in the human prostate (Colinot *et al.*, 2017). Some Nigerian plants have been found to be effective in preventing the build-up of pro-inflammatory kinins and ameliorating the deleterious effects of chronic inflammation.

12.2 Nigerian Plants with Anti-Inflammatory Properties

Simple physiological processes, like eating, can also trigger an inflammatory response that activates the immune systems of healthy individuals and has a protective effect, because an average diet contains not only nutrients but is also laden with significant quantities of bacteria. In healthy individuals, short-term inflammatory responses play an important role in sugar uptake and the activation of the immune system. In overweight individuals, however, this inflammatory response fails so dramatically that it can lead to diabetes. It has been shown that the number of macrophages (a type of immune cell) around the intestines increases during mealtimes. These so-called "scavenger cells" produce the messenger substance IL-1beta in varying amounts, depending on the concentration of glucose in the blood (Konrad *et al.*, 2017). This, in turn, stimulates insulin production in pancreatic beta cells. The insulin then causes the macrophages to increase IL-1beta production. Insulin and IL-1beta work together to regulate blood sugar levels, while the messenger substance IL-1beta ensures that the immune system is supplied with glucose and thus remains active (Dror *et al.*, 2017). The study showed that the mechanism of the metabolism and immune system is dependent on the bacteria and nutrients that are ingested during meals. With sufficient nutrients, the immune system is able to adequately combat foreign bacteria. Conversely, when there is a lack of nutrients, the few remaining calories must be conserved for important life functions at the expense of an immune response. Thus, infectious diseases occur more frequently in times of famine or low nutritional level.

Studies have established that inflammation is a key contributory factor in the pathogenesis of most common kinds of arthritis, especially osteoarthritis (OA) and rheumatoid arthritis (RA). It is established that various cytokines are involved in RA and OA pathology. TNF-a, IL-1b, and interferon-g (IFN-g), produced by macrophages, dendritic cells, and T cells, are the most important cytokines stimulating matrix metalloproteinase expression (MMP) and synovial inflammation under inflammatory conditions (Hall and Bravo-Clouzet, 2013). The characteristic joint swelling and the cartilage and bone erosion through osteoclast formation are caused by the action of the pro-inflammatory cytokines (Ritchilin, 2000). Accordingly, an effective therapeutic strategy for the management of RA and OA is the blockade of these cytokines and their downstream effectors.

It is now established that transcription factors and signalling molecules regulate the inflammatory cascade and that most chronic diseases are caused by dysregulation of inflammatory pathways, leading in turn to chronic inflammation. Identification of such molecular targets has led to the discovery that many herbs and spices used as part of African cuisine possess anti-inflammatory properties, as evidenced by *in vitro* and *in vivo* effects, as well as clinical reports. These include ginger, turmeric, alligator pepper, onions, scent leaves, sage, coco, red and black peppers, and carrots. The classes of compounds responsible for the biological activities of these anti-inflammatory foods cover a very wide range, including I) aromatic acids, II) polyphenols, III) alkaloids, IV) terpenoids, V) amino acids and peptides, and VI) organosulphides. The anti-inflammatory functional foods, which contain many bioactive molecules, are more suitable for the management of chronic diseases than pharmaceutical-based agents, by being active on multiple molecular targets, since it is well known that chronic diseases are caused by dysregulation of multiple genes, whereas many modern medicines are based on the modulation of a single target and therefore are less likely to be as effective.

The observed anti-inflammatory health benefits of phytochemicals have been attributed to the following mechanisms of action: (1) direct antioxidant activity or increase in the expression of antioxidant proteins; (2) attenuation of endoplasmic reticulum stress signalling; (3) blockade of pro-inflammatory cytokines; (4) blockade of transcription factors related to metabolic diseases; (5) induction of metabolic gene expression; and (6) activation of transcription factors that antagonise inflammation (Joven *et al.*, 2013).

12.3 Profile of Selected Nigerian Plants Used in the Treatment of Inflammatory Diseases

The plants listed in Table 12.1 above are those frequently reported in ethnobotanical surveys, research publications and clinical reports as being useful in the treatment of various inflammatory disorders (Sagnia *et al.*, 2014). A few of the plants have been selected for this chapter, to illustrate the biological properties of these medicinal plants, their varied mode of application, and probable mechanism of action.

12.3.1 *Bridelia ferruginea*

Bridelia ferruginea Benth. (Phyllanthaceae, Formerly Euphorbiaceae) is a shrub used extensively in traditional medicine for the treatment of diabetes, arthritis, kidney diseases, and fungal infections, especially for the management of opportunistic thrush in HIV/AIDS patients. It is also used externally as an embrocation for the treatment of bruises, boils, dislocation, and burns (Iwu, 2014). The anti-inflammatory activity of the aqueous extract of *Bridelia ferruginea* stem bark has been evaluated in several *in-vivo* models (Olajide *et al.*, 2000). *B. ferruginea* extract (BFE) at a dose of 25–200 μg inhibited the production of PGE2, nitrite, tumour necrosis factor-α (TNFα), and interleukin-6 (IL6), as well as COX-2 and iNOS protein expression in lipopolysaccharide (LPS)-activated microglial cells. Studies to elucidate the mechanisms of anti-inflammatory action of BFE revealed interference with nuclear translocation of NF-κBp65 through to mechanisms involving inhibition of IκB degradation. BFE prevented phosphorylation of p38, but not p42/44 or JNK MAPK. It is suggested that *B. ferruginea* exhibits anti-inflammatory action through mechanisms involving p38 MAPK and NF-κB signalling (Olajide *et al.*, 2012).

One of the probable mechanisms of the anti-inflammatory effects of *B. ferruginea* involves the suppression of TNFα up-regulation, as laboratory studies have shown that pre-treatment with *B. ferruginea* extract (10–80 mg/kg bodyweight) produced a dose-dependent inhibition of the septic shock syndrome in mice, with 80 mg/kg dosage of the extract exhibiting comparable activity to pentoxifylline (100 mg/kg). The LPS-induced dye leakage in the skin of mice was also suppressed by the extract at a dose of 10–80 mg/kg (Olajide *et al.*, 2003).

B. ferruginea extracts appear to be safe and well tolerated. A toxicity study on the aqueous extract of the stem bark reported that the half-maximal lethal dose (LD_{50}) of the extract was estimated to be >4,000 mg/kg orally, and that neither significant visible signs of toxicity nor mortality were observed (Awodele *et al.*, 2015). There were also no significant differences in the animal or organ weights, nor in haematological or biochemical parameters in the treated groups, compared with the control group. Surprisingly, it was observed that there was a significant increase ($P<0.05$) in the level of lipid peroxidation and a significant ($P<0.05$) decrease in sperm count in the treated animals compared with the control group. The observed potential of the aqueous extract of *B. ferruginea* stem bark to cause lipid peroxidation and damage sperm quality in animals calls for caution in long-term use of the herb (Awodele *et al.*, 2015). The phenolic-rich extracts of *B. ferruginea* leaves have been shown to possess strong antioxidant activity, based on evaluation using assays for ferric-reducing antioxidant power, total antioxidant activity (phosphomolybdenum-reducing ability), 1,1-diphenyl-2-picrylhydrazyl, and thiobarbituric acid reactive species. Also, the leaf extracts inhibited alpha-amylase and alpha-glucosidase activities *in vitro* (Afolabi *et al.*, 2018)

Although *B. ferruginea* is the plant used most among the members of the genus *Bridelia*, several species are used interchangeably by traditional medical practitioners for the preparation of remedies. They include *Bridelia atroviridis, Bridelia speciosa,* and *Bridelia micrantha*. Others, such as *Bridelia micrantha* var.

Table 12.1 Nigerian Plants with Anti-Inflammatory Activities

Plants	Part Used
Abelmoschus esculentus	Fruits
Adansonia digitata	Fruits
Aframomum melegueta	Seeds
Allium cepa	Bulbs, rhizome
Allium sativum	Bulbs, rhizome
Anchomanes difformis	Leaves, roots
Andrographis paniculata	Leaves, aerial parts
Ananas comosus	Fruits
Anona muricata	Fruits
Antrocaryon klaineanum	Stem bark
Artocarpus heterophyllus	Fruits, seeds
Asparagus africanus	Leaves, underground parts
Azadirachta indica	Leaves, stem bark, oil
Balanitas aegyptica	Fruits
Bridelia ferruginea	Leaves, stem bark, root
Bulbine frutescens	Leaves
Cajanus cajan	Leaves, seeds
Capsicum frutencens	Fruits
Carica papaya	Leaves, seeds, fruits, root
Cocos nucifera	Nuts, fruit, oil
Colocasia esculenta	Leaves, tubers
Commelina nudiflora	Leaves
Costus afer	Aerial parts
Garcinia kola	Seeds, fruit rind
Cleome gynandra	Leaves
Delonix regia	Aerial parts
Dichrostachys cinerea	Leaves
Dracaena manii	Aerial parts
Elaeis guineensis	Fruits, fronds
Eremomastax spectiosa	Leaves, stem
Ficus iteophylla	Stem bark
Gardenia ternifolia	Leaves
Hibiscus sabdariffa	Calyces
Indigofera pulchra	Leaves
Ipomoea pes-caprae	Leaves
Moringa oleifera	Leaves, seeds, oil
Ocimum basilicurn	Leaves
Ocimum gratissimum	Leaves
Palisota hirsuta	Aerial parts
Piper guineense	Fruits
Polyscias fulva	Aerial parts
Pseudospondias longifolia	Stem bark
Schwenkia americana	Leaves
Securidaca longepedumcula	Aerial parts

(*Continued*)

Table 12.1 (Continued) Nigerian Plants with Anti-Inflammatory Activities

Plants	Part Used
Solenostemon rotundifolius	Leaves, tubers
Sorghum bicolor	Seeds, leaves
Telfairia occidentalis	Seeds, fruit pulp
Terminalia ivorensis	Fruits, leaves
Tetrapleura tetraptera	Fruits
Theobroma cacao	Seeds, pod
Vernonia amygdalina	Leaves
Vigna unguiculata	Beans
Withania somnifera	Leaves
Zingiber officinale	Rhizome

ferruginea, *Bridelia speciosa* var. *kourousensis*, and *Gentilia chevalieri* are synonyms used in literature that refer to *B. ferriginea*. All members of the genus *Bridelia* have been used in the preparation of folk remedies for the treatment of a variety of illnesses, such as fevers, diabetes, rheumatism, gonorrhoea, and diarrhoea. They feature significantly in the preparation of many Yoruba "Agbo" recipes, especially as a key ingredient for the preparation of a traditional gargle called "Ogun Efu," used in the treatment of 'Efu', a pathological disorder characterised by a furred tongue caused by overgrowth of papillae, that produces a creamy curd-like coating on the tongue (Iwu, 2014).

12.3.2 Turmeric

Turmeric is used in medicine and food for its strong antioxidant, anti-inflammatory, antiproliferative, anticancer and antiangiogenic activities. It contains the polyphenol curcumin (diferuloylmethane), which has antioxidant effects greater than those of vitamins, and induces enzymes of the glutathione-linked detoxification pathways. The curcuminoids modulate signalling molecules, including inflammatory molecules, transcription factors, enzymes, protein kinases, protein reductases, carrier proteins, cell survival proteins, drug resistance proteins, adhesion molecules, growth factors, receptors, cell-cycle regulatory proteins, chemokines, DNA, RNA and metal ions, important in oxidation, obesity, neurologic and psychiatric disorders, and cancers. Curcumin ameliorates various chronic illnesses, linked to free radical damage, affecting the eyes, lungs, liver, kidneys, and the gastrointestinal and cardiovascular systems (Gupta *et al.*, 2012). Studies have shown that turmeric and curcumin prevent oxidative stress in diabetic mammals without changing blood sugar levels (Suryanarayana *et al.*, 2005). Positive results have been reported in the use of turmeric for the treatment of acute and chronic diseases linked to enhanced pro-inflammatory kinins and/or chronic inflammation (such as arteriosclerosis, cancer, respiratory, hepatic, pancreatic, intestinal and gastric diseases, neurodegeneration, and degenerative diseases). The probable mechanism of action of turmeric and its most abundant bioactive constituent, curcumin, is through the inhibition of inflammation mediators such as NFKB, COX-2, lipooxygenase (LOX) and iNOS (Bengmark, 2006). Its remarkable anti-inflammatory activity and effect in the prevention of degenerative disease is ascribed to its ability to modulate (i) several cell-signalling pathways at multiple levels (Chen *et al.*, 2010); (ii) cellular enzymes, e.g. COX-2 and glutathione S-transferases; (iii) the immune system; (iv) angiogenesis; (v) cell–cell adhesion; (vi) gene transcription; and (vii) apoptosis in preclinical cancer models (Shehzad and Lee, 2010).

Turmeric and curcumin are well tolerated and safe even at high doses (12 g/day) of turmeric because of its low bioavailability, poor absorption, rapid metabolism and rapid systemic elimination (Mohamed, 2014). In comparative evaluation, turmeric was found to be more effective than curcumin in countering oxidative stress when comparing the rates for changes in (i) lipid peroxidation, (ii) reduced glutathione, (iii) protein carbonyl content, (iv) antioxidant enzyme activities, and (v) osmotic stress (Anand *et al.*, 2007).

12.3.3 *Prunus africanum*

The anti-inflammatory activity of *Prunus (Pygeum) africana* extract has been shown to be due mainly to its ability to inhibit leukotriene synthesis (Marconi *et al.*, 1986; Paubert-Braquet *et al.*, 1994a). Extracts of the stem bark of *P. africana* are the herbal components frequently recommended by both traditional healers and orthodox physicians for the treatment of micturitional difficulties due to benign prostatic hyperplasia (BPH). The extract has also been shown in animal study to ameliorate the viscoelastic properties of bladder muscle, which has degenerated due to aging (Riffaud and Lacolle, 1990). Although its mechanism of action is not completely understood, it is generally associated with the anti-inflammatory activity of the herb.

It is believed that the basic fibroblast growth factor (bFOF) probably plays a role in the development of BPH and a study has been published that examined the effects of *P africana* extract on basal cell proliferation and on the proliferation induced by bFOF, epidermal growth factor (EOF) and insulin-like growth factor-l (lOF-I) (Paubert-Braquet *et al.*, 1994b). The proliferation of 3T3 fibroblasts was measured by the incorporation of tritiated methyl thymidine and the staining of nuclei with crystal violet. It was found that the extract slightly inhibited the basal growth of fibroblasts, with a much larger inhibitory effect on cell proliferation induced by bFOF at 0.5 μg/mL, and the effect was significant at 1 μg/mL. The *Prunus* extract also inhibited cell proliferation induced by EOF, but to a lesser extent. This suggests that the therapeutic effect of the *P africana* extract may be partly due to inhibition of cell growth induced by certain growth factors (Paubert-Braquet *et al.*, 1994b).

12.3.4 *Tinospora cordifolia*

The leaves, stem bark and roots of *T. cordifolia* are ingredients in the preparation of many polyherbal remedies used for the treatment of diseases including diabetes, liver damage, free radical-mediated injury, infections, stress and cancer, as well as rheumatoid arthritis, dermatological diseases, cancer, gout, jaundice, asthma, leprosy, and in the treatment of bone fractures. It is a major ingredient in the formulation of the gynaecological traditional medicine known as Aju Mbaise. Studies have provided evidence for its immunomodulatory, diuretic, anti-inflammatory, analgesic, anticholinesterase and gastrointestinal protective effects (Panchabhai *et al.*, 2008). The therapeutic application of the plant is attributed to its diverse chemical constituents, including berberine, tinosporin, tinosporal, tinosporaside, tinosporic acid, tinocordiofolioside, columbin and related alkaloids, diterpenes, glycosides and polysaccharides (Panchabhai *et al.*, 2008). A bioassay-guided extraction has led to the isolation of a new clerodane furano diterpene glycoside along with five known compounds, namely cordifolioside A (β-D-glucopyranoside,4-(3-hydroxy-1-propenyl)-2,6-dimethoxyphenyl 3-*O*-D-apio-β-D-furanosyl), β-sitosterol (,2β,3β: 15,16-diepoxy-4α,6 β-dihydroxy-13(16),14-clerodadiene-17,12:18,1-diolide), ecdysterone and tinosporoside, that exhibit anticancer activity *via* induction of mitochondrion-mediated apoptosis and autophagy in HCT116 cells, with potential benefit for the treatment of colon cancer (Sharma *et al.*, 2018).

12.3.5 *Garcinia kola*; Bitter Kola (Kolaviron) (Family: Clusiaceae)

The anti-inflammatory properties of extracts of bitter kola, *G. kola*, and its major isolates, kolaviron, kolanone and garcinone, have been demonstrated in several experimental models (Iwu, 1985; Farombi, 2011). Literature (involving 68 published studies) has shown that constituents of *G. kola* are effective blockers of pro-inflammatory endochemicals. *G. kola* protects against carcinogen-induced hepatotoxicity by free radical scavenging, metal chelation, up-regulation of the detoxification system, inhibition of stress-response proteins, and down-regulation of NF-kB and AP-1 (Farombi, 2011). The ability of kolaviron to modulate the expression of inflammatory genes in a lipopolysaccharide (LPS)-stimulated Sertoli cell line 93RS2 has been investigated (Agbariukwu, 2015). Kolaviron decreased the expression of inflammatory genes encoding TNF-α, Tlr-4, and Nfκb1, and had a synergistic effect on LPS-induced COX-2 and iNOS expression at high concentrations (25–100 μM). At lower concentrations (5–15 μM), the expressions of TNF-α, IL-6, and IL-1α were down-regulated by kolaviron, except for Tgfβ1, which was up-regulated. The LPS-induced decrease in the expression of the genes encoding anti-inflammatory IL-3, IL-4, and IL-10 was blocked by kolaviron at all concentrations of kolaviron tested. The LPS-induced phosphorylation of mitogen-activated protein kinase family members (ERK1/2, and phospho-JNK) decreased IκBα expression, and decreased Akt phosphorylation was blocked by kolaviron (Agbariukwu, 2015). The study highlighted the potential for kolaviron at lower concentrations to ameliorate cellular damage caused by local inflammation (Agbariukwu, 2015).

12.3.6 Ginger (*Zingiber officinalis*)

Ginger is a well-known anti-inflammatory herb that has been studied extensively for its ability to ameliorate various inflammatory conditions (Baliga *et al.*, 2013), both in animal studies (Srivastava and Mustafa, 1992) and in clinical evaluations (Kasahara *et al.*, 1983). Ginger exerts anti-inflammatory properties through multiple mechanisms and probably suppresses the activation of the NF-κB signalling pathway. For the treatment of osteoarthritis (OA), ginger has been shown to reduce significantly the pain in patients with OA in a randomized, double-blind, placebo-controlled, multicentre, parallel group, over a six-week study period. At the end of the study, when compared to the placebo cohorts, administering a ginger capsule (255 mg of extract drawn from 2500–4000 mg of dried ginger rhizomes) twice daily, morning and evening, reduced the pain on standing and after walking 50 feet (Altman and Marcussen, 2001). The activity of ginger in the management of OA has been corroborated in another randomized double-blind, placebo-controlled, crossover study of six months duration. The study reported that, although administration of ginger extract (250 mg per capsule, four times a day) was only as effective as the placebo during the first three months of the study, by the end of six months, three months after crossover, ginger showed a significant superiority over the placebo group in reducing the pain and discomfort associated with OA (Wigler *et al.*, 2003).

12.3.7 The Oil Palm (*Elaeis guineensis*) – Palm Fruit Bioactives

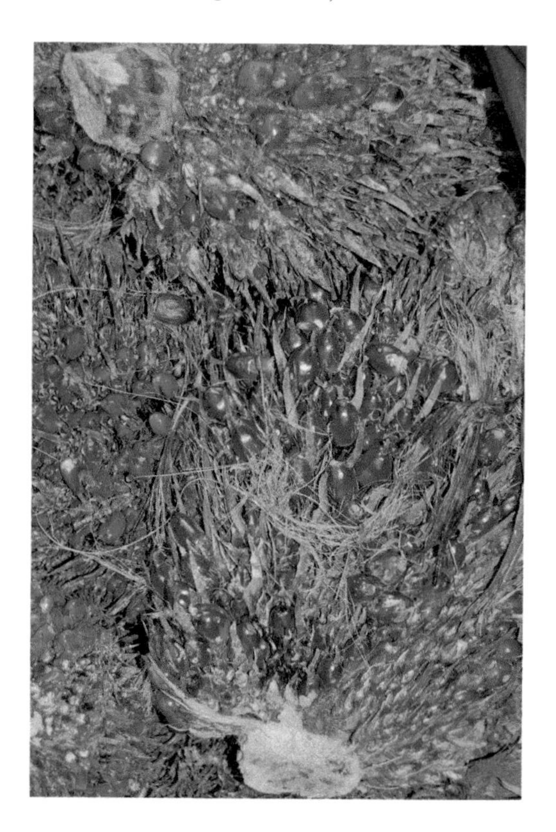

The leaves and fruits of the oil palm are commonly used topically for the treatment of inflammatory conditions. The therapeutic value of palm fruit is well recognised among the communities in the oil-palm-growing southern region of Nigeria. For example, the Igbo prepare a special healing soup with palm fruit called "Ofe akwu", which is similar to the famous "Banga soup" prepared by the Itsekiri, Urhobo, and other Niger Delta tribes. The palm fruit contains mainly lipid-soluble compounds such as carotenoids, tocopherols and tocotrienols, which are known to have antioxidant activity. Perhaps what is less well known is the use of water-soluble palm fruit bioactives (PFB), usually produced as a byproduct of palm oil processing, in the oral treatment of inflammatory conditions.

During the production of palm oil, large amounts of aqueous liquor are generated, which are typically discarded in the waste stream as palm oil mill effluent, which, over time, constitutes an environmental nuisance. The liquor can also be processed to yield a water-soluble product rich in polyphenolics, referred to as "water-soluble palm fruit bioactives" or Oil Palm Phenolics (OPP). The palm fruit bioactives fraction is known to have strong antioxidant, radical-scavenging, hydrogen-donating and reducing properties in *in-vitro* assays (Balasundram *et al.*, 2005). Animal studies have also shown that the product exhibits potential beneficial effects on degenerative diseases (Sekaran *et al.*, 2010; Lynch *et al.*, 2017). In various animal models, OPP reduced blood pressure in a NO-deficient rat model, protected against ischaemia-induced cardiac arrhythmia in rats and reduced plaque formation in rabbits fed an atherogenic diet. In Nile rats, a spontaneous model of metabolic syndrome and type 2 diabetes, OPP protected against multiple aspects of metabolic syndrome and diabetes progression. In tumour-inoculated mice, OPP protected against cancer progression. Microarray studies on the tumours showed differential transcriptome profiles that suggested anti-tumour molecular mechanisms involved in OPP action. Thus, initial studies suggest that OPP may have potential against several chronic disease outcomes in mammals (Sambanthamurthi *et al.*, 2011).

Inflammation of the brain is often the first phase of many neurodegenerative diseases and reactive astrocytes are key regulators in the development of neuroinflammation. A study has been published which showed the effects of Palm Fruit Bioactives (PFB) on the behaviour of human astrocytes, which have been activated by IL-1β (Weinberg *et al.*, 2018). When activated, the astrocytes proliferate, release numerous cytokines/chemokines, including TNFα, RANTES (CCL5), and IP-10 (CXCL10), generate reactive oxygen species (ROS), and express specific cell surface biomarkers, such as the Intercellular Adhesion Molecule (ICAM), Vascular Cellular Adhesion Molecule (VCAM) and the Neuronal Cellular Adhesion Molecule (NCAM). Interleukin 1-beta (IL-1β) causes activation of human astrocytes with marked up-regulation of pro-inflammatory genes. It was found that PFB caused significant inhibition of these pro-inflammatory processes when IL-1β-activated were exposed to PFB. It causes a dose-dependent and time-dependent reduction in specific cytokines, namely TNFα, RANTES, and IP-10 (Weinberg *et al.*, 2018). The study also showed that PFB significantly reduced ROS production by IL-1β-activated astrocytes. PFB additionally reduced the expression of ICAM and VCAM, both in activated and naïve human astrocytes *in vitro*; reactive astrocytes play an essential role in the neuroinflammatory state preceding neurodegenerative diseases (Weinberg *et al.*, 2018).

OPP has been found to be safe even at very high doses; in a phase one single-blind clinical trial conducted to evaluate the safety and effects of OPP, 25 healthy human volunteers showed no major adverse effects (AE) or serious adverse effects (SAE). The volunteers were supplemented with 450 mg gallic acid equivalent (GAE)/day of OPP or control treatments for a 60-day period. Fasting blood and urine samples were collected at days 1, 30 and 60. Medical examination was performed during these trial interventions. All clinical biochemistry profiles observed throughout the control and OPP treatment periods were within the normal range. Additionally, OPP supplementation resulted in improvement of total cholesterol and low-density lipoprotein-cholesterol (LDL-C) levels, compared to the control treatment (Fairus *et al.*, 2018).

12.3.8 *Delonix regia*; Flame of the Forest, Flamboyant Bark, Royal Poinciana (Family: Fabaceae)

D. regia, a semi-deciduous tree distributed widely throughout Africa and Asia, is used extensively in many traditional systems of medicine for the treatment of various diseases, including diabetes, inflammation, rheumatism, bronchitis, anaemia, fever, gynaecological disorders, and infectious diseases. The major application in Nigeria is as a potherb for the treatment of chronic inflammatory conditions and asthma, and in polyherbal preparations for the treatment of malaria, viral infections, and diabetes. Pharmacological studies have indicated that parts of the plant showed antioxidant, hepatoprotective, gastroprotective, wound-healing, antiarthritic, larvicidal, antimalarial, antiemetic, antibacterial, antifungal, anti-inflammatory, analgesic, antidiarrhoeal, antihaemolytic, diuretic, and anthelmintic activities (Modi *et al.*, 2016). *D. regia* has been the subject of extensive phytochemical investigations, which showed that the plant contains many polyphenolic compounds, viz. flavonols, anthocyanins, and phenolic acids, as bioactive secondary metabolites. The flowers, which are often used in folk medicine, contain tannins, saponins, flavonoids, steroids, alkaloids, and carotenoids (such as phytoene, phytofluene, β-carotene, pigment X, δ-carotene, γ-carotene, prolycopene, neolycopene, and lycopene). The leaves contain a complex mixture of flavonoids (kaempferol-3-rhamnoside, quercetin-3-rhamnoside, kaempferol-3-glucoside, kaempferol-3-rutinoside, kaempferol-3-neohesperidoside, quercetin

3-rutinoside, and quercetin 3-glucoside), alkaloids, glycosides, tannins, and carbohydrates, whereas the bark contains saponins, tannins, terpenoids, alkaloids, glycosides, carbohydrates, and phytosterols.

The anti-inflammatory activity of the leaf extracts have been evaluated using carrageenan-induced rat paw oedema and cotton pellet granuloma at three different dosages (100, 200, or 400 mg/kg bodyweight (b.w.) p.o.(per os, by mouth) of an ethanol extract. The ethanol extract exhibited significant anti-inflammatory activity at the dose of 400 mg/kg in both models, when compared with the positive control group treated with indomethacin (10 mg/kg b.w. p.o) (Shewale *et al.*, 2012).

The ethanol extract of the stem and one of its bioactive constituents, β-elemene, have been studied for their anti-inflammatory activity, using standard *in-vivo* anti-inflammatory models: carrageenan-induced paw oedema, cotton pellet granuloma, and acetic acid-induced vascular permeability. It was found that the *D. regia* stem ethanol extract or β-elemene efficiently inhibited inflammation in laboratory studies (Patra *et al.*, 2016). It has been demonstrated that the probable mechanism of the anti-inflammatory activity of *D. regia* involves (at least in part) the modulation of major inflammatory mediators (NO, iNOS, and COX-2-dependent prostaglandin E2 or PGE2), and pro-inflammatory cytokines (TNF-a, IL-1b, IL-6, and IL-12). Additionally, we quantified expression of the Toll-like receptor 4 (TLR4), myeloid differentiation primary response gene 88 (MyD88), and bymRNA expression in a drug-treated LPS-induced murine macrophage model (Patra *et al.*, 2016). The significant anti-inflammatory activity of *D. regia* extract and β-elemene was observed, along with the inhibition of TNF- α, IL-1b, IL-6 and IL-12 expression. Furthermore, the expression of TLR4, NF-kBp65, MyD88, iNOS and COX-2 were reduced in the drug-treated groups, but not in the LPS-stimulated untreated or control groups (Patra *et al.*, 2016).

The leaf extract has also been shown to possess cardioprotective effects by activating vasodilation through the NO pathway and preventing the myocyte injury *via* inhibition of the TNF- α pathway (Wang *et al.*, 2016).

12.3.9 *Carica papaya*; Paw-Paw, Papaya (Family: Caricaceae)

C. papaya is consumed as a fruit all over the world. Since every part of papaya tree possesses economic value, it is cultivated on a commercial scale in Nigeria. The leaves, unripe fruits, stems and roots are employed in traditional medicine for the treatment of different ailments, including malaria, diabetes, cancer (brain, colorectal, prostate and benign), and parasitic infections, as well as being a major component of topical dressings used for treating ulcers and dermatitis, as an anti-fertility agent, and in gastrointestinal uses such as antihelmintic (worms) and antibacterial activity. Its modern phytomedical applications include the treatment of cardiovascular diseases, dengue fever, digestion disorders, inflammatory diseases, wound healing, malaria, hypoglycaemia, hyperlipidaemia (Nafiu *et al.*, 2019), bacterial and fungal diseases (Iwu *et al.*, 1999) and as a male contraceptive (Nwaehujor *et al.*, 2014). Many reports have also demonstrated its protective effects against cancers of the colorectum, prostate, cervix, breast, and gall bladder. The fruit and seed extracts of papaya displayed significant cytotoxic and anti-proliferative activities against breast (MCF-7), liver (HepG2) and leukaemia (HL-60) cancer cells. The plant contains many biologically active compounds. The orange or red flesh is an excellent source of provitamin A and ascorbic acid. *C. papaya* yields two important digestive enzymes, chymopapain and papain. Papain is also used to treat arthritis. The yield of the enzymes varies among the fruit, latex, leaves, and roots. The leaves contain alkaloids as well as quercetin and kaempferol as the main phenolic compounds. In addition, the leaves of papaya have been shown to contain antioxidant compounds such ascystatin, α-tocopherol, ascorbic acid, flavonoids, cyanogenic glucosides and glucosinolates (Iwu, 2016).

Ethanol extracts of *C. papaya* leaves showed strong antioxidant properties against both hydrogen peroxide and the superoxide anion. The effect of the plant extract on cd T cells and imDC was evidenced by the dosage-dependent reduction in TNF-α production in the presence of the extract. *C. papaya* leaf ethanol extract has been shown to significantly reduce carrageenan-induced paw oedema and cotton granuloma tissue formation. Similarly, the extract reduced oedema in a formaldehyde-induced arthritis model (Owoyele *et al.*, 2008). The anti-inflammatory activity of the seeds has also been demonstrated on laboratory animals. At all doses, the extracts showed a dose- and time-dependent inhibition effects of oedema ($P<0.05$), which was indicative of the anti-inflammatory activity of the seeds of *C. papaya* (Amazu *et al.*, 2010).

Pawpaw is often used in combination with other plants for the treatment of inflammatory conditions. The fallen dry papaya leaf, along with leaves of *Azadirachta indica*,

Cymbopogon citratus, *Psidium guava* and stem bark of *Alstonia boonei* are boiled together, cooled and drunk for the treatment of arthritis and rheumatism (Gill, 1992). A double-blind, randomized placebo-controlled trial reported the effects of a complex fruit/vegetable concentrate prepared from *C. papaya* and 24 other fruits and vegetables, that reduced systemic inflammation in humans. The subjects were given six capsules of the concentrate daily for eight weeks. The concentrate markedly reduced plasma TNF-α without a change in plasma C-reactive protein (CRP), especially in subjects with higher baseline systemic inflammation (serum CRP ≥3.0 mg/mL). The concentrate also altered the expression pattern of several genes: NF-κB, 5′-adenosine monophosphate–activated protein kinase (AMPK) signalling, phosphomevalonate kinase (PMVK), zinc finger AN1-type-containing 5 (ZFAND5), and calcium-binding protein-39 (CAB39). It was shown that the polyherbal concentrate improved blood lipid and metabolic profiles through reduction of systemic inflammation, to alleviate the risk of obesity-induced chronic disease (Williams *et al.*, 2017). Papain from papaya latex or unripe pulp has been found to be beneficial to patients with inflammatory disorders and was effective in controlling oedema and inflammation associated with surgical or accidental trauma (Rakhimov, 2000).

The widely acclaimed anti-ageing properties of *C. papaya* fruit is due largely to its general anti-inflammatory effect. Clinical observational studies suggest the potential efficacy of *C. papaya* as an anti-ageing agent in humans. In a reported double-blind, randomized clinical trial, the anti-ageing effects of oral supplementation with a fermented papaya preparation (FPP) were evaluated (Bertuccelli *et al.*, 2016). The FPP caused marked improvement in skin quality and integrity in treated patients compared with those without FPP. This anti-ageing effect was further corroborated by significant changes in the gene expression of proteins involved in skin integrity (Bertuccelli *et al.*, 2016). Clinical studies have shown that human subjects (male and female) fed 100 g fresh papaya fruit per day for two days exhibited decreased Th1 T-cell phenotypes, but increased Th2 and Tregs T-cell phenotypes in their PBMCs (peripheral blood mononuclear cells) (Abdullah *et al.*, 2011). It was also established that the increased Tregs phenotype of the male sample was markedly associated with IL-1β level in the *in vitro* culture supernatant (Abdullah *et al.*, 2011). This finding is indicative of an immune system modulatory role of *C. papaya*.

Pawpaw extracts are generally non-toxic; however, there are many reports of interaction between *C. papaya* extract and certain medications, other herbal preparations and some food ingredients (Rodríguez-Fragoso *et al.*, 2011; Oluduro and Omoboye, 2010; Giordani *et al.*, 1997; Nafiu *et al.*, 2019). These interactions may cause potentiation of the effect of certain drugs and herbal supplements or interfere with their bioavailability, absorption and excretion.

12.3.10 *Spilanthes africana*

Spilanthes is a genus comprising over 60 species, which is used in traditional medicine in different parts of the world (Prachayasittikul *et al.*, 2013; Paulraj *et al.*, 2013). *S. africana* is one of the African species used in Nigeria, with a folk reputation as an effective remedy for toothache and inflammation of the gums, hence it is known as the "toothache plant." In traditional medicine, it is used for the treatment of rheumatism, fever, snakebite, fractures, dysentery, and malaria. Studies have shown that the anti-inflammatory activity of the herb is due to the presence of spilanthol, which exerts its anti-inflammatory action via inhibition of the NF-κB pathway, reduction in the mRNA level and protein expression of COX-2 and iNOS, and induced free radical-scavenging activity (Wu *et al.*, 2008). An herbal remedy containing a mixture of aqueous extracts of *S. africana*,

Portulaca oleracea and *Sida rhombifolia* is used in the management of diabetes. The mixture has been shown to significantly ($P<0.05$) decrease the level of glycaemia, and the total cholesterol, triglyceride, and LDL-cholesterol concentrations as well as MDA (Malondialdehyde), AST (Aspartate transaminase), ALT (Alanine transaminase), and creatinine (CT) levels. It also significantly increases the concentration of high-density lipoprotein (HDL)-cholesterol, glutathione, and TAOS. A considerable reduction of the atherogenic indexes CT/HDL and LDL/HDL of the treated groups was also observed. The major phytochemicals present are saturated and unsaturated alkyl ketones, alkamides, hydrocarbons, acetylenes, lactones, alkaloids, terpenoids, flavonoids, and coumarins.

12.3.11 Pterocarpus soyauxii (Oha)

Three species of *Pterocarpus, P. soyauxii, Pterocarpus santalinoides* and *Pterocarpus osun* are used medicinally in Nigeria. The bark, roots, and leaves of these species are commonly used in the preparation of herbal remedies for various ailments. The leaves are valued as a vegetable for preparation of special healing soups and as a tonic. Decoctions of *P. santalinoides* are administered externally to wounds to promote healing, and to treat haemorrhoids and fever. They are taken internally to treat bronchial complaints, amoebic dysentery, stomach ache and sleeping sickness, to prevent abortion and to ease childbirth. *P. soyauxii* is consumed in Southeast Nigeria as a vegetable for the preparation of the highly cherished Oha soup, although the leaves are not tender even after cooking. The plant has been shown to be relatively non-toxic (Tchamadeu *et al.*, 2011). The heartwood of *Pterocarpus* species (especially *P. soyaxii*) is rich in pterostilbenes, a class of phenylpropanoid that possesses remarkable and varied pharmacological activities (McCormack and McFadden, 2013). For example, the heartwood of *P. santalinus* has been reported to possess antioxidative, antidiabetic, antimicrobial, anticancer, and anti-inflammatory properties (Kumar, 2011), and protective effects on the liver, gastric mucosa, and nervous system (Bulle *et al.*, 2016). These health-promoting effects were attributed to the presence of the following bioactive compounds in the plant: santalin A and B, savinin, calocedrin, pterolinus K and L, and pterostilbenes (Bulle *et al.*, 2016).

Studies on the related *P. erinaceus* showed that the methanol extract and its dichloromethane and ethyl acetate fractions possessed significant anti-inflammatory and analgesic activities, as well as strong antioxidant properties (Noufou *et al.*, 2016). Phytochemical analysis indicated the presence of friedelin, lupeol, and epicatechin in the methanol extract and the dichloromethane fraction. The anti-inflammatory effect was established using the croton oil- induced ear oedema assay. The analgesic effect of the extract was evaluated using acetic acid-induced writhing, and the antioxidant properties of the methanol extract and its dichloromethane and ethyl acetate fractions were assessed using the 1,1-diphenyl-2-picrylhydrazyl method. The methanol extract showed a stronger radical-scavenging activity than did its dichloromethane and ethyl acetate fractions (Noufou *et al.*, 2016). Another species, *P. santalinus*, has exhibited significant anti-inflammatory and mild analgesic activity following topical application as a gel in a rat model of the chronic inflammation assay. *P. santalinus* gel also showed significant reduction (P=0.03) in paw volume of rats compared to the other groups. There was significant reduction in the pain threshold (g/s) due to chronic inflammation (Dhande *et al.*, 2017).

12.3.12 *Andrographis paniculata*; King of Bitters (Family: Acanthaceae)

A. paniculata is a very bitter herb used in Nigeria for the preparation of bitter tonic for the treatment of malaria and hypertension. A decoction of the aerial parts of the plant is often used with pawpaw leaves for the treatment of arthritis and other inflammatory conditions. It is also a major ingredient in IHP-Detox Tea, which is used for the management of tissue inflammation, among other indications. *Andrographis* and its major constituent, andrographolide (a diterpenoid lactone), have been extensively studied and their biological properties are well reported in the literature (Iwu, 2014).

Andrographolide-mediated inhibition of enzymes/transcription factors is associated with inflammatory processes. It inhibited the production of NO and TNF-α in lipopolysaccharide-activated PMф in a dosage-related manner. Immunoblot analyses revealed that andrographolide suppressed activation of both iNOS and COX-2 by directly targeting nuclear transcription factor (NF)-κB. Complete Freund's Adjuvant-induced paw oedema in rats was also significantly inhibited by andrographolide treatment. From the data, we conclude that andrographolide exhibited anti-inflammatory effects by suppressing two key inflammatory enzymes and a signalling pathway that mediates expression of a variety of inflammatory cytokines/agents *in situ* (Gupta *et al.*, 2018). Both the crude drug and its major bioactive, andrographolide, are useful as effective phytotherapeutic agents for the clinical treatment of various inflammatory diseases, like rheumatoid arthritis or diseases associated with joint pain.

References

Abarikwu, S.O. (2015) Anti-inflammatory effects of kolaviron modulate the expressions of inflammatory marker genes, inhibit transcription factors ERK1/2, p-JNK, NF-κB, and activate Akt expressions in the 93RS2 Sertoli cell lines. *Molecular and Cellular Biochemistry*, 401, 197–208.

Abdullah, M., Chai, P.S., Loh, C.Y., Chong, M.Y., Quay, H.W., Vidyadaran, S., Seman, Z., Kandiah, M. and Seow, H.F. (2011) *Carica papaya* increases regulatory T cells and reduces IFN-γ+ CD4+ T cells in healthy human subjects. *Molecular Nutrition and Food Research*, 55(5), 803–806.

Afolabi, O.B., Oloyede, O.I. and Agunbiade, S.O. (2018) Inhibitory potentials of phenolic-rich extracts from *Bridelia ferruginea* on two key carbohydrate-metabolizing enzymes and Fe2+-induced pancreatic oxidative stress. *Journal of Integrative Medicine*, 16(3), 192–198.

Altman, R.D. and Marcussen, K.C. (2001) Effects of a ginger extract on knee pain in patients with osteoarthritis. *Arthritis and Rheumatism*, 44: 2531–2538.

Amazu, L.U., Azikiwe, C.C.A., Njoku, C.J., Osuala, F.N. and Enye, J.C. (2010) Antiinflammatory activity of the methanolic extract of the seeds of *Carica papaya* in experimental animals. *Asian Pacific Journal of Tropical Medicine*, 3(11), 884–886.

Anand, P., Kunnumakkara, A.B., Newman, R.A. and Aggarwal, B.B. (2007) Bioavailability of curcumin: problems and promises. *Molecular Pharmaceutics*, 4(6), 807–818.

Awodele, O., Amagon, K.I., Agbo, J. and Prasad, M.N. (2015) Toxicological evaluation of the aqueous stem bark extract of *Bridelia ferruginea* (Euphorbiaceae) in rodents. *Interdisciplinary Toxicology*, 8(2), 89–98.

Balasundram, N., Ai, T.Y., Sambanthamurthi, R., Sundram, K. and Samman, S. (2005) Antioxidant properties of palm fruit extracts. *Asia Pacific Journal of Clinical Nutrition*, 14(4), 319–324.

Baliga, M.K., Latheef, L., Haniadka, R., Fazal, F., Chacko, J. and Arora, R. (2013) *Bioactive Food as Dietary Interventions for Arthritis and Related Inflammatory Diseases*, 41, 529.

Bengmark, S. (2006) Curcumin, an atoxic antioxidant and natural NFkappaB, cyclooxygenase-2, lipooxygenase, and inducible nitric oxide synthase inhibitor: a shield against acute and chronic diseases. *Journal of Parenteral and Enteral Nutrition*, 30(1), 45–51.

Bertuccelli, G., Zerbinati, N., Marcellino, M., Kumar, N., Shanmugam, N., He, F., Tsepakolenko, V., Cervi, J., Lorenzetti, A. and Marotta, F. (2016) Effect of a quality-controlled fermented nutraceutical on skin aging markers: an antioxidant-control, double-blind study. *Experimental and Therapeutic Medicine*, 11(3), 909–916.

Bulle, S., Reddyvari, H., Nallanchakravarthula, V. and Vaddi, D.R. (2016) Therapeutic potential of *Pterocarpus santalinus* L.: an update. *Pharmacognosy Reviews*, 10(19), 43–49.

Chen, M., Hu, D.N., Pan, Z., Lu, C.W., Xue, C.Y. and Aass, I. (2010) Curcumin protects against hyperosmoticity-induced IL-1beta elevation in human corneal epithelial cell via MAPK pathways. *Experimental Eye Research*, 90(3), 437–443.

Colinot, D.L., Garbuz, T., Bosland, M., Rice, S., Wang, L., Sullivan, W.J., Arrizabalaga, G. and Jerde, T. (2017) The common parasite *Toxoplasma gondii* causes prostatic inflammation and microglandular hyperplasia in a mouse model. *The Prostate*, 77(10), 1066–1075.

Dhande, P.P., Gupta, A.O., Jain, S. and Dawane, J.S. (2017) Anti-inflammatory and analgesic activities of topical formulations of *Pterocarpus* santalinus powder in rat model of chronic inflammation. *Journal of Clinical and Diagnostic Research*, 11(7), FF01–FF04.

Dror, E., Dalmas, E., Meier, D.T., Wueest, S., Thévenet, J., Thienel, C., Timper, K., Nordmann, T.M., Traub, S., Schulze, F., Item, F., Vallois, D., Pattou, F., Kerr-Conte, J., Lavallard, V., Berney, T., Thorens, B., Konrad, D., Böni-Schnetzler, M. and Donath, M.Y. (2017) Postprandial macrophage-derived IL-1β stimulates insulin, and both synergistically promote glucose disposal and inflammation. *Nature Immunology*, 18(3), 283–292.

Fairus, S., Leow, S.S., Mohamed, I.N., Tan, Y.A., Sundram, K. and Sambanthamurthi, R. (2018) A phase I single-blind clinical trial to evaluate the safety of oil palm phenolics (OPP) supplementation in healthy volunteers. *Scientific Reports*, 8(1), 8217.

Farombi, E.O. (2011) Bitter Kola (Garcinia kola) Seeds and Hepatoprotection. In: V.R. Preed, R.R. Watson, and V.B. Patel (eds.), *Nuts and seeds in health and disease prevention*, Chapter 26. New York: Elsevier, p. 221.

Gill, L. (1992) *Ethnomedicinal uses of plants in Nigeria*. Benin City, Nigeria: University of Benin Press.

Giordani, R., Gachon, C., Moulin-Traffort, J. and Regli, P. (1997) A synergistic effect of *Carica papaya* latex sap and fluconazole on *Candida albicans* growth. *Mycoses*, 40(11–12), 429–437.

Gupta, S.C., Patchva, S., Koh, W. and Aggarwal, B.B. (2012) Discovery of curcumin, a component of golden spice, and its miraculous biological activities. *Clinical and Experimental Pharmacology and Physiology*, 39(3), 283–299.

Gupta, S., Mishra, K. P., Singh, S. B. and Ganju, L. (2018) Inhibitory effects of andrographolide on activated macrophages and adjuvant-induced arthritis. *Inflammopharmacology*. 26(2), 447–456.

Hall, J. and Bravo-Clouzet, R. (2013) Anti-inflammatory herbs for arthritis. In: Ronald Ross Watson, R. S. and Preedy, V. R. (Eds.) *Bioactive food as dietary interventions for arthritis and related inflammatory diseases*, Elsevier Press, Amsterdam. Chapter 46, pp. 619–631.

Iwu, M.M. (1985) Antihepatoxic constituents of *Garcinia kola* seeds. *Experientia* 41(5), 699–700.

Iwu, M.M., Duncan, A.R. and Okunji, C.O. (1999) New antimicrobials of plant origin. In: J. Janick (ed.), *Perspectives in new crops and new uses*. Alexandria, VA: ASHS Press, pp. 457–462.

Iwu, M.M. (2014) Handbook of African medicinal plants. Second ed. Boca Raton, FL: CRC Press/Taylor and Francis Group, pp. 476.

Iwu, M.M. (2016) Food as medicine – functional food plants of Africa. Boca Raton, FL: CRC Press/Taylor and Francis Group, pp. 384.

Joven, J., Rull, A., Rodriguez-Gallego, E., Camps ,J., Riera-Borrull, M., Hernández-Aguilera, A. et al. (2013) Multifunctional targets of dietary polyphenols in disease: a case for the chemokine network and energy metabolism. *Food and Chemical Toxicology*, 51, 267–279.

Kasahara, Y., Saito, E. and Hikino, H. (1983) Pharmacological actions of Pinellia tubers and Zingiber rhizomes. *Shoyakugaku Zasshi* 1983; 37(1): 73–83.

Konrad, D., Böni-Schnetzler, M. and Donath, M.Y. (2017) Postprandial macrophage-derived IL-1β stimulates insulin, and both synergistically promote glucose disposal and inflammation. *Nature Immunology*, 18(3), 283–292.

Kumar, D. (2011) Anti-inflammatory, analgesic, and antioxidant activities of methanolic wood extract of *Pterocarpus santalinus* L. *Journal of Pharmacology and Pharmacotherapeutics*, 2(3), 200–202.

Lynch, B.S., West, S. and Roberts, A. (2017) Safety evaluation of water-soluble palm fruit bioactives. *Regulatory Toxicology and Pharmacology*, 88, 96–105.

Marconi, M., D'Angelo, L., Del Vecchio, A., Caravaggi, M., Lecchini, S., Frigo, G.M. and Crema, A. (1986) Anti-inflammatory action of *Pygeum africanum* extract in the rat. *Farmaci and Terapia*, 3, 135–137.

McCormack, D. and McFadden, D. (2013) A review of pterostilbene antioxidant activity and disease modification. *Oxidative Medicine and Cellular Longevity*, 2013, 575482.

Modi, A., Mishra, V., Bhatt, A., Jain, A. and Kumar, V. (2016) *Delonix regia*: historic perspectives and modern phytochemical and pharmacological researches. *Chinese Journal of Natural Medicines*, 14(1), 31–39.

Mohamed, S. (2014) Herbs and spices in aging. In: V. Preedy (ed.), *Aging*, 1st ed. *Oxidative Stress and Dietary Antioxidants*. New York: Academic Press, pp. 316.

Nafiu, A.B., Alli-Oluwafuyi, A., Haleemat, A., Olalekan, I.S. and Rahman, M.T. (2019) Papaya (*Carica papaya* L., Pawpaw). In: S. Nabavi and A. Silva (eds.), *Nonvitamin and nonmineral nutritional supplements*, Chapter 3.32. Amsterdam: Elsevier Press, pp. 335–359.

Noufou, O., Anne-Emmanuelle, H., Claude W.O.J., Richard, S.W., André, T., Marius, L., Jean-Baptiste, N., Jean, K., Marie-Genevieve, D.F. and Pierre, G.I. (2016) Biological and phytochemical investigations of extracts from *Pterocarpus erinaceus* poir (fabaceae) root barks. *African Journal of Traditional, Complementary, and Alternative Medicines*, 14(1), 187–195. doi:10.21010/ajtcam.v14i1.2

Nwaehujor, C.O., Ode, J.O., Ekwere, M.R. and Udegbunam, R.I. (2014) Anti-fertility effects of fractions from *Carica papaya* (Pawpaw) Linn. methanol root extract in male wistar rats. *Arabian Journal of Chemistry*, 7(5): 805–810.

Olajide, O.A., Makinde, J.M., Okpako, D.T. and Awe, S.O. (2000) Studies on the anti-inflammatory and related pharmacological properties of the aqueous extract of *Bridelia ferruginea* stem bark. *Journal of Ethnopharmacology*, 71, 153–160.

Olajide, O.A., Okpako, D.T. and Makinde, J.M. (2003) Antiinflammatory properties of *Bridelia ferruginea* stem bark: inhibition of lipopolysaccaride-induced septic shock and vascular permeability. *Journal of Ethnopharmacology*, 88(2–3), 221–224.

Olajide, O.A., Aderogba, M.A., Okorji, U.P. and Fiebich, B.L. (2012) *Bridelia ferruginea* produces antineuroinflammatory activity through inhibition of nuclear factor-kappa B and p38 MAPK signalling. *Evidence-Based Complementary and Alternative Medicine*, 2012, 546873.

Oluduro, A. and Omoboye, O. (2010) In vitro antibacterial potentials and synergistic effect of south-western Nigerian plant parts used in folklore remedy for *Salmonella typhi* infection. *Nature and Science*, 8(9), 52–59.

Owoyele, B.V., Adebukola, O.M., Funmilayo, A.A. and Soladoye, A.O. (2008) Anti-inflammatory activities of ethanolic extract of *Carica papaya* leaves. *Inflammopharmacology*, 16(4), 168–173.

Panchabhai, T.S., Kulkarni, U.P. and Regge, N.N. (2008) Validation of therapeutic claims of *Tinospora cordifolia*: a review. *Phytotherapy Research*, 22, 425–441.

Patra, S., Muthuraman, M.S., Meenu, S., Priya, P. and Pemaiah, B. (2016) Anti-inflammatory effects of royal poinciana through inhibition of toll-like receptor 4 signaling pathway. *International Immunopharmacology*, 34, 199–211.

Paubert-Braquet, M., Cave, A., Hocquemiller, R., Delacroix, D., Dupont, C., Hedef, N. and Borgeat, P. (1994a) Effects of *Pygeum africanum* extract on A23187-stimulated production of lipoxygenase metabolites from human polymorphonuclear cells. *Journal of Lipid Mediators and Cell Signalling*, 9, 285–290.

Paubert-Braquet, M., Monboisse, J.C., Servent-Saez, N., Serikoff, A., Cave, A., Hocquemiller, R., Dupont, C., Foumeau, C. and Borel, J.P. (1994b) Inhibition of bFGF and EGF-induced proliferation of 3T3 fibroblasts by extract of *Pygeum africanum* (Tadenanv) *Biomedicine and Pharmacotherapy*, 48(Suppl. I), 43s–47s.

Paulraj, J., Govindarajan, R. and Palpu, P. (2013) The genus spilanthes ethnopharmacology, phytochemistry, and pharmacological properties: a review. *Advances in Pharmacological Sciences*, 2013, 510298.

Prachayasittikul, V., Prachayasittikul, S., Ruchirawat, S. and Prachayasittikul, V. (2013) High therapeutic potential of *Spilanthes acmella*: a review. *EXCLI Journal*, 12, 291–312.

Rakhimov, M.R. (2000) Pharmacological study of papain from the papaya plant cultivated in Uzbekistan. *Eksperimental'naia i Klinicheskaia Farmakologiia*, 63(3), 55–57.

Riffaud, J.P. and Lacolle, J.Y. (1990) Effects of Tadenan on the detrusor smooth muscle of young and old rats. *European Urology*, 18(SI), 309.

Ritchilin, C. (2000) Fibroblast biology. Effector signal released by the synovial fibroblast in arthritis. *Arthritis Research*, 2(5), 356–360.

Rodríguez-Fragoso, L., Martínez-Arismendi, J.L., Orozco-Bustos, D., Reyes-Esparza, J., Torres, E. and Burchiel, S.W. (2011) Potential risks resulting from fruit/vegetable–drug interactions: effects on drug-metabolizing enzymes and drug transporters. *Journal of Food Science*, 76(4), R112–R124.

Sagnia, B., Fedeli, D., Casetti, R., Montesano, C., Falcioni, G. and Colizzi, V. (2014) Antioxidant and anti-inflammatory activities of extracts from *Cassia alata*, *Eleusine indica*, *Eremomastax speciosa*, *Carica papaya* and *Polyscias fulva* medicinal plants collected in Cameroon. *PLoS ONE*, 9(8), e103999.

Sambanthamurthi, R., Tan, Y., Sundram, K., Hayes, K.C., Abeywardena, M., Leow, S.S., Sekaran, S.D., Sambandan, T.G., Rha, C., Sinskey, A.J., Subramaniam, K., Fairus, S. and Wahid, M.B. (2011) Positive outcomes of oil palm phenolics on degenerative diseases in animal models. *The British Journal of Nutrition*, 106(11), 1664–1675.

Sekaran, S.D., Leow, S.S., Abobaker, N., Tee, K.K., Sundram, K., Sambanthamurthi, R. and Wahid, M.B. (2010) Effects of palm oil phenolics on tumor cells in vitro and in vivo. *African Journal of Food Science*, 4(8), 495–502.

Sharma, N., Kumar, A., Sharma, P.R., Qayum, A., Singh, S.K., Dutta, P., Paul, S., Gupta, V., Verma, M.K., Satti, N.K. and Vishwakarma, R. (2018) A new clerodane furano diterpene glycoside from *Tinospora cordifolia* triggers autophagy and apoptosis in HCT-116 colon cancer cells. *Journal of Ethnopharmacology*, 211, 295–310.

Shehzad, A. and Lee, Y.S. (2010) Curcumin: multiple molecular targets mediate multiple pharmacological actions - a review. *Drugs Future*, 35(2), 113–119.

Shewale, V.D., Deshmukh, T.A., Patil, L.S. and Patil, V.R. (2011) Anti-inflammatory activity of *Delonix regia* (Boj. Ex. Hook). *Advances in Pharmacological Sciences*, 2012, 789713.

Srivastava, K.C. and Mustafa, T. (1992) Ginger (Zingiber officinale) in rheumatism and musculoskeletal disorders. *Medical Hypotheses*, 39, 342–348.

Suryanarayana, P., Saraswat, M., Mrudula, T., Krishna, T.P., Krishnaswamy, K., Reddy, G.B. (2005) Curcumin and turmeric delay streptozotocin-induced diabetic cataract in rats. *Investigative Ophthalmology and Visual Science*, 46(6), 2092–2099.

Tchamadeu, M.C., Dzeufiet, P.D., Nana, P., Kouambou-Nouga, C.C., Ngueguim-Tsofack, F., Allard, J., et al. (2011) Acute and sub-chronic oral toxicity studies of an aqueous stem bark extract of *Pterocarpus soyauxii* Taub (Papilionaceae) in rodents. *Journal of Ethnopharmacology*, 133, 329–335.

Wang, L.-S., Lee, C.T., Su, W.L., Huang, S.C. and Wang, S.C. (2016) *Delonix regia* leaf extract (DRLE): a potential therapeutic agent for cardioprotection. *PloS one*, 11(12), e0167768.

Weinberg, R.P., Koledova, V.V., Schneider, K., Sambandan, T.G., Grayson, A., Zeidman, G., Artamonova, A., Sambanthamurthi, R., Fairus, S., Sinskey, A.J. and Rha, C. (2018) Palm fruit bioactives modulate human astrocyte activity in vitro altering the cytokine secretome reducing levels of TNFα, RANTES and IP-10. *Scientific Reports*, 8(1), 16423.

Wigler, I., Grotto, I., Caspi, D. and Yaron, M. (2003) The effects of Zintona EC (a ginger extract) on symptomatic gonarthritis .*Osteoarthritis and Cartilage*, 11: 783–789.

Williams, E.J., Baines, K.J., Berthon, B.S. and Wood, L.G. (2017) Effects of an encapsulated fruit and vegetable juice concentrate on obesity-induced systemic inflammation: a randomised controlled trial. *Nutrients*, 9(2), 116.

Wu, L. C., Fan, N. C., Lin, M.H., Chu, I.R., Huang, S.J., Hu CY et al. (2008) Antiinflammatory effect of spilanthol from Spilanthes acmella on murine macrophage by down-regulating LPS-induced inflammatory mediators. J Agric Food Chem. 56: 2341-2349.

13 Skin Care, Dental, Oral Care and Cosmeceuticals from Nigerian Plants

13.1 Introduction – The Skin

The skin is the largest organ of the body and serves as a protective shield from the external environment. It is the soft outer covering of the body, that acts as a waterproof, insulating buffer, guarding the body against extremes of temperature, damaging sunlight, and harmful chemicals. It also exudes antibacterial substances, that prevent infection, manufactures vitamin D for converting calcium into healthy bones, and permits the sensations of touch, heat, and cold. In an average adult, the area of skin is about 22 square feet (2 square meters) and weighs about 3.5 to 10 kilograms (7.5 to 22 pounds). It is an amazingly versatile organ that envelops the whole body and, at the same time, allows the free movement of the body because of its elasticity.

There are three layers that constitute the skin structure, namely the epidermis, dermis, and hypodermis. The epidermis is the outermost layer, which consists mainly of cells called keratinocytes, made from the tough protein keratin (the same material as in hair and nails). Keratinocytes form several layers that constantly grow outwards as the exterior cells die and flake off. It takes roughly five weeks for newly created cells to work their way to the surface. This covering of dead skin is known as the stratum corneum, or horny layer, and its thickness varies considerably, being more than ten times thicker on the soles of the feet than around the eyes. The epidermis harbors defensive Langerhans cells, which alert the body's immune system to viruses and other infectious agents.

The dermis is the deeper skin layer below the epidermis, which gives the organ its strength and elasticity due to its content of collagen and elastin fibres. The regulation of body temperature is carried out by the dermis, by increasing blood flow to the skin, to allow heat to escape through a network of blood vessels, or by restricting the flow when it is cold. An intricate system of nerve fibres and receptors that abound in the dermis are responsible for the transmission to the brain of feelings, such as touch, temperature, and pain. The dermis also houses hair follicles, the apocrine glands which develop during puberty and produce a scented sweat linked to sexual attraction, that can also cause body odor, especially around the armpits, with ducts that pass up through the skin. The dermis also contains sweat glands that bring down internal temperature through perspiration, while ridding the body of waste fluids containing urea and lactate, as well as the sebaceous glands, which secrete oil-like sebum that lubricates the hair and skin.

The third layer is the hypodermis, also known as the subcutis, subcutaneous tissue, hypoderm, or superficial fascia. It is an irregular layer of adipose and connective tissue, stroma, and membranes, that is found beneath the dermis. The types of cells that occur in the hypodermis are fibroblasts, adipose cells, and macrophages. The hypodermis is the skin's base layer, that works as insulation and cushions the body from knocks and falls. It includes a seam of fat laid down as a fuel reserve in case of food shortage, although, in some parts of the body, it is nearly devoid of fat (as in the auricles, eyelids, scrotum, and penis). This layer of the skin varies most between sexes and among different nutritional states. The subcutis is penetrated by, and gains support from, skin ligaments [TA] (retinacula cutis [TA]) extending between the dermis and the deep fascia; cutaneous nerves and superficial vessels course within the subcutaneous tissue, with only their terminal branches passing to the skin.

Nigeria is situated in the tropics, with enormous climatic variation within the country, ranging from the dry semi-desert areas of the northern regions to the dry savannah belt in the North central area, the deciduous forests of the middle-belt, and the rainforest jungles of the South. Each geographical region poses its own peculiar challenges to the skin of the inhabitants. The people living in these climatic zones have developed natural adaptations, as evidenced by the different skin types that are best suited to each particular environment. They have also developed a range of strategies to cope with their harsh environment by developing skin care products for nurturing and maintaining the skin, as well as natural remedies to treat cutaneous diseases. The skin is naturally colonized by a large number of microorganisms (the skin microbiota), most of which are harmless and even beneficial, as some of the microorganisms maintain symbiotic relationship with the skin surface by providing protection against invasion by pathogenic microorganisms. In a healthy skin, a balance is maintained between the host and the microorganisms, with disruptions in this balance resulting in skin disorders or infections (Grice and Segre, 2011).

13.2 Natural Personal Care

Many Nigerian plants are used in the preparation of cosmetics or personal care products that are aimed at maintaining the integrity of the skin against the external environment or at enhancing the cultural perspective of beauty, aesthetics, or appearance. Several oil-bearing plants and herbs are used as skin lubricants, moisturisers, or colourants. Many of these plants are used as ingredients in the formulation of a large variety of products, such as creams, lotions, soaps, shampoos, shower products, sunscreens, hair care products, hair dyes, makeup, lipstick, toothpaste, dental care products, deodorants, personal hygiene products, and many others (Alade *et al.*, 2015; Oyedemi *et al.*, 2018).

13.2.1 Skin Conditions and Chronic Skin Diseases

The topical application of drugs for the treatment of various skin conditions is a well-known aspect of medical treatment. Sometimes, certain medicines for systemic treatment are administered as intra-dermal patches. The semi-permeable property of the skin has been successfully exploited for administration of many active pharmaceutical products that require delayed release or time-dependent bioavailability, such as insulin (for diabetics), contraceptives, painkillers and cardiotonic agents. It is generally believed that substances that are safe for oral consumption can be assumed to be safe when applied topically, and this had led to the introduction of many therapeutic agents for intra-dermal applications (Tirant *et al.*, 2018).

The same thinking has led to a renewed interest in herbal preparations to address skin conditions and diseases by deliberately including functional ingredients into cosmetic products, that are now classified as cosmeceuticals (Charles – Dorni *et al.*, 2017).

Chronic skin conditions are typically not curable, but they can be managed using medicines and by modifications in lifestyle. Outlined below are the major skin conditions, their description, symptoms, and herbal products used for the treatment.

Condition	Description	Plants Used
Acne also known as acne vulgaris	The most common skin condition, that occurs especially during adolescence, acne affects over 85% of people at some time in life. It is a long-term skin disease that occurs when hair follicles are clogged with dead skin cells and oil from the skin. It is characterized by blackheads or whiteheads, pimples, oily skin, and sometimes skin scarring.	*Ocimum gratissimum*, black seed oil, turmeric; lemongrass, and pawpaw leaves (as cleansers)
Cellulitis	Inflammation of the dermis and subcutaneous tissues, usually due to an infection. A red, warm, often painful skin rash generally results.	*Garcinia kola*, *Andrographis paniculata*, pineapple fruit peel

(*Continued*)

Condition	Description	Plants Used
Dandruff	A scaly condition of the scalp that is characterized by flaking and sometimes mild itchiness. It may be caused by seborrhoeic dermatitis, psoriasis, or eczema.	*Dostenia mannii*, black seed oil, *Bryophyllum pinnatum*
Dermatitis	A general term for inflammation of the skin. Atopic dermatitis (a type of eczema) is the most common form.	*Bryophyllum pinnatum*, ginger, turmeric, shea butter
Eczema	Skin inflammation (dermatitis), causing an itchy rash. Most often, it iss due to an overactive immune system.	*Bryophyllum pinnatum, Garcinia kola, Psidium guaja* leaves
Herpes	The herpes viruses HSV-1 and HSV-2 can cause periodic blisters or skin irritation around the lips or the genitals.	*Andrographis paniculata, Garcinia kola, Picralima nitida* seeds (extracted with kerosene)
Hives	Raised, red, itchy patches on the skin that arise suddenly. Hives usually result from an allergic reaction.	*Carica papaya* leaves, Camwood, shea butter
Psoriasis	An autoimmune condition that can cause a variety of skin rashes. Silver, scaly plaques on the skin are the most common form.	*Rothmannia longiflora, Pterocarpus* spp., *Baphia nitida*, black soap from palm husk; internal oral medication: turmeric, bitter kola, *Spathodea campanulata, Tinospora* spp., *Arnica montana*
Skin abscess (boil or furuncle)	A localized skin infection creates a collection of pus under the skin. Some abscesses must be opened and drained by a doctor in order to be cured.	*Andrographis paniculata*, Black seed, *Garcinia kola, Chamomilla recutita, Carica papaya leaves, Psidium guajava, Piper guineense, Allium cepa*
Rash	Change in the skin's appearance can be called a rash. Most rashes are from simple skin irritation; others result from medical conditions	*Ocimum gratissimum, Chromolaena odorata*, coconut oil, palm kernel oil, shea butter
Ringworm	A fungal skin infection (also called tinea). The characteristic rings it creates are not due to worms.	*Andrographis paniculata*, Black seed, *Garcinia kola, Carica papaya leaves, Cassia alata*, black soap, *Acalypha hispida*
Rosacea	A chronic skin condition causing a red rash on the face, pimples, swelling, and small and superficial dilated blood vessels. Often the nose, cheeks, forehead, and chin are most involved. A red enlarged nose may occur in severe cases, a condition known as rhinophyma. Rosacea may look like acne, and is poorly understood.	*Corchorus olitorius, Garcinia kola* root, *Carica papaya* leaves, *Cola nitida, Citrus x aurantifolia* (lime), *Aloe vera*

(*Continued*)

Condition	Description	Plants Used
Scabies	Tiny mites that burrow into the skin cause scabies. An intensely itchy rash in the webs of fingers, wrists, elbows, and buttocks is typical of scabies.	*Butyrospermum paradoxum* (shea) oils, Black seed, *Garcinia kola, Carica papaya leaves,* black soap, *Calpurnia aurea, Ageratum conyzoides, Azadirachta indica*
Tinea versicolor	A benign fungal skin infection creates pale areas of low pigmentation on the skin.	*Acalypha wikesiana, Aframomum melegueta,* shea butter, *Argemone mexicana*
Warts	A virus infects the skin and causes the skin to grow excessively, creating a wart. Warts may be treated at home or removed through a minor surgical operation.	*Azadirachta indica, Aloe vera, Ricinus communis, Dorstenia scaphigera, Solanum nigrum.*

13.3 Major Nigerian Plants Used in the Preparation of Polyherbal Skin-Care or Cosmeceutical Products

Traditional healers, that are specialists in treating skin diseases, use certain plants, usually mixed with vegetable oils or added to black soap preparations, as cosmeceuticals. They are used to achieve a specific therapeutic effect, such as alleviation of topical pain, or antiallergic, antioxidant, or antiageing effects, removal of wrinkles, warts and stretch marks, skin-tone enhancement, hair darkening, etc. As a result of a better understanding of the physiology of the skin, including the peculiarity and resistance of the black African skin texture and the ageing process, many herbal preparations are currently in use, with potential to repair damaged skin, or to enable skin to retain and/or remain in a healthy state. The phytochemicals found in such plants have been shown by studies to possess the ability to alter or recover the original healthy skin and external appearance.

The most frequently prescribed herbs for the management of aberrant skin conditions include:

Abelmoschus esculentus
Acalypha ornate
Adansonia digitatA
Albizia ferruginea
Aloe vera
Capsicum frutescens
Cardiospermum africana
Cardiospermum halicacabum
Carica papaya
Cola nitida
Combretum micranthum
Crabbea velutina

Malva sylvestris
Mangifera indica
Pentas longiflora
Polygala erio
Pseudovigna argentea
Rhus vulgaris

Croton macrostachys
Cyphostremma spp.
Dorstenia scaphigera
Eleusine coracana
Erigeron floribunda
Euphorbia hirta
Galium spp.
Hibiscus sabdariffa
Ludwigia erecta
Rhynchosia viscosa
Ricinus communis
Solanum nigrum
Spathodea campanulata
Synaedenium spp.
Theobroma cacao
Vismia orientalis
Withania somnifera

The plants listed below are used in various formulations to protect the integrity of the healthy skin and/or to enhance its beauty, aesthetics, or appearance. They may possess specific physiological properties but they are rarely used for the treatment of skin diseases.

Functional Use: Plant Examples

1. **Antioxidants**: Tomatoes, turmeric, bitter kola, cocoa extracts, *Andrographis paniculata*, black seed oil, *Cajanus cajan*, *Pterocarpus* spp., camwood (*Baphia nitida*), carrot, and essential oils from various flowers, fruits and leaves.

2. **Anti-inflammatory/Anti-irritant**: *Rothmannia longiflora*, *Andrographis paniculata*, black seed, bitter kola, pawpaw leaves, palm oil, and *Bryophyllum pinnatum* leaves

3. **Natural Colourants:** Henna, indigo, chamomile, tomatoes, vegetable dyes, and camwood

4. **Fragrances, Essential Oils**: Lemongrass oil, *Eucalyptus globulus*, *Ocimum* spp., queen of the night (*Selenicereus grandiflorus*), cloves, as well as extracts containing terpenes, terpenoids, aldehydes, alcohols, esters, ketones, phenols, or methoxyphenols from aromatic leaves, flowers, and barks.

5. **Cleansers, Preservatives, Antiseptics:** Soaps prepared from palm husk, vegetable oils and leaves, and saponin-bearing plants (usually identified from foaming in water), plants containing essential oils, benzoic/salicylic acids, and plants containing organic acids and esters or phenols; *Camellia sinensis*, *Capsicum annuum* (red pepper, totoshi), *Cinnamomum* spp. (cinnamon), *Citrus limon* (lemon), and *Citrus paradisi* (grapefruit)

6. **Hydration/Moisturising**: Palm kernel oil, coconut oil, various seed oils, *Citrullus vulgaris*, moringa oil, *Aloe vera*, camwood, black seed oil, and *Bryophyllum pinnatum*

7. **Skin Toning/Whitening**: Cashew nut juice, lime, pawpaw fruits, soy proteins, cocoa pod fruit pulp extracts, bitter kola fruit pulp and water

extract of fresh bitter kola seeds, *Mucuna pruriens*, *Prunus domestica*, *Artemisia afra*, *Cichorium intybus*, *Cynara scolymus*, and leaves of *Helichrysum stoechas*

8. **Anti-ageing/Free Radical Scavenging:** Grape seed extract, bitter kola lipid extract, palm fruit bioactive fraction, carrot and coconut oil, tomatoes, guava leaves, *Pterocarpus* species, *Tamarindus indica*, *Medicago sativa*, *Pueraria mirifica*, *Prunus* spp., and turmeric.

In addition to the above-listed plants several Nigerian plants have been used in skin care preparations that help to stimulate and regenerate skin repair after injury. The conceptual framework is that, since plant regeneration at the cellular and tissue level is a process that can be similar to that in animals, such undifferentiated stem cells in plants have properties that are beneficial in skin therapy. The unique properties of such plant stem cells have been a recent area of interest and focus, both in developing new cosmetics and in studying how these plant extracts will influence animal skin. Available evidence-based trends in plant stem cell-based cosmetics shed light on the challenges that we need to overcome in order to see meaningful changes in human skin using topical cosmetics derived from plant stem cells.

13.4 Dental and Oral Care Plant Products

The oral cavity is heavily colonized by many microbes such as bacteria (Eubacteria), viruses, protozoa, fungi, and members of Archaea. Most of the micro-organisms isolated from the mouth are innocuous, whereas a few are potentially disease-causing and could be harmful if not controlled. Two common diseases of the mouth, dental caries and periodontal diseases, are caused by these microbes. Dental caries and periodontal diseases (including gingivitis) are among the most common infectious diseases affecting humans (Chinsembu, 2016). Dental caries is a medical term used to describe tooth cavities or decay. It is a polymicrobial disease caused by specific bacteria that produce acid, which destroys the tooth's enamel and the layer under it (dentine). The disease process involves acidogenic plaque bacteria, including *Streptococcus mutans*, *Streptococcus sobrinus* and *Lactobacillus* spp. (Maeda *et al.*, 2013). The human tooth is a hard, calcified structure found in the jaws, with roots that are covered by gums. They are used primarily to break down food, whereas some animals, particularly carnivores, also use teeth for hunting or for defensive purposes. It is important to note that teeth are not made of bone, but rather of multiple tissues of varying density and hardness. The cellular tissues that ultimately become teeth originate from the embryonic germ layer, the ectoderm. The care of the teeth has to take into account the characteristics of the tooth structure and its susceptibility to dental caries-causing microbes. Although dental caries can be prevented and treated by good oral hygiene, if not properly managed, it can progress to inflammation and death of vital pulp tissue, with eventual spread of infection to the periapical area of the tooth and beyond (Allaker and Douglas, 2009).

Periodontitis involves inflammation around the tooth. It is a serious gum infection that damages the soft tissue and bone that supports the tooth. Periodontitis is associated with oral microbiota including *Aggregatibacter actinomycetemcomitans, Eikenella corrodens, Fusobacterium nucleatum, Porphyromonas gingivalis, Prevotella intermedia, Tannerella forsythia, Enterococcus faecalis* and *Treponema denticola* (Alireza et al., 2014). *P. gingivalis, P. inter-media* and *A. actinomycetemcomitans* are regarded as the major pathogens in causing periodontitis (Allaker and Douglas, 2009). Archaea, including *Methanobrevibacter oralis, Methanobacterium* spp., *Methanosarcina* spp., and *Thermoplasmata* spp., are also responsible for oral diseases (Maeda *et al.*, 2013). In most cases of sub-gingival plaque, the dominant microorganism is the archaeal species *M. oralis.* Gingivitis is a mild and reversible form of periodontal disease, but periodontitis causes permanent damage to tooth-supporting tissues and may lead to tooth loss (Bonifait *et al.*, 2012). Whereas as many as 700 different bacterial species may be present in sub-gingival plaque samples, available evidence indicates that *P. gingivalis*, a Gram-negative anaerobic bacterium, is often found in association with *T. forsythia* and *T. denticola* as the key pathogens in periodontitis (Azelmat *et al.*, 2015; Chinsembu, 2016). The major Nigerian plants used for the treatment of periodontal diseases include *Garcinia kola, Allium cepa, Acalypha hispida, Distemonanthus benthamianus, Bridelia ferruginea, Anogeissus leiocarpus. Terminalia glaucescens, Baphia nitida, Macrolobium macrophyllum, Fagara rubescens, Cassia alata, Breynia nivosa, Eleusine indica*, and *Chromolaena odorata.*

13.4.1 Chewing Sticks

Chewing sticks are dried stems or roots from plants, which are used in Nigeria and other parts of Africa and the Middle East as a means of maintaining good oral hygiene (Lewis and Elvin-Lewis, 1977). In using the stick, the end of the stick is chewed into a fibrous brush, which is rubbed or brushed against the teeth and gum. It imparts positive effects on the tooth and oral cavity by the mechanical cleansing action and the antimicrobial properties of many plants used as chewing sticks (Wolinsky and Okoye and Sote, 1983; Akinyemi *et al.*, 2005; Xu *et al.*, 2013; Rao *et al.*, 2014; Anyanwu, 2017). Many studies have established the fact that chewing stick users have a lower rate of dental caries and a better general oral health than non-users (Emslie, 1966; Enwonwu, 1974; Gazi *et al.*, 1990; Malik *et al.*, 2014).

The major plants used as chewing sticks in Nigeria include *Butyrospermum paradoxum, Garcinia kola, Distemonanthus benthamianus, Bridelia ferruginea, Anogeissus leiocarpus. Terminalia glaucescens* and *Fagara rubescens.* It is remarkable that analyses of the constituents of these plants, using cyclic neutron activation analysis (CNAA) (Adesanmi, 1989), PIXE (Proton Induced X-ray Emission) or PIGE (Proton Induced Gamma-ray Emission) techniques (Olabanji *et al.*, 1996), showed the presence of fluorine and other minerals in all the plant samples examined. The concentration of fluorine in the chewing sticks was specifically determined using the $^{19}F(p, p'y)^{19}F$ reaction. PIXE and PIGE results show the presence of twenty elements in the chewing sticks studied but of particular interest is the detection of fluorine in all the samples, and calcium at relatively high concentrations because of their direct useful impact on teeth.

Other plants used as chewing sticks but at a lower level of use include *Vernonia amygdalina, Anacardium occidentale, Cocos nucifera, Citrus sinensis, Canarium schweinfurthii, Carica papaya, Ricinodendron heudelotii, Mangifera indica*, and the comb of *Zea mays*.

13.4.2 Plants Used for Other Oral Diseases and Conditions

Apart from tooth problems, the oral cavity has other medical conditions and diseases, that can be treated or managed with medicinal plant products. Examples of oral health problems treated with plant-based traditional medicines include mouth sores, fungal infections, xerostomia, sore throat, mouth sores, abscess, broken tooth and jaw, tooth sensitivity, mouth thrush, tonsillitis, oral syphilis, oral cancer, halitosis, and dental abscess. Herbal remedies for these conditions are prepared with one or more of the following plants as the key ingredients: *Irvingia gabonensis, Ricinodendron heudoletti, Pterocarpus soyauxii, Alchornea cordifolia, Garcinia kola* and *Piptadeniastrum africanum*.

Xerostomia, for example, which is “dry mouth”, resulting from reduced or absent saliva flow due to salivary gland hypofunction, is usually treated with an oral rinse prepared with *Alchornia cordifolia* leaves and *Aframomum melegueta* seeds. Xerostomia is not actually a disease, but occurs as a symptom of various medical conditions, such as a side-effect of irradiation to the head and neck during cancer therapy, or side-effects of certain medications. The condition can cause serious discomfort by creating difficulties eating, speaking, and sleeping, thereby impairing the patient’s quality of life. Xerostomia may increase the risk of oral infections.

Halitosis, which is the emission of unpleasant mouth odour, may be caused by several oral conditions, including poor oral hygiene, and periodontal disease, as well as by respiratory tract conditions, such as chronic sinusitis, tonsillitis, and bronchiectasis (Scully and Porter, 2008). Simple halitosis is treated with oral rinses and herbal decoctions, but, in some cases, halitosis can be caused by underlying systemic disease that would require disease-specific treatment. Tooth brushing and antiseptic mouthwashes are often not enough to treat halitosis. Use of herbal remedies, tongue cleaning, artificial saliva, and lifestyle modifications, such as dietary changes, switch to sugar-free chewing gums, and the use of zinc and herbal toothpastes have been shown to reduce or eliminate halitosis The plants used in the preparation of remedies to treat halitosis include *Ocimum viride* leaves, *Garcinia kola* seeds, *Xylopia aethiopica, Alchornia cordifolia* leaves and *Aframomum melegueta* seeds.

References

Adesanmi, C.A. (1989) Multielement analysis of Nigerian chewing sticks by cyclic neutron activation analysis (CNAA). *International Journal of Radiation Applications and Instrumentation. Part A. Applied Radiation and Isotopes*, 40(3), 263–264.

Alade, G.O., Okpako, E., Ajibesin, K.K. and Omobuwajo, O.R. (2015) Indigenous knowledge of herbal medicines among adolescents in Amassoma, Bayelsa State, Nigeria. *Global Journal of Health Science*, 8(1), 217–237.

Alireza, R.G.A., Afsaneh, R., Hosein, M.S.S., Siamak, Y., Afshin, K., Zeinab, K., Reza, R.A. (2014). Inhibitory activity of Salvadora persica extracts against oral bacterial strains associated with periodontitis: an in-vitro study. J. Oral Biol. Craniofac.Res. 4 (1), 19–23.

Allaker, R.P. and Douglas, C.W. (2009) Novel anti-microbial therapies for dental plaque-related diseases. *International Journal of Antimicrobial Agents* 33(1), 8–13.

Akinyemi, K.O., Oladapo, O., Okwara, C.E., Ibe, C.C. and Fasure, K.A. (2005) Screening of crude extracts of six medicinal plants used in South-West Nigerian unorthodox medicine for anti-methicillin resistant *Staphylococcus aureus* activity. *BMC Complementary and Alternative Medicine*, 2005: 5–6.

Anyanwu, M.U. and Okoye, R.C. (2017) Antimicrobial activity of Nigerian medicinal plants. *Journal of Intercultural Ethnopharmacology*, 6(2), 240–259.

Azelmat, J., Larente, J.F. and Grenier, D. (2015) The anthraquinone rhein exhibits synergistic antibacterial activity in association with metronidazole or natural compounds and attenuates virulence gene expression in *Porphyromonas gingivalis*. *Archives of Oral Biology*, 60(2), 342–346.

Bonifait, L., Marquis, A., Genovese, S., Epifano, F. and Grenier, D. (2012) Synthesis and antimicrobial activity of geranyloxy-and farnesyloxy-acetophenone derivatives against oral pathogens. *Fitoterapia*, 83(6), 996–999.

Charles - Dorni, A.I., Amalraj, A., Gopi, S., Varma, K. and Anjana, S.N. (2017) Novel cosmeceuticals from plants — an industry guided review. *Journal of Applied Research on Medicinal and Aromatic Plants*, 7, 1–26.

Chinsembu, K.C. (2016) Plants and other natural products used in the management of oral infections and improvement of oral health. *Acta Tropica*, 154, 6–18.

Emslie, R.D. (1966) A dental health survey in the Republic of the Sudan. *British Dental Journal*, 120, 167–178.

Enwonwu, C.O. (1974) Socio-economic factors in dental caries prevalence and frequency in Nigerians: an epidemiological study. *Caries Research*, 8, 155–171.

Gazi, M., Saini, T., Ashri, N. and Lambourne, A. (1990). Meswak chewing stick versus conventional tooth brush as an oral hygiene aid. *Clinical Preventive Dentistry*, 12, 19–23.

Grice, E.A. and Segre, J.A. (2011) The skin microbiome. *Nature Reviews: Microbiology*, 9, 244–253.

Lewis, W.H. and Elvin-Lewis, M.P.F. (1977) *Oral hygiene, in medical botany plants affecting man's health*. New York: John Wiley & Sons.

Maeda, H., Hirai, K., Mineshiba, J., Yamamoto, T., Kokeguchi, S. and Takashiba, S. (2013) Medical microbiological approach to Archaea in oral infectious diseases. *Japanese Dental Science Review*, 49(2), 72–78.

Malik, A.S., Shaukat, M.S., Qureshi, A.A. and Abdur, R. (2014) Comparative effectiveness of chewing stick and toothbrush: a randomized clinical trial. *North American Journal of Medical Sciences*, 6(7), 333–337.

Olabanji, S.O., Makanju, O.V., Haque, A.M.I., Buoso, M.C., Ceccato, D., Cherubini, R. and Moschini, G. (1996) PIGE-PIXE analysis of chewing sticks of pharmacological importance. *Nuclear Instruments and Methods in Physics Research Section B: Beam Interactions with Materials and Atoms*, 113(1–4), 368–372.

Oyedemi, B.O., Oyedemi, S.O., Chibuzor, J.V., Ijeh, I.I., Coopoosamy, R.M. and Aiyegoro, A.O. (2018) Pharmacological evaluation of selected medicinal plants used in the management of oral and skin infections in Ebem-Ohafia District, Abia State, Nigeria. *The Scientific World Journal*, 2018, 4757458.

Rao, D.S., Penmatsa, T., Kumar, A.K., Reddy, M.N., Gautam, N.S. and Gautam, N.R. (2014) Antibacterial activity of aqueous extracts of Indian chewing sticks on dental plaque: an *in vitro* study. *Journal of Pharmacy and Bioallied Sciences*, 6(Suppl 1), S140–S145.

Scully, C. C. and Porter, S. (2008) Halitosis. *BMJ Clinical Evidence*, 2008, 1305 (12 pages).

Tirant, M., Lotti, T., Gianfaldoni, S., Tchernev, G., Wollina, U. and Bayer, P. (2018) Integrative dermatology – The use of herbals and nutritional supplements to treat dermatological conditions. *Open Access Macedonian Journal of Medical Sciences*, 6(1), 185–202.

Wolinsky, L.E. and Sote, E.O. (1983). Inhibiting effect of aqueous extracts of eight Nigerian chewing sticks on bacterial properties favoring plaque formation. *Caries Research*, 17, 253–257.

Xu, H.-X., Mughal, S., Taiwo, O. and Lee, S.F. (2013) Isolation and characterization of an antibacterial biflavonoid from an African chewing stick *Garcinia kola* Heckel (Clusiaceae). *Journal of Ethnopharmacology*, 147(2), 497–502.

14 Nigerian Healing Plants in Global Trade

A major development in healthcare delivery over the past two decades is the paradigm shift in people's understanding of what constitutes good health and the clear distinction between diseases and medicine on one hand and healthy living on the other. Another observation is that, contrary to popular literature, which claims that more than 80% of the population in traditional societies depend almost exclusively on traditional medicines and herbal remedies for their health needs (Bannerman, 1982), traditional medicines are used in most developing countries only as an alternative to western allopathic medicine or as a treatment of last resort when modern medicine fails. This often-cited statement by World Health Organization (WHO), which estimates that 80% of the world's population relies solely or largely on traditional remedies for health care, can only apply to isolated hunter-gatherer communities and is no longer tenable in most developing countries. It is noteworthy that, as traditional societies embrace modern medicine, the global desire for herbal medicines has moved in the opposite direction. It is also remarkable that, whereas the poor, who are often the custodians of traditional medical secrets, are demanding better access to modern medicine, the urban rich are increasingly choosing traditional and alternative medicines in place of modern therapeutics. Traditional medicine is increasingly consumed in western countries, where it is commonly called "alternative/complementary/holistic/herbal/indigenous/integrative/native/natural/non-toxic/oriental/unconventional" and "fringe/non-traditional/unproven/

unscientific" medicines. These types of medical care are sometimes referred to collectively as Complementary and Alternative Medicines (CAM). This term excludes what has been termed "allopathic/conventional/mainstream/modern/orthodox/western" medicine (Smith-Hall *et al.*, 2012). The medicinal plants themselves have largely moved out of the classical drug category to become part of a new class of safe but active botanical agents called functional foods, dietary supplements, nutraceuticals, etc.

These developments have fueled the growth in the global herbal medicine business (Mohammed-Abubakar *et al.*, 2017). Annual sales of these ethnic-based products account for a large proportion of the wider market for nutritional supplements that is valued at US$50 billion and seems impervious to the economic instability that has affected other business sectors (Harrison-Dunn, 2014). This profitable trade provides a powerful incentive for investment in the cultivation, processing and development of various types of natural products. According to WHO (2013), the key drivers of the trade in medicinal plants and its increasing demand include:

- Concern about the adverse effects of pharmaceutical drugs.
- The perceived safety and efficacy of herbal medicines.
- Questioning of orthodox medicine and concern about physician and prescription costs.
- Increased risk of long-term chronic illness, for which pharmaceutical drugs are seen to be ineffective.
- Increased access to health information and herbal medicinal products, through the internet (Picking, 2017).

Nigerian medicinal plants have been used in the prevention, diagnosis, and treatment of diseases based on the practical experience of thousands of years, but their introduction into the global trade on herbal medicine, nutraceuticals, and phytotherapeutics has been slow and limited. A critical and pivotal gap is the low level of practical translational research that will transform laboratory findings into fruitful formulations, leading to the development of newer products for the cure of diseases that are not adequately treated by existing allopathic drugs. Most of the scientific investigations on the pharmacology, phytochemistry, and formulations of Nigerian medicinal plants rarely go beyond the proof of concept stage. The potential role of Nigerian plants in the treatment of diseases such as AIDS, cancers, obesity, diabetes, cardiovascular diseases, and metabolic syndrome has been highlighted in previous chapters, and the focus of this chapter is on those plants that are of commercial significance and can be possible commodities in global trade. In the local markets in Nigeria, medicinal plants are sold as fresh or dried whole plants, and also as plant parts, such as roots, rhizomes and bulbs, stem bark, wood, shredded leaves, and fruits. Most of the herbal medicinal products are consumed as generally unrefined plant products with limited processing, apart from grinding, chopping, and mixing. Others are used in the preparation of infusions, concentrates, smoke inhalants, and burned plants (Dzoyema *et al.*, 2013).

Medicinal plants are also exported as items of commerce in many forms, viz. crude or processed whole plants, extracts and exudates, and sometimes as purified or isolated characterised compounds. The first type are medicinal plants whose pharmacological activity is based on ethnomedical leads and often with several established medicinal properties used to treat a variety of disease conditions by Nigerian immigrant populations in Europe, North America, and Asia. Such plants are imported by wholesalers and distributed through African and Asian food stores and supermarkets. The second type consists of extracts and other liquid preparations sold directly to drug producers for the preparation of galenicals and formulated phytomedicines. Furthermore, a single plant may contain many chemical constituents such as phenols, glycosides, polysaccharides, alkaloids, resins, and terpenoids, which can exhibit therapeutic activities towards more than one medical condition. Herbal medicines are also formulated in dosage forms, such as tablets and capsules, powders, extracts, pastes, gels, and oils. Tablets and capsules provide dose accuracy compared to other dosage forms. Hence, this market is expected to record the fastest growth over the period forecast. Poor regulatory framework and low numbers of institutes providing knowledge of herbal therapeutics due to a lack of relevant research evidence have hindered the market growth of Nigerian plants in global trade.

14.1 Herb and Medicinal Plant Value Chain in Nigeria

The global herbal medicine market size was valued at US$71.19 billion in 2016 (Hexa, 2017), with a projected 2017 value of US$107 billion. The remarkable increase in the market for herbal medicines is attributed to the increasing preference of western consumers for traditional medicines, which have been proved to be safer than conventional drugs, do not cause overdose toxicity, and have fewer side-effects. The extracts segment alone generated a revenue of US$27.1 billion in 2016 and is expected to reach US$44.6 billion by 2024. Higher absorption rates from extracts, in comparison with other dosage forms, is supposed to drive the market over the forecast period. With the demand growing for natural medicine at the rate of 15–25% annually, it is projected that, by 2050, the trade will be worth up to US$5 trillion. It has been noted that commercial herbal medicine has been dominated to date by northern hemisphere plants, mainly from the traditional medicine systems of Europe and Asia (especially Ayurveda and Chinese Traditional Medicine) and, to a lesser extent, plants derived from Latin American and Australian traditional medicines (Van Wyk, 2015).

It has been suggested that local commercialisation of natural products can contribute to reduced poverty and vulnerability of the rural poor (Shackleton *et al.*, 2008). In a study on rural communities in the Ndoki and Kalabari areas of the Niger Delta region of Nigeria, it was reported that 32.8% of households in the communities have, as their major occupation, the extraction and trade in non-timber forest products, including medicinal plants (Chukwuone *et al.*, 2013). The economic value of medicinal plants goes beyond the tradeable market value to have a direct impact on sustainable livelihoods of rural communities. From a socio-economic perceptive, benefits are derived by multiple stakeholders involved at several levels in the supply, distribution, and consumption channels. While the trade in herbs and spices may play a key role in enhancing the livelihood security of the poorest households, such products were unlikely to provide a route out of poverty for most of the farmers, herb collectors and sellers because the products are sold without much added value. A key element of this review is an assessment of the relative benefits of the herbal medicinal plant trade to the primary producers, as opposed to middlemen, wholesalers, and retailers. Often, it was found that farmers or gatherers obtained only a minimal share of the benefits of such products. Medicinal herbs, and the products derived from them, also seem to have very varied value chains. However, despite the size of the trade in medicinal herbs and herbal products, surprisingly, only a minuscule part of the value chain is retained in Nigeria because most of the processing activities are carried out offshore. Five major levels of medicinal plant transactions in the market can be identified: direct consumers, street traders, shop traders, traditional healers, and manufacturing companies (Dzoyema *et al.*, 2013).

Major factors that are responsible for the low participation of Nigerian medicinal plants in the global trade include the absence of large-scale medicinal plant farms, lack of bio-processing facilities, that would enable speedy drying and processing of farm produce, poor quality control, and difficulty in obtaining credible information on Nigerian plants. These factors are also responsible for the low impact of the medicinal plants trade to the Nigerian economy (Muhammad and Awaisu 2008). Fortunately, it has been established that value chains are not fixed in terms of composition, relationships, or market positioning, and that there is a competitive need to alter and improve the value chain in light of strategic choices that businesses can make, regarding the markets in which they compete. Whereas a value chain's purpose is to link production to the target market advantageously, it is the private sector that decides which markets and where to compete — and alters the value chain accordingly (Martin-Webber and Labaste, 2010).

14.2 Market Structure and Segmentation

There is a variety of stakeholders in the medicinal plants business in Nigeria, as outlined above. The market itself exits in three clusters, namely (Adegboyega and Oluwalana 2011): 1. Plant products that are used as food and in the preparation of traditional medicine remedies; 2. Plants used in the preparation of formulated and packaged medicines, which are usually registered or listed by the National Agency for Food and Drug Administration and Control (NAFDAC); and 3. Speciality crops that are often in pharmacopoeias or official monographs, that are used for the production of standard herbal medicinal products.

Within the second category of the medicinal plants trade clusters, two distinct segments are evident: the imported finished products from Asia, America and Europe, and the crude unprocessed herbs from other African countries

Table 14.1. Nigerian Medicinal Plants Traded in African/Asian Food Stores in North East USA

Species	Plant part	Percentage
Vernonia amygdalina	Leaves (processed or raw, dried)	92%
Gnetum africanum	Sliced leaves (frozen or dried)	84%
Elaeis guineensis	Oil from fruits	84%
Colocasia esculenta	Starchy tuber	84%
Capsicum annuum	Fresh and dried fruits	79%
Manihot esculenta	Starchy tuber and frozen leaves	79%
Abelmoschus esculentus	Fresh fruits	79%
Ipomoea batatas	Fresh leaves	74%
Musa × paradisiaca	Fresh fruits	68%
Hibiscus acetosella	Fresh and dried leaves	32%
Hibiscus sabdariffa	Fresh leaves, dried calyces	26%
Aloe vera	Fresh leaves	26%
Zingiber officinale	Fresh rhizomes	26%
Dacryodes edulis	Fresh fruits	21%
Cola nitida/Cola acuminata	Fresh seeds	21%

constituting one segment, while the other segment comprises locally produced plants in both crude and processed forms. Based on volume alone, the imported herbal medicinal products segment has been estimated to account for 12.3% of the medicinal plants' products, while local products accounted for the remaining 87.7%. Most of the export trade in Nigerian medicinal plants is mainly focused on serving the demands for culinary plants by Nigerian immigrants in Europe and America (Van Andel and Fundiko, 2016). As shown in Table 14.1, a survey of some African and Asian stores in North East USA (Maryland, New York, and Washington D.C.) revealed that culinary plants are more readily available than medicinal plants, based on the customers' demands. However, some sizeable quantities of Nigerian medicinal plants have been recorded in the trade of globally established medicinal plants, such as *Rauwolfia vomitoria*, *Zingiber officinale*, *Andrographis paniculta*, *Aloe vera*, *Cola nitida/Cola acuminata*, *Moringa oleifera*, etc.

14.3 Major Nigerian Plants in Global Trade

14.3.1 *Aframomum melegutta*, Alligator Pepper or Grains of Paradise (Family: Zingiberaceae)

All parts of the plant are used in traditional medicine in Nigeria but only the fruits and rhizome are traded in both local markets and as items of global commerce. The seeds (also known in commerce as mbongo spice, hepper pepper) are used in the food industry as essence to flavour spirits, wines, and beer. Laboratory studies and clinical outcome evaluations support its use for the treatment of male sexual dysfunction. The extract of the seeds and rhizome have been shown to have antioxidant and anti-inflammatory activities. The pungent and peppery taste of the seeds is caused by the presence of aromatic ketones, namely 1-(4-hydroxy-3-methoxyphenyl)-decan-3-one ((6)-paradol), zingiberone, 6-gingerol and 6-shogaol. Alligator pepper seed enjoys a GRAS ('generally recognized as safe') classification in USA, and no side-effects have been reported from its consumption or usage over the years. It is relatively easy to cultivate and has been incorporated into many dietary supplements and herbal medicines. The root, when boiled, is taken by nursing mothers to check excessive lactation and to control postpartum haemorrhage. The rhizome serves as an ingredient used in the preparation of a remedy to promote conception. The export price for the dried seeds ranges from US$650.00–US$700.00/100 kg bag.

14.3.2 *Azanza garckeana*; Silky Kola, Goron Tula (Tula Kola nut), Kayan Mata (Family: Malvaceae)

Azanza garckeana (syn. *Thespesia garckeana*) is a semi-deciduous tree, which can grow up to 10 m tall, with a roundish crown. Silky kola is a valuable edible indigenous fruit which is widely distributed in northern Nigeria, and is a major traded plant in Tula, Kaltungo Local Government Area of Gombe State and in Michika, Adamawa State. This indigenous fruit tree species has been semi-domesticated by the people of Tula because of the increased demand. The fruit is an important food in Tula and it is used widely as a fertility medicine (Ochokwu *et al.*, 2015). Its major commercial value is associated with its acclaimed ability to increase libido in women, hence its name, *Kayan Mata*. The tree also provides fodder for animals, timber for firewood, shade, and achieves soil conservation, whereas the bark and root are utilised for health purposes. The major constituents identified in the pulp and seed include ascorbic acid, magnesium, calcium, sodium, iron, potassium, phosphorus, carotenoids, tannins, saponins, and alkaloids, as well as flavonoids, phenols, C-glycosides and xanthones, and they are a source of crude fibre. Pharmacological studies on the fruits of *Azanza* indicate potential for the treatment of female reproductive problems, diabetes, cancer, sexual dysfunction, fertility, and urinary tract infections. The seeds are highly priced with 20 × 2 g sachets selling for up to US$16 per pack.

14.3.3 *Tetrapleura tetraptera*; Aridan (Yoruba: Prekese (Prɛkɛsɛ); Igbo: Ohikogho) (Family: Fabaceae)

This is a highly prized Nigerian spice used in the preparation of the popular fish soup, 'Banga soup'. It produces a very aromatic volatile oil with a pleasant and characteristic taste. It is also used in the preparation of post-partum soup for nursing mothers to promote healing and restore tonicity of the womb and reduce weight gain. Alcoholic extracts of the fruit exhibited sedative effects in laboratory studies. One of the major constituents of Aridan, scopoletin, has been shown to possess *in-vitro* and *in-vivo* anti-bronchoconstrictor and antiarrhythmic effects. It has clinical antimalarial activity that is comparable with that of the standard antimalarial drug, chloroquine. It is also used as an anti-inflammatory, antihypercholesterolaemic and antidiabetic phytomedicine, as oral formulations in the form of tablets and teas. Its commercial value is mainly among the immigrant communities from West Africa in Europe and North America. It is available also on eBay and Amazon, with prices ranging from US$30.00 for 8 oz dried fruits to US$50.00 for the dried powder.

14.3.4 *Parkia biglobosa*; Locust Beans (Dawa-Dawa) (Family: Fabaceae)

The parts of locust beans used in medicine are the seeds and leaves. The seeds and leaves have antihypertensive properties. Fermented seeds are used locally as a seasoning or condiment for flavouring soups and stews, known as Dawa-Dawa (Hausa), Ogiri (Igbo) and iru (Yoruba). It is becoming a popular soup ingredient globally and it is increasingly traded as a commodity alongside the unfermented beans. Another commercially important product from locust beans is the gum. Locust bean gum is a low-cost food additive, which is used as a gelling agent, thickener, stabilizer and fat replicator in food technology. Its sweet flavour makes it an acceptable chocolate substitute. The locust bean gum is relatively cheap, and the market can be segmented on the basis of application, function and geography. The application segment is further separated into food and beverages, pharmaceutical, textile, cosmetics and others. Its other functions include stabilising, texturing, and coating functions.

Laboratory studies on the boiled and fermented seeds showed that they contain 35% protein, 29% lipid, and 16% carbohydrate, as well as tannins, flavonoids, and coumarins, and have good organoleptic properties with a positive (prebiotic) effect on intestinal flora. The seeds also contain lupeol, 4-*O*-methyl-epi-gallocatechin, epi-gallocatechin, epi-catechin 3-*O*-gallate, and epi-gallocatechin 3-*O*-gallate. It has a rich pool of essential amino acids. The seeds exhibit anti-platelet, antioxidant and antihypertensive activities. It has been shown to possess chemoprevention activity against different strains of human cancer cell lines. The powdered beans sell for about US$5 per kg and the gum can be purchased for US$10–50/kg to US$25/kg.

14.3.5 *Cajanus cajan*; Pigeon Pea (Family: Fabaceae)

Pigeon pea is a legume cultivated mainly in the northern parts of Nigeria, including parts of the middle-belt and the drier parts of southern regions of the country. The leaves and the seeds are used in traditional medicine. It contains more than 20% protein, many essential amino acids, non-protein amino acids, fats, carbohydrates, saponins, stilbenes, flavonoids, and isoflavones, as well as minerals, such as calcium, phosphorus, and iron. The leaves contain the phytoalexins, pinostrobin and cajaninstilbene acid, and the coumarin, cajanuslactone (7-hydroxy-5-*O*-methyl-8-(3-methyl-2-butylene)-4-phenyl-9,10-dihydro-benzopyran-2-one). Cajanus is a strong antioxidant and anti-inflammatory agent. It possesses antifungal, antidiabetic and anti-cholesterolaemic activities. A standardized aqueous formulation called Cyclavit (Neimeth International Pharmaceuticals) is marketed in West Africa for the management of sickle cell anaemia. The export price for pigeon pea seeds ranges from US$800 to US$1320 per tonne.

14.3.6 *Curcuma longa*; Turmeric (Family: Zingiberaceae)

Turmeric is a spice that is widely cultivated in northern parts of Nigeria as an introduced crop, and consists of the primary and secondary rhizomes of *C. longa*. It is used for the treatment of arthritis and gastrointestinal system dysfunction. It is also an immunostimulant and antidepressant. It has remarkable antioxidant and anti-inflammatory properties. It has been used in clinical medicine for the treatment of various inflammatory diseases. Turmeric contains a group of compounds called curcuminoids, which include curcumin (diferuloylmethane), demethoxycurcumin, and bisdemethoxycurcumin. The best-studied compound is curcumin, which constitutes 3.14% (on average) of powdered turmeric dry weight. It contains other important compounds in the volatile oil, including turmerone, atlantone, and zingiberene. It also contains proteins, resins, and carbohydrates. Because of the many applications of turmeric, there is enormous price variation, depending on the intended use and quality of the raw material. Unprocessed turmeric sells for about US$1000–1400/tonne, while the powdered material fetches US$1.50 to US$3.00 per kg.

14.3.7 *Ocimum gratissimum*; Scent Leaf (Family: Lamiaceae)

Scent leaf is an aromatic herb used as culinary herb. It is called Nchawun in Igbo, Ntong in Efik, and Effirin in Yoruba. Like other members of the basil genus *Ocimum*, it is used as a vegetable, herb, and spice, as well as in the preparation of traditional medicines in Nigeria. It yields a volatile oil that is easily distinguishable from those of related *Ocimum* species, such as *Ocimum basilicum, Ocimum canum,* and *Ocimum suave.* Scent leaf has diuretic, antioxidant, anti-inflammatory, antidiabetic and antianxiety properties. A proprietary formulation of *Ocimum gratissimum* and *Vernonia amygdalina* (IHP) is used in Nigeria for the treatment of diabetes and in cancer chemoprevention. The herb contains terpenes, tannins, flavonoids, and oligosaccharides. It is an ingredient in formulations used for the management of high blood sugar, hyperlipidaemia and body weight. The essential oil from scent leaf is custom produced and no prices are available in the literature. The dried leaves are sold at food stores at about US$2.5 per kg.

14.3.8 *Garcinia kola*; Bitter Kola (Family: Clusiaceae)

Bitter kola is a commonly used remedy for the treatment of sore throat, coughs and respiratory tract infections. It is also used in the treatment of impotence by increasing blood flow to the core area in men, and to treat the flu virus and laryngitis. It is also antidiabetic. It is not only the seeds that are used in traditional medicine, but also the fruit pulp and rind which are used for the treatment of various diseases, including diabetes and tumours. Extracts of bitter kola and some of the constituents have been found to be antidiabetic, anti-cholesterolaemic and antiviral.

There are three groups of chemical constituents in *G. kola*: the polyphenols, including alkylated-benzophenones; xanthones, biflavonoids and related compounds; and tocotrienols and the triterpenoids. Kolaviron, a mixture of biflavonoids, benzophenones and chromanols, as well as other phenolic compounds in *G. kola*, exhibits strong antioxidant activity. Several reports on the experimental validation of *G. kola* indicate that the antioxidant and anti-inflammatory properties of *G. kola* are due essentially to kolaviron. An evaluation of the antioxidant properties of garcinoic acid (structurally similar to vitamin E) and its congeners, including investigation of the structure-antioxidant activity relationships, identified a semisynthetic derivative with antioxidant activity that was 18.7 times stronger than dl-alpha-tocopherol. It has also been shown to inhibit aflatoxin B1 genotoxicity in human liver-derived HepG2 cells. The *Garcinia* benzophenones have the ability to modulate epigenetic histone methylation by xenobiotics and therefore have a positive role in the management of chronic inflammation in the aged. Bitter kola is well tolerated but some of the reported side-effects include hives, upset stomach, and nausea. The seeds sell for about US$2.5 per kg. The processed powder and extracts fetch higher prices.

14.3.9 *Vernonia amygdalina*; Bitter Leaf (Family: Asteraceae)

V. amygdalina, bitter leaf, is used extensively in traditional medicine and as a leafy vegetable in the preparation of the West African bitter-leaf soup. The parts of the plant used are the leaves, bark, roots, and fruits. The leaf is taken as a digestive tonic, laxative and for fever. It is also used to treat diabetes. The bark and root are used to treat rheumatism, parasitic infection, ringworm, smallpox, chickenpox, and measles. When boiled, the root is used to treat coughs, colds and general weakness of the body. The fruit is recommended for piles, dysentery, and menstrual disorders. The leaves are chewed raw in most parts of Nigeria as an effective remedy for gastrointestinal disorders and as a general tonic. The plant contains the saponin vernonin, the sesquiterpenes vernoleptin and vernodalin, and the ubiquitous flavonoid kaempferol. Laboratory and clinical studies have established its antimalarial, antidiabetic, antioxidant, antihypercholestrolaemic, prebiotic, anthelmintic, anticancer, immune boosting activity in HIV infections, and anti-inflammatory properties, as well as its effect on uterine contractility.

Several proprietary formulations exist in which *Vernonia* is either the active component or the main ingredient (for example, *Vernonia-Ocimum*®, manufactured by IHP, for diabetes and immunostimulation; and Edotite®, which contains bitter leaf as the only active pharmaceutical ingredient and is promoted as an anticancer agent. The dried leaves are sold in ethnic food shops in Europe and America in packages of up to 2 kg at US$5 each.

14.3.10 *Irvingia gabonensis*; Ogbono seeds (Family: Irvingiaceae)

This tree is commonly known as African bush mango. The Ogbono seeds, used as a condiment for the preparation of the West African 'draw soup', is a highly prized weight-loss medicine in Europe and America. The parts used in traditional medicine are the leaves, twigs, seeds and the fruit. The seeds are a major item of commerce and are sold in European and American ethnic food shops. However, a random survey of 17 proprietary products from American dietary supplements shops by the US Department of Agriculture showed that none of the products that claimed to contain African bush mango contained any *Irvingia*. These cases of adulterations can be attributed partly to the scarcity of the authentic plant species from Africa. It is sold at US$ 95/kg for small quantities of less than twenty kilograms and US$200 to US$570 per tonne for large supplies, depending on quality.

14.3.11 *Moringa oleifera*; Tree of Life (Family: Moringaceae)

This plant is commonly known as the "tree of life" because of its many uses and the near- perfect mixture of vitamins and minerals. It has seven times the vitamin C concentration of oranges, four times the vitamin A concentration of carrots, four times the calcium concentration of milk, three times the potassium concentration of bananas, and twice the protein concentration of yogurt. The parts used are the leaves, flowers, seeds, roots and bark. The leaves contain all the essential amino acids. It is used to treat poor vision and goitre. The leaves contain amino acids, including aspartic acid, glutamic acid, serine, glycine, threonine, α-alanine, valine, leucine, isoleucine, histidine, lysine, cysteine, methionine, arginine, and tryptophan. The flowers and the fruits also contain amino acids. The seeds yield an almost colourless fixed oil, known in commerce as Beni or Moringa oil. The oil consists of a 60% liquid olein fraction and 40% solid fat. The major constituents of the oil are oleic acid (65%), stearic acid (10.8%), behenic acid (8.9%), myristic acid (7.3%), palmitic acid (4.2%), and lignoacetic acid (3.0%). *Moringa* has remarkable antioxidant, anti-inflammatory, and immunomodulatory activities. It has been shown to have chemoprotective and antifertility properties. There is much scientific and clinical evidence to support the use of the leaves as dietary supplements for the treatment of hyperlipidaemia, hypertension, and diabetes, diseases caused by oxidative stress, anaemia, and chronic arthritis. In commerce, *Moringa* consists of three distinct products: the highly nutritious leaves, usually sold in powdered form, the seeds, and the oil. The powdered leaves are sold at US$20–25 per kg, whereas the seeds have a wholesale price of US$ 3,000 to 4,000 per tonne.

14.3.12 *Hibiscus sabdariffa*; Zobo (Family: Malvaceae)

Zobo is used in Nigeria as a vegetable and for the preparation of hot and cold beverages, as a flavouring agent in the food industry and as an herbal medicine for the treatment of high blood pressure. Zobo is derived from the calyx and flowers of *H. sabdariffa*. Laboratory studies have shown that the extracts exhibited antibacterial, antioxidant, nephro- and hepatoprotective, renal/diuretic effects, effects on lipid metabolism (anticholesterol), antidiabetic and antihypertensive effects, among others (Da-Costa-Rocha *et al.*, 2014) The main constituents found in the plant include phenolic acids (especially protocatechuic acid), organic acids (hydroxycitric acid and hibiscus acid) and anthocyanins (delphinidin-3-sambubioside and cyanidin-3-sambubioside).

A variety of fruit drinks made from zobo and other herbs, such as ginger, turmeric, and alligator pepper, are sold in markets in West Africa as general health tonics for the improvement of general body function, blood circulation, and hypertension. It is a considered to be a safe caffeine-free, low-tannin tea and beverage. Zobo is traded in commerce mainly as the dried calyx. It has a global market as a colourant for herbal teas, with a rising profile and application as a nutraceutical. The Nigerian zobo is sold mainly in African food markets in Europe and America. Although *H. sabdariffa* was first domesticated in Africa sometime before 4000 BC, it is now widely cultivated in both tropical and subtropical regions.

It is an important industrial crop also valued for its fibres, and the demand for roselle fibre is likely to increase as a result of the rising interest in natural, biodegradable fibres for making jute bags and for other uses. This multipurpose crop provides farmers with food and cash income when other vegetables have become scarce. Processing generates additional family income, from which women in particular benefit. Use of roselle as a vegetable or as a beverage should be promoted through research to improve cultivars, husbandry, and post-harvest technologies. The market competition is stiff and the application of rigorous on-farm quality standards for grading, processing, and packaging will boost competitiveness in the international market. The current market price for *H. sabdariffa* HA in the international market is about US$2350–2450 per tonne MT/CIF.

14.3.13 *Xylopia aethiopica*; Ethiopian Pepper (Family: Annonaceae)

X. aethiopica or Ethiopian pepper is a spice with a very characteristic aromatic flavour. *Xylopia* is employed extensively in African traditional medicine for the treatment of a plethora of illnesses. The essential oils, alcoholic extract, and aqueous extracts of *X. aethiopica* have been shown to exhibit significant antimicrobial, antioxidant, cardiovascular and antidiuretic effects. When extracts were subjected to bioassay-directed phytochemical examination, the diterpene kaurenoids were shown to be responsible for the significant systemic hypotensive and coronary vasodilatory effect accompanied with brachycardia (Iwu, 2016). These effects were attributed to calcium antagonistic mechanism. The diuretic and natriuretic effects found were similar to the effects of chlorothiazide, suggesting inhibition of Na^+ and K^+ reabsorption in the early portion of the distal tubule. The extract also has the ability to reduce the intra-ocular pressure in glaucoma patients. The essential oil contains a complex mixture of diterpenes and phenolics, including α- and β-pinenes, carrageenan, cymene, α-phellandrene, limonene and terpinoline, cineole, bisabolone, and linalool, terpinen-4-ol, terpineal, cuminyl alcohol, and cuminic aldehyde. The fruit also contains the diterpenic acid, xylopic acid [15β-acetoxy (-) 16ene-19-oic acid], and its analogues, kaurenoic, 15-oxo-kaurenoic acid, and kauran-16-α-ol. The ground raw pepper is sold for about US$20/kg. The oils are usually custom produced to meet specific manufacturing needs at prices of up to US$1,000 per kg, depending on the processing. Blended oils for aromatherapy sell for about US$25 per 5 g.

14.3.14 *Zingiber officinale*; Ginger (Family: Zingiberaceae)

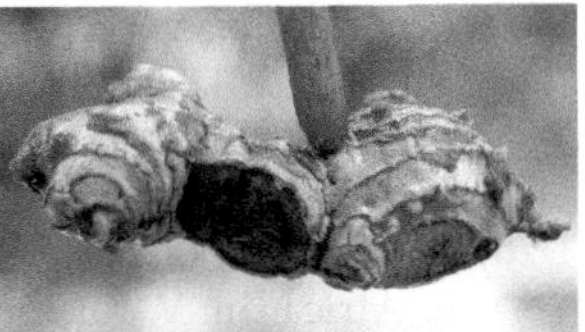

Ginger has a distinctive odour and aromatic, pungent, and agreeable taste. Ginger yields a volatile oil, which consists mainly of camphene, citral, cineol, linalool, zingiberene, bisbolene, zingiberol, zingiberenol, and methylheptenone. The plant also yields pungent phenolic principles, known as gingerols and shogaols. The anti-inflammatory, antioxidant, and antinausea properties of ginger has been the subject of many laboratory studies. Other biological activities of ginger and its constituents, that have been validated by laboratory studies, include antihepatotoxic, antiviral, antiarthritic, and chemoprevention, and in the treatment of morphine dependence. In human clinical studies, ginger has been shown to suppress gastric secretion and to reduce vomiting. It was effective in reducing the pain and immobility caused by osteoarthritis. Capsules containing 940 mg of the dried rhizome were found to be superior to the antihistamine dimenhydrinate (100 mg) in preventing motion sickness. Proprietary products containing ginger are available in China and Japan for the treatment of elevated blood pressure and degenerative heart diseases. In some formulations, it is claimed that steamed ginger has better therapeutic outcomes than the fresh rhizome. Two varieties are known in commerce, the mild tasting and often peeled West Indian variety called "Jamaican Ginger", and the more pungent and usually unpeeled variety called "Nigerian" or "African ginger". The Nigerian ginger, therefore, has an exclusive market position, which has not yet been fully explored by the growers. Peeled Nigerian ginger sells for US$800–1500/tonne.

References

Adegboyega, E.O. and Oluwalana, I.E. (2011) Structure, control and regulation of the formal market for medicinal plants' products in Nigeria. *African Journal of Traditional, Complementary and Alternative Medicines*, 8(3), 370–376.

Bannerman, R.H. (1982) Traditional medicine in modern health care. *World Health Forum*, 3, 8–13.

Chukwuone, N., Odele., M., Onugu, A., Iwu, M., Schopp, D., Pendergrass, J., Czebiniak, R. and Otegbulu, A. (2013) Natural resource valuation and damage assessment in Nigeria: a case study of the Niger Delta. Toward a legal framework for improved natural resource decision-making in the Niger Delta: conserving biodiversity through enhanced natural resource valuation practice. *MacArthur Foundation Report*. Abuja. pp. 109.

Da-Costa-Rocha, I., Bonnlaender, B., Hartwig Sievers, H., Pischel, I. and Heinrich, M. (2014) *Hibiscus sabdariffa* L. – A phytochemical and pharmacological review. *Food Chemistry*, 165, 424–443.

Dzoyema, J.P., Tshikalange, E. and Kuete, V. (2013) Medicinal plants market and industry in Africa. In: V. Kuete (ed.), *Medicinal plant research in Africa*. Amsterdam and New York: Elsevier.

Harrison-Dunn, A.-R. (2014) *A global look at supplements on the rise*. Assessed 10 March 2014. Available online at: www.nutraingredients.com.

Hexa Research. (2017) Herbal medicine market size and forecast, by product (tablets & capsules, powders, extracts), by indication (digestive disorders, respiratory disorders, blood disorders), and trend analysis, 2014–2024. Assessed May, 2019. Available at https://www.hexaresearch.com/research-report/global-herbal-medicine-market.

Iwu, M.M. (2016) *Food as medicine – functional food plants of Africa*. Boca Raton, FL: CRC Press/Taylor and Francis Group, pp. 384.

Martin-Webber, C. and Labaste, P. (2010) Building competitiveness in Africa's agriculture: a guide to value chain concepts and applications. Pub. *The International Bank for Reconstruction and Development / The World Bank*, Washington DC. pp. 187.

Mohammed Abubakar, B., Mohd Salleh, F., Shamsir Omar, M.S. and Wagiran, A. (2017) Review: DNA barcoding and chromatography fingerprints for the authentication of botanicals in herbal medicinal products. *Evidence-Based Complementary and Alternative Medicine*, 2017, 1352948.

Muhammad, B.Y. and Awaisu, A. (2008) The need for enhancement of research, development, and commercialization of natural medicinal products in Nigeria: lessons from the Malaysian experience. *African Journal of Traditional, Complementary, and Alternative Medicines*, 5(2), 120–130.

Ochokwu, I.J., Dasuki, A. and Oshoke, J.O. (2015) Azanza garckeana (Goron Tula) as an edible indigenous fruit in north eastern part of Nigeria. *Journal of Biology, Agriculture and Healthcare*, 5(15), 26–31.

Picking, D. (2017) The global regulatory framework for medicinal plants. Chapter 35. In: S. Badal and R. Delgoda (eds.), *Pharmacognosy - fundamentals, applications and strategy*. New York: Elsevier Press, pp. 663–674.

Shackleton, S., Campbell, B., Lotz-Sisitka, H. and Shackleton, C. (2008) Links between the local trade in natural products, livelihoods and poverty alleviation in a semi-arid region of South Africa. *World Development*, 36(3), 505–526.

Smith-Hall, C., Larsen, H.O. and Pouliot, M. (2012) People, plants and health: a conceptual framework for assessing changes in medicinal plant consumption. *Journal of Ethnobiology and Ethnomedicine*, 8, 43. Available at http://www.ethnobiomed.com/content/8/1/43

Van Andel, T. and Fundiko, M.C. (2016) The trade in African medicinal plants in Matonge-Ixelles, Brussels (Belgium). *Economic Botany*, 70(4), 405–415.

Van Wyk, B.-E. (2015) A review of commercially important African medicinal plants. *Journal of Ethnopharmacology*, 176, 118–134.

World Health Organization (WHO). (2013) Traditional medicine strategy 2014_2023. Retrieved 10 October 2014, available from http://www.who.int/medicines/publications/traditional/trm_strategy14_23/en/

Index

B

C

D

E

F

G

H

N

O

P

U

V

W

X

Y

Z

Taxonomic Index

D

S

T